An Introduction to the Mechan
Properties of Ceramics

This book is a comprehensive introduction to the mechanical properties of ceramics, and is designed primarily as a textbook for undergraduate and graduate students in materials science and engineering.

Over the past 25 years ceramics have become key materials in the development of many new technologies as scientists have been able to design these materials with new structures and properties. An understanding of the factors that influence their mechanical behavior and reliability is essential. Some of these new applications are structural, and for these it is important to understand the factors that control their mechanical behavior. Non-structural applications are also being developed, but in each case it is necessary to design mechanically reliable materials. This is a particular challenge for materials that are inherently brittle. This book will introduce the reader to current concepts in the field. It contains problems and exercises to help readers develop their skills.

Although designed principally as a textbook for advanced undergraduates and graduate students, this book will also be of value as a supplementary text for more general courses and to industrial scientists and engineers involved in the development of ceramic-based products, materials selection and mechanical design.

An Introduction to the Mechanical Properties of Ceramics

Cambridge Solid State Science Series

Editors

Professor D.R. Clarke
Department of Materials,
University of California, Santa Barbara

Professor S. Suresh
Department of Materials Science and Engineering,
Massachusetts Institute of Technology

Professor I.M. Ward FRS
IRC in Polymer Science and Technology,
University of Leeds

Titles in print in this series

S. W. S. McKeever
Thermoluminescence of Solids

P. L. Rossiter
The Electrical Resistivity of Metals and Alloys

D. I. Bower and W. F. Maddams
The Vibrational Spectroscopy of Polymers

S. Suresh
Fatigue of Materials

J. Zarzycki
Glasses and the Vitreous State

R. A. Street
Hydrogenated Amorphous Silicon

T–W. Chou
Microstructural Design of Fiber Composites

A. M. Donald and A. H. Windle
Liquid Crystalline Polymers

B. R. Lawn
Fracture of Brittle Solids – Second Edition

T. W. Clyne and P. J. Withers
An Introduction to Metal Matrix Composites

V. J. McBrierty and K. J. Packer
Nuclear Magnetic Resonance in Solid Polymers

R. H. Boyd and P. J. Phillips
The Science of Polymer Molecules

D. P. Woodruff and T. A. Delchar
Modern Techniques of Surface Science – Second Edition

J. S. Dugdale
Electrical Properties of Metallic Glasses

M. Nastasi, J. Mayer and J. K. Hirvonen
Ion–Solid Interactions: Fundamentals and Applications

D. Hull and T. W. Clyne
An Introduction to Composite Materials – Second Edition

J.W. Martin, B. H. Doherty and B. Cantor
Stability of Microstructure in Metallic Systems – Second Edition

T. G. Nieh, J. Wadsworth and O. D. Sherby
Superplasticity in Metals and Ceramics

L. J. Gibson and M. F. Ashby
Cellular Solids – Second Edition

K. K. Chawla
Fibrous Materials

The book is dedicated to the memory of
Christina Cushing Green

An Introduction to the Mechanical Properties of Ceramics

David J. Green
The Pennsylvania State University

CAMBRIDGE
UNIVERSITY PRESS

PUBLISHED BY THE PRESS SYNDICATE OF THE UNIVERSITY OF CAMBRIDGE
The Pitt Building, Trumpington Street, Cambridge CB2 1RP, United Kingdom

CAMBRIDGE UNIVERSITY PRESS
The Edinburgh Building, Cambridge CB2 2RU, United Kingdom
40 West 20th Street, New York, NY 10011–4211, USA
10 Stamford Road, Oakleigh, Melbourne 3166, Australia

First published 1998

Printed in the United Kingdom at the University Press, Cambridge

Typeset in 10¼ on 13½pt Monotype Times   [SE]

A catalogue record for this book is available from the British Library

Library of Congress Cataloguing in Publication data

Green, D. J. (David J.)
   An introduction to the mechanical properties of ceramics / David
J. Green
      p.   cm. – (Cambridge solid state science series)
   Includes index.
   ISBN 0 521 59087 6 (hardcover) – ISBN 0 521 59913 X (paperback)
   1. Ceramic materials – Mechanical properties   I. Title.
II. Series.
   TA455.C43G738   1998
   620.1'40423 – dc21   97-18018 CIP

ISBN 0 521 59087 6 hardback
ISBN 0 521 59913 X paperback

Contents

Preface   xi

Chapter 1   **Introduction   1**

References   12

Chapter 2   **Elastic behavior   13**

2.1   Elastic deformation of atomic bonds   14
2.2   Failure of Hooke's Law   17
2.3   Engineering elastic constants   18
2.4   Strain at a point   24
2.5   Transformation of strains   30
2.6   Dilatational and deviatoric strains   34
2.7   Strain compatibility   35
2.8   Tensors   36
2.9   Coefficients of thermal expansion   36
2.10   Definition of stress   40
2.11   General version of Hooke's Law   44
2.12   Elastic behavior of anisotropic materials   47
2.13   Elastic behavior of isotropic materials   55
2.14   Miscellaneous effects on the elastic constants   57
2.15   Propagation of mechanical disturbances   58
2.16   Resonant vibrations   60

2.17   Measurement of elastic constants   62

            Problems   65

            References   69

Chapter 3   **Effect of structure on elastic behavior   70**

3.1   Relationship of elastic constants to interatomic potential   70

3.2   Elastic anisotropy and atomic structure   75

3.3   Elastic behavior of particulate composites   78

3.4††   Advanced constitutive relationships for composites   83

3.5   Constitutive relations for random polycrystals   87

3.6   Effects of porosity and microcracking on elastic constants   88

3.7   Thermal expansion behavior of polycrystalline ceramics   94

3.8   Elastic behavior of sandwich panels   98

            Problems   99

            References   103

Chapter 4   **Elastic stress distributions   105**

4.1   Statically determinate and indeterminate problems   106

4.2   Thin-walled pressure vessels   107

4.3   Bending of beams   108

4.4   Elastic stability and buckling   113

4.5   Plane stress and plane strain   114

4.6   Cylindrical polar coordinates   117

4.7   Pressurized thick-walled cylinders   118

4.8   Residual stresses in composites   120

4.9   Stress concentrations due to pores and inclusions   124

4.10   Contact forces   127

            Problems   129

            References   133

Chapter 5   **Viscosity and viscoelasticity   134**

5.1   Newton's Law of viscosity   134

5.2   Temperature dependence of viscosity   137

5.3   Simple problems of viscous flow   139

5.4††   General equations for slow viscous flow   142

5.5   Non-linear viscous flow   145

5.6   Dispersion of solid particles in a fluid   146

5.7   Viscoelastic models   148

5.8   Anelasticity in ceramics and glasses   156
            Problems   158
            References   161

Chapter 6   **Plastic deformation   162**

6.1   Theoretical shear strength   162
6.2   Dislocations   164
6.3   Stress fields of dislocations   166
6.4   Attributes of dislocations   169
6.5   The geometry of slip   172
6.6   Partial dislocations   176
6.7   Plasticity in single crystals and polycrystalline materials   179
6.8   Obstacles to dislocation motion   183
6.9   Plasticity mechanics   186
6.10   Hardness   188
            Problems   189
            References   191

Chapter 7   **Creep deformation   193**

7.1   Creep in single crystals   195
7.2   Creep in polycrystals   196
7.3   Deformation mechanism maps   201
7.4   Measurement of creep mechanisms   202
            Problems   204
            References   208

Chapter 8   **Brittle fracture   210**

8.1   Theoretical cleavage strength   210
8.2   Stress concentrations at cracks   212
8.3   The Griffith concept   213
8.4   Nucleation and formation of cracks   216
8.5   Linear elastic fracture mechanics   218
8.6   Stress intensity factor solutions   224
8.7††   Methods of determining stress intensity factors   231
8.8   Indentation fracture   243
8.9 R curves   245
8.10   Mixed mode fracture   247
8.11   Microstructural aspects of crack propagation   248

8.12   Sub-critical crack growth   264

8.13   Fractography   266

8.14   Contact-damage processes   269

8.15††   *J*-integral   278

            Problems   280

            References   283

Chapter 9   **Strength and engineering design   285**

9.1   Strength testing   285

9.2   Failure statistics   286

9.3   Time dependence of strength   291

9.4   Determination of sub-critical crack growth parameters   293

9.5   SPT diagrams   295

9.6   Improving strength and reliability   296

9.7   Temperature dependence of strength   298

9.8   Thermal stresses and thermal shock   298

9.9   Thermal shock resistance parameters   301

9.10   Residual stresses   305

            Problems   306

            References   314

Comprehension exercises   316

Appendices

1   Explicit relations between the stiffness and compliance constants
            for selected crystal classes   325

2   Young's modulus as a function of direction for various single
            crystals   326

3   Relationship between engineering elastic constants for isotropic
            materials   327

4   Madelung constants for various crystal types   328

5   Stress and deflection for common testing geometries   329

            Index   331

Preface

The aim of this book is to provide a text for a senior undergraduate course on the mechanical behavior of ceramics. There are, however, some advanced sections that would allow the book to be used at the graduate level (marked ††). The format of the book owes much to the text, *Mechanical Properties of Matter*, by A. H. Cottrell, which helped me through graduate school. In teaching a course in this area, it has always been frustrating that there are so few texts aimed primarily at ceramics. There is often the concern of discerning whether ideas applied to other materials could also be used to understand ceramic materials. I have also been fortunate in being involved in the field of structural ceramics at a time it has undergone remarkable developments and I have tried to incorporate my interpretation of these recent advances into the text.

I would be amiss in not acknowledging the support I have received in undertaking this project. I owe much to Pat Nicholson, Dave Embury and Dick Hoagland, who patiently introduced me to this field of research and to Tom Wheat, who taught me about the processing of ceramics. I am particularly grateful to Fred Lange, who took a chance on me and became my mentor. His enthusiastic, intuitive advice and sense of fun encouraged me to pursue many new ideas. I also appreciate the interaction with my other colleagues at Rockwell International Science Center. The undergraduates at Penn State in Ceramic Science and Engineering should be acknowledged for suffering through the various versions of this book. I should particularly thank Fred Fitch for patiently pointing out many typographical errors in an earlier version of the book. Thanks are also owed to Brian Watts and Patty Phillips for their patient proof-reading skills. I would also thank George Scherer, David Clarke, David

Wilkinson and John Ritter for reading some of the chapters and giving insightful advice. I owe much to the graduate students I have advised in the last 12 years and to my faculty colleagues at Penn State.

Finally, I should mention the encouragement I have received from my extended family; Mel and Vera Smith, the Knapps, Marc, Tina and Tony. The emotional support from Chris Cushing Green, Cyndi Asmus and Patty Phillips was also essential in completing this project.

David J. Green
State College, Pennsylvania

Chapter 1

Introduction

In the last 25 years there has been a strong movement to use ceramics in new technological applications and a key facet of this work has been directed at understanding the mechanical behavior of these materials. First, let us consider the various **technological functions** of ceramics as shown in Table 1.1. The diverse properties of ceramics are not always appreciated. For structural functions, adequate mechanical properties are of prime importance. Ceramic materials that are considered for these applications are termed **structural ceramics**. In some cases, such as engine parts, the choice is based on their high-temperature stability and corrosion resistance. These factors imply the engine temperature could be raised, making the overall performance more efficient. Unfortunately, ceramics can be brittle, failing in a sudden and catastrophic manner. Consequently, there has been a strong emphasis on understanding the mechanical properties of ceramics and on improving their strength, toughness and contact-damage resistance. Indeed, it is appropriate to state that there has been a revolution in the understanding of these properties and the associated research has led to the discovery of new classes of structural ceramic materials.

It is important to realize that mechanical properties can also be critical in non-structural applications. For example, in the design of the thermal protection system of the space shuttle, highly porous, fibrous silica tiles are used. The microstructure of these materials, shown in Fig. 1.1, consists of a bonded array of fibers, usually based on silica glass. Clearly, the prime reason for using these materials was their low thermal conductivity but the resistance to thermal and structural stresses was a key item in the final design. In some non-structural applications, mechanical properties can be important in determining the lifetime

Table 1.1 *Functions and technological applications of ceramics*
(Adapted from Kenney and Bowen, 1983, reproduced courtesy of The American Ceramic Society.)

Function	Primary characteristic	Examples of applications
Electrical	Electrical insulation (e.g., Al_2O_3, BeO)	Electronic substrates and packages, wiring, power-line insulators
	Ferroelectricity (e.g., $BaTiO_3$, $SrTiO_3$)	Capacitors
	Piezoelectricity (e.g., PZT)	Vibrators, oscillators, filters, transducers, actuators, spark generators
	Semiconductivity (e.g., $BaTiO_3$, SiC, ZnO–Bi_2O_3, CdS, V_2O_5)	NTC thermistor (temperature sensor)
		PTC thermistor (heater element, switch)
		CTR thermistor (heat sensor)
		Thick-film thermistor (IR sensor)
		Varistor (noise elimination, surge arrestors)
		Solar cells, furnace elements
	Ionic conductivity (β-alumina, ZrO_2)	Solid state electrolytes (batteries, fuel cells, oxygen sensors)
	Superconductivity (YBCO)	Magnets, electronic components
Magnetic	Soft magnets (ferrites)	Magnetic recording heads
	Hard magnets (ferrites)	Magnets, electric motors
Optical	Translucency (Al_2O_3, MgO, mullite, Y_2O_3, PLZT)	High-pressure sodium-vapor lamps, IR windows, lighting tubes and lamps, laser materials, light memory, video display and storage, light modulation and shutters.
	Transparency (silicate glasses)	Optical fibers, containers, windows
Chemical	Chemical sensors (ZnO, Fe_2O_3, SnO_2)	Gas sensors and alarms, hydrocarbon and fluorocarbon detectors, humidity sensors

Table 1.1 (*cont.*)

(Adapted from Kenney and Bowen, 1983, reproduced courtesy of The American Ceramic Society.)

Function	Primary characteristic	Examples of applications
	Catalyst carriers (cordierite, Al_2O_3)	Emission control, enzyme carriers, zeolites
	Electrodes (titanates, sulfides, borides)	Electrowinning, photo-chemical processes
Thermal	Thermal insulation (fiberglass, aluminosilicate fibers)	IR radiators, thermal protection systems for aerospace vehicles
	Thermal conduction (diamond films, AlN)	Heat sinks in electronic devices
	Thermal stability (AZS, Al_2O_3)	Refractories
Structural	Hardness (SiC, TiC, TiN, Al_2O_3)	Cutting tools, wear-resistant materials, mechanical seals, abrasives, armor, bearings
	Stiffness and thermal stability (SiC, Si_3N_4)	Ceramic engine parts, turbine parts, burner nozzles, radiant tubes, crucibles.
Biological	Chemical stability (hydroxyapatite, Al_2O_3)	Artificial teeth, bones and joints
Nuclear	Nuclear fission (UO_2, PuO_2)	Nuclear fuels, power sources,
	Neutron absorption (C, SiC, B_4C)	Cladding and shielding

Notes:

PZT – lead zirconium titanate	NTC – negative temperature coefficient
YBCO – yttrium barium copper oxide	PTC – positive temperature coefficient
PLZT – lead lanthanum zirconium titanate	CTR – critical temperature resistance
AZS – Alumina zirconium silicate	IR – infra-red

of the component. For example, the sodium–sulfur battery, that has high energy and power density, has been developed for transportation and energy-storage applications. This battery is based on the use of a solid ceramic electrolyte known as β-alumina, but cracking during recharging can lead to a limited lifetime. Clearly, improvements in the mechanical properties of the electrolyte could significantly impact the economic viability of the battery.

Before launching into the details of the various aspects of mechanical properties, it is worthwhile considering the overall philosophy of **materials science and engineering**. Figure 1.2 shows an overview of the way this discipline is used in developing technological applications. An understanding of the possible properties of materials allows a particular application to be identified and a design put forward. This is usually in the realm of engineering and it is important to be able to identify and measure all of the critical properties required in the design. In materials science one is also concerned with properties, but here it is usually the optimization of properties, structure and processing that is the key item. The aim is often to adjust the processing so that a particular structure, that gives the best set of properties, is obtained. *The search for new or improved materials is an*

Figure 1.1 Microstructure of space shuttle tile material, a high-porosity fibrous silica; secondary electron image using the scanning electron microscope (SEM). (Reproduced courtesy of Plenum Press, New York.)

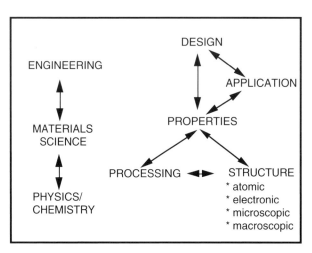

Figure 1.2 Overview of scientific approach for the development of materials for technological applications.

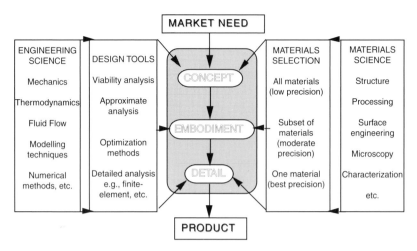

Figure 1.3 A flow chart showing how design tools and materials selection enter the design process. (Adapted from Ashby, 1992. Reproduced courtesy of Heineman Publishers, Oxford, UK.)

essential aspect of the field. The structure of the material that controls a particular property can occur at different scale levels: from the electron and atomic structure through to the macroscopic structure. For example, elastic properties are basically dependent on the chemical bonding, which determines the stiffness of the atomic bonds. Thus, one approach to changing the elastic behavior would be to adjust the composition by adding a solute. It is, however, also possible to change the elastic properties of a material by adjusting the microstructure, i.e., at a larger scale. For example, adding other phases or changing the amount of porosity at the microscopic level will also change the elastic properties. The introduction of ceramic fibers into glass, polymers or metals can substantially increase stiffness and has led to the development of **fiber composites**. At the macroscopic level, the use of different materials in particular configurations and geometries can be important. For example, in laminated structures the elastic behavior can be controlled by placing a stiffer component in the region of highest stress, such as in **sandwich structures**. As indicated in Fig. 1.2, materials science often plays an intermediary role between engineering and the basic sciences of physics and chemistry.

Figure 1.3 shows a flow chart that describes the overall design process. As the design moves from concept to product, design tools of increasing sophistication are needed from engineering science. Concurrently, materials science needs data on the properties of materials with increasing degrees of precision, in order to select the best candidate. In the early stages, approximate data are useful to identify the best group of materials but, later, standard test procedures and in-house testing are often required. The material selection process is becoming more sophisticated as computer databases are being developed. Such databases are expected to increase in sophistication and 'intelligence'. This latter aspect is particularly important in being able to check and correct any input errors. Clearly this process will be influenced by many other factors, such as the economics and the aesthetics of the design. It is important to realize that there are often many different processing routes to produce a particular material. Shifting economic patterns and the development of new processing techniques can strongly impact the final decisions. There is also greater concern over 'greener' processing and industrial processes are now being studied from an ecological viewpoint (**industrial ecology**).

To illustrate techniques that have been introduced to improve **mechanical reliability**, it is useful to consider some examples for a particular material, say alumina (Al_2O_3). The microstructure of hot-pressed alumina, which consists of fine equiaxed grains (~5 μm), is shown in Fig. 1.4. Polycrystalline alumina has found applications from electronic circuit substrates through to armor plating. It is now established, however, that the strength and toughness of alumina can be substantially improved by several techniques. For example, adding zirconia as a second phase gives microstructures similar to that shown in Fig. 1.5. In this approach, a mechanism known as **transformation toughening** can be introduced which increases toughness and strength. The toughness of alumina can also be

Figure 1.4 Microstructure of hot-pressed polycrystalline aluminum oxide; secondary electron image using the SEM.

increased by incorporating ceramic fibers, whiskers or platelets into the micro-structure. For example, Fig. 1.6 shows the microstructure of Al_2O_3 containing SiC platelets. As shown in Chapter 8, **crack bridging** and the frictional pull-out of the reinforcements is often the source for the improved properties of these **ceramic-matrix composites**. The structure of an alumina product can also be manipulated at the macroscopic level. For example, Fig. 1.7 shows the loading of an alumina ceramic sandwich panel. The incorporation of a porous alumina core between dense alumina plates can be used to produce materials with maximum strength or stiffness at minimum weight, especially in flexural loading modes. There is also a current interest in producing ceramic hybrid laminates in which layers of different compositions are interspersed. For example, Fig. 1.8 shows a laminate consisting of alternating alumina and zirconia layers and such a structure can significantly influence the crack-propagation behavior. The above approaches emphasize the use of composites in controlling the structure. There is, however, a push to produce materials in which **self- reinforcement** is 'grown' into the microstructure. For example, Fig. 1.9 shows the microstructure of alumina, in which grain shape and texture are used to control the physical prop-erties and Fig. 1.10 shows the microstructure of self-reinforced silicon nitride in which the production of 'fibrous' grains is found to increase fracture toughness.

The way a particular material is processed is also very important in deter-

Figure 1.5 Microstructure of a zirconia-toughened alumina; back-scattered electron image using the SEM. The bright phase is zirconia.

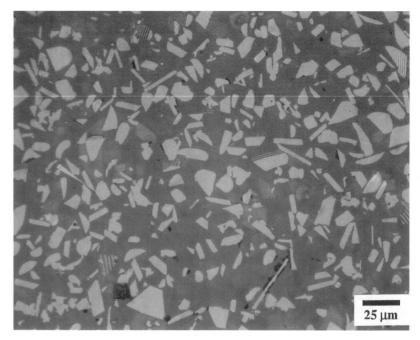

Figure 1.6 Microstructure of a SiC platelet-reinforced alumina; optical micrograph. (Courtesy of Matt Chou.)

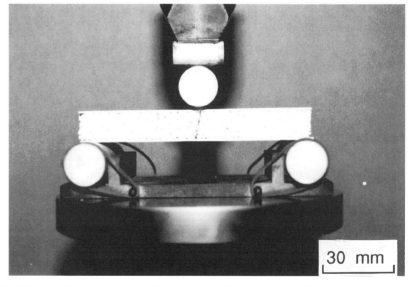

Figure 1.7 Three-point bending of an alumina sandwich panel, consisting of a porous, cellular core and dense faceplates. (Reproduced courtesy of The American Ceramic Society, Westerville OH.)

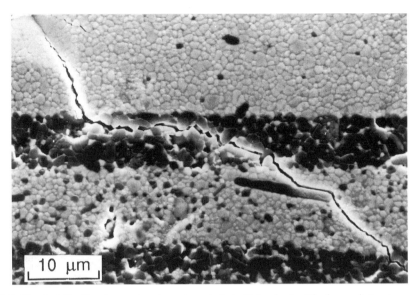

Figure 1.8 Crack deflection in an alumina–zirconia hybrid laminate. The layered structure was produced by electrophoretic deposition; scanning electron micrograph. (Courtesy of P. Sarkar and P. S. Nicholson.)

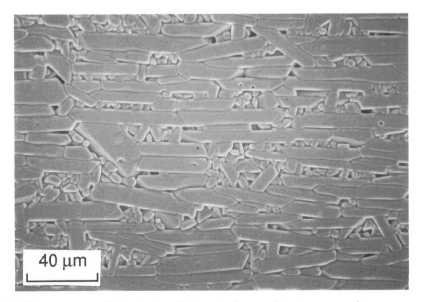

Figure 1.9 Microstructure of alumina in which grain shape and texture are used to control physical properties. (Courtesy of Matthew Seabaugh and Gary L. Messing.)

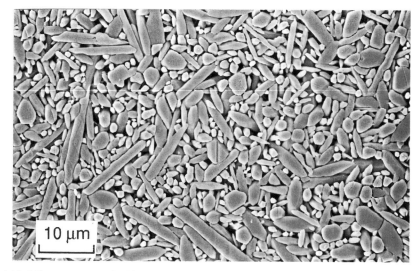

Figure 1.10 Microstructure of self-reinforced silicon nitride in which grain shape is used to control mechanical reliability. (Courtesy of Chien-Wei Li, Allied–Signal Corp.)

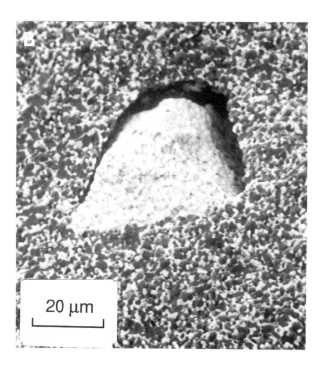

Figure 1.11 Failure origin in a zirconia-toughened alumina, caused by the presence of a zirconia aggregate in the starting powder. (Reproduced courtesy of CRC Press, Boca Raton, FL.)

mining mechanical behavior, as failure is often initiated at **microstructural imperfections**. A useful philosophy in improving the strength of ceramics is to identify the nature of the failure origins in the material. The processing is then adjusted to eliminate the particular source of failure. Figure 1.11 shows an example of a failure origin in zirconia-toughened alumina that was caused by poor mixing of the two phases. In this case, a zirconia agglomerate densified more extensively than the surrounding material and led to a 'crack-like' void. Clearly, more attention to the mixing of the components would eliminate this type of defect and thereby lead to an improvement in strength. Techniques are also being developed to identify microstructural imperfections in a non-destructive fashion. This approach can also be used to 'improve' strength by identifying and eliminating components that contain large defects.

Figure 1.12 summarizes the tremendous improvements that have been made in the strength of ceramics over the last 25 years. This is a direct result of improved processing, the detection of new toughening mechanisms, and the refined understanding of the relationships between microstructure and mechanical properties. In the remainder of this book, the various mechanical properties of ceramics will be discussed, both in terms of a formal description and the way

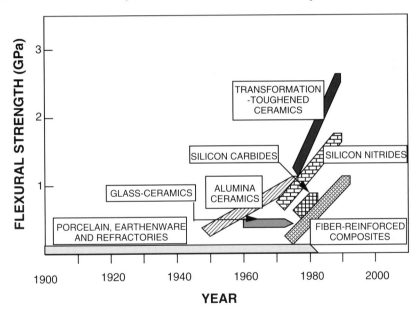

Figure 1.12 The dramatic increase in the strength of bulk ceramics in the last 25 years.

these properties are linked to the structure. The initial chapters will emphasize the elastic behavior of ceramics as this is a fundamental mode of deformation in ceramics. Indeed, engineering structural design is primarily based on the elastic properties of materials. At high temperatures, however, ceramics may undergo permanent deformation by processes such as viscous flow, plastic deformation and creep, and these are covered in Chapters 5 to 7. Inevitably, if stresses are increased further, fracture will occur, and Chapter 8 considers the fundamental basis for understanding fracture in brittle materials such as ceramics. Chapter 9 is concerned with the methodologies needed in the engineering design of ceramic components to ensure reliability under the action of mechanical or thermal stresses.

References

M. F. Ashby, *Materials Selection in Mechanical Design*, Pergamon Press, Oxford, UK, 1992.

D. J. Green, Industrial applications of ceramics, pp. 81–122 in *Industrial Materials Science and Engineering,* edited by L. E. Murr, Marcel Dekker, New York, 1984.

D. J. Green, Mechanical behavior of space shuttle thermal protection tiles, pp. 123–43 in *Industrial Materials Science and Engineering,* edited by L. E. Murr, Marcel Dekker, New York, 1984.

G. B. Kenney and H. K. Bowen, High tech ceramics in Japan; current and future markets, *Bull. Am. Ceram. Soc.,* **62** (1983) 590–96.

Chapter 2

Elastic behavior

In describing the mechanical properties of materials, one is interested in understanding the response of the materials to force. For example, consider the forces that are exerted on materials as we walk around. The forces arise because our bodies are being acted on by gravity and this force is acting on each particle of our body. A force that acts on every particle of a body, animate or not, is known as a **body force**. The force produced by our bodies is then transmitted to the floor through our feet. As the force is being transmitted via a surface, it is known as a **surface force**. Now, let us consider what is happening to the floor as we transmit this force. In general, we do not notice much of a reaction but, following Newton's Third Law, we know for every action there is an equal and opposite reaction. In a way this is rather remarkable, as it indicates the floor is pressing back on our feet with exactly the same force as that caused by our weight. If the reactive force was less, we would sink and if it was too high, we would rise. To understand the mechanical properties of materials, it is important to understand how this reaction arises and as materials scientists we are interested in determining whether this reaction can be controlled.

The reaction of a material becomes clearer if one walks across a wooden plank. The force exerted by our body causes the plank to bend and the only logical explanation is that the atoms in the plank have moved to create the reactive force. Once one has completed the journey across the plank (unless fracture intervenes), the plank usually returns to its original position. If a deformation is reversible, it is termed **elastic** and this phenomenon is very common in ceramics and glasses, except at high temperatures. Clearly, many materials are not elastic (**inelastic**) and the forces can create a permanent change in size or shape. For example, if we walk through mud or wet sand we often create, at least temporarily, a lasting impression. There are various types of inelastic responses in materials, such as **plasticity** and **viscosity** and, in some cases, time can be important in the deformation. For example, hanging a weight on a string of chewing gum can cause the **permanent deformation** to increase with time. This type of behavior is known as **creep** and this process often occurs in ceramics and glasses at high temperatures. Ultimately, if the force on a material is high enough, whether the deformation is elastic or inelastic, **fracture** will occur. In the current chapter, we will be concerned with the formal description of elasticity. Inelastic deformation and fracture will be the subject of later chapters.

2.1   Elastic deformation of atomic bonds

An indirect way of demonstrating that atoms move in a material as one applies a force is shown schematically in Fig. 2.1. A polycrystalline material is being bent by a set of forces, while x-ray diffraction is used to determine the spacing of atomic planes inclined to the surface. The interplanar spacing d of a particular set of planes can be determined from the angle θ associated with a particular diffraction peak and the wavelength of the radiation λ using Bragg's Law. i.e., $d=\lambda/(2\sin\theta)$. As the force on the bending specimen increases, the diffraction peak changes its angular position, indicating an increase in interplanar spacing, i.e., the atoms have moved. If the atoms move back to their original positions after the stress is removed, the material is termed **elastic**. As shown in Fig. 2.2, one can plot the form of the relationship between the applied force F and the interplanar spacing d. If this relationship is linear and the material is elastic, the material or the deformation is termed **linear elastic**.

If elastic behavior is associated with the movement of atoms, consider a force

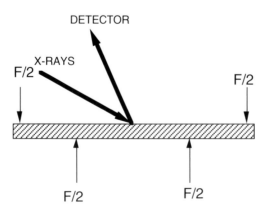

Figure 2.1 Schematic of experimental arrangement for the study of atomic displacements under the application of stress.

f being applied to a pair of atoms, as shown in Fig. 2.3. Ignoring atomic vibrations, the atoms would be at their equilibrium spacing a_0, if $f=0$. In a simple sense, one can consider the atomic bond to be a sort of spring. As f increases, the atoms will move apart or, if the force direction is reversed, the atoms should move closer. If the force increases the interatomic spacing, it is termed **tensile** and if it decreases the spacing, **compressive**. One would expect the ease of this deformation to depend on the chemical nature of the atomic bond and it is of interest to determine how the deformation relates to the interatomic potential ϕ. This potential consists of two important components, an attractive term and a repulsive term, as shown in Fig. 2.4(a). The summation of these two terms leads to the minimum in the potential that is associated with the equilibrium atomic spacing. Clearly, the interatomic potential changes if one displaces the atoms, the displacement u being defined as $(a-a_0)$. In order to transform the potential-displacement relationship to force and displacement, one needs to use $f=d\phi(u)/du$. If one performs this differentiation, a curve such as that illustrated in Fig. 2.4(b) is obtained. In the vicinity of a_0, the slope of the force–displacement function is expected to be approximately linear and it is in this region that one expects to obtain linear elastic behavior.

In order to move the atoms depicted in Fig. 2.3, work has to be performed on the system. For example, if one atom is moved by an amount δu, the work done is $f\,\delta u$, using the **principle of virtual work**. In doing this work, the bond energy has changed by an amount $\Delta\phi=\phi(u+\delta u)-\phi(u)$. As indicated earlier, the equilibrium condition for the deformation is given by $f=d\phi(u)/du$ and thus as δu approaches zero, $\Delta\phi=f\delta u$. Thus, for small displacements, the work done on the atoms is equivalent to the change in bond energy. It is now possible to define a

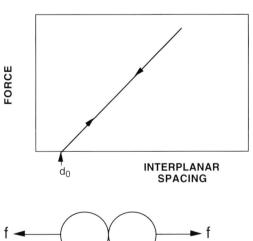

Figure 2.2 Example of possible output from experimental arrangement shown in Fig. 2.1 for a linear elastic material.

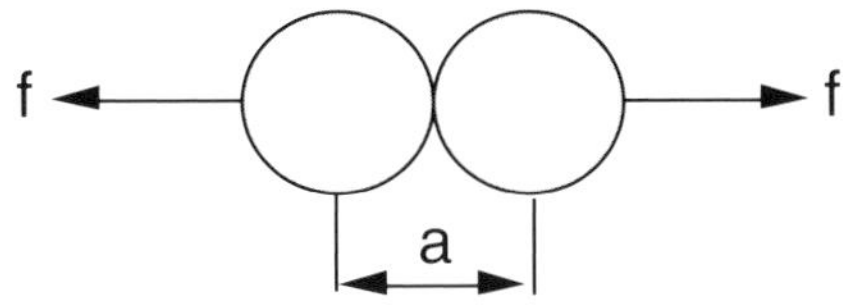

Figure 2.3 Tensile force acting on a pair of atoms.

perfect elastic deformation; it is a deformation that has the following qualities: a) the force is a function of displacement; b) for each displacement there is a corresponding force; and c) it is reversible.

Let us consider the force–displacement relationship for the two atoms in Fig. 2.3 in more detail. First, note that $\phi(u)$ is a continuous function and has a minimum value when $u=0$, implying that $d\phi(u)/du=0$ at the equilibrium spacing. As $\phi(u)$ is continuous, one can expand this function around the equilibrium position as a Taylor series, i.e.,

$$\phi(u)=\phi+\left(\frac{d\phi}{du}\right)_0 u+\left(\frac{d^2\phi}{du^2}\right)_0\left[\frac{u^2}{2}\right]+\ldots \tag{2.1}$$

where the subscript zero is used to represent the value at $u=0$. It has already been indicated that $(d\phi/du)_0=0$ so the second term on the right-hand side vanishes. In addition, for small values of u, one can neglect the term in u^3 and the higher terms. Thus, using $f=d\phi/du$, one obtains

$$f=\left(\frac{d^2\phi}{du^2}\right)_0 u=ku \tag{2.2}$$

i.e., a linear relationship between force and displacement is expected. Thus, for small displacements one predicts **linear elastic** behavior in a material. Moreover,

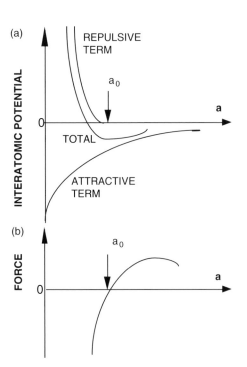

Figure 2.4 Schematic of (a) interatomic potential and (b) force relationships for a pair of atoms as a function of their separation, a. The potential is composed of attractive and repulsive terms.

one can define a proportionality constant, $k = (d^2\phi/du^2)_0$, which is an **elastic constant** for a given material. This constant represents the stiffness of the atomic bond and depends on the curvature of the interatomic potential at $u=0$. For interatomic potentials with a small radius of curvature at $u=0$, k will be large and hence the bond will exhibit a high stiffness. Equation (2.2) is an atomic version of the law known as **Hooke's Law,** which is used to describe linear elastic deformations. An important property of linear elastic deformations is that displacements produced by more than one force can be superposed, provided the forces do not strain the bonds beyond the range of constant curvature in ϕ near $u=0$. This property is known as the **principle of superposition**. The link between elastic behavior and atomic structure has now been established and these ideas will be extended further in Chapter 3.

2.2   Failure of Hooke's Law

Linear elastic behavior is common in macroscopic bodies, especially in ceramics. Indeed, one expects all materials to obey Hooke's Law over some range of displacements. When the deformation of the bonds becomes too large, however, one expects that Hooke's Law will fail to describe the behavior. Figure 2.5 shows the wide range of deformation behaviors that can be found in ceramic materials. Many ceramics, such as Al_2O_3, are linear elastic but ceramic whiskers, such as SiC, can often be so strong that one can no longer assume the displacements are small. Higher-order terms can no longer be neglected in Eq. (2.1) and the force–displacement behavior becomes non-linear but still elastic. If one considers the slope of the force–displacement function in Fig. 2.4(b), one can see the bonds become more compliant for large tensile deformations and stiffer for compressive deformations. **Non-linear elastic** behavior is also found

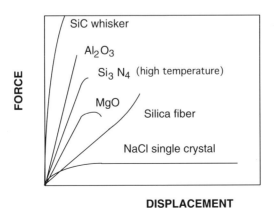

Figure 2.5 Schematic of load–displacement behavior for different ceramics, showing various types of deformation behavior.

in high-strength, silica-glass fibers but in this case tensile deformations lead to an increase in stiffness. This effect has been related to the rotation of the $(SiO_4)^{4-}$ tetrahedra that form the glass structure, giving an increase in stiffness once the rotation has occurred. An important distinction in the various types of non-linear behavior is whether the material remains entirely elastic or whether **non-elastic** or **inelastic** processes intervene. For example, in poly-crystalline MgO at high temperatures (>1800 °C), the initial deformation behavior is linear elastic but once a critical force is passed, termed the **yield point**, the material becomes inelastic. The implication is that inelastic materials have the ability to 'flow'. In the case of MgO and the alkali halides, atoms can slide past each other and this is termed **plastic deformation**. The process by which this occurs will be discussed in more detail in Chapter 6. Inelastic deformation is also found in many covalent ceramics, such as Si_3N_4, at high temperatures. In these cases, the inelasticity is caused by **viscous flow** of a glassy phase, which is often present in the grain boundaries of these materials. Viscous flow will be discussed in Chapter 5. It is clearly important to be able to distinguish non-linear elasticity from inelasticity and this is accomplished by unloading the material. This is shown in Fig. 2.6 for an inelastic material and one finds there is no longer a single functional dependence on load. Some of the deformation is still elastic but there is also a permanent displacement that remains at zero force. If a material was non-linear elastic, the displacement would return to zero.

For the remainder of this chapter, we will return to the subject of linear elasticity and concentrate on the way it is formally described for macroscopic bodies. At first, uniform deformations in a body will be considered but then we will move to using continuum mechanics for the description of non-uniform deformations.

2.3   **Engineering elastic constants**

For large bodies that are linear elastic, the version of Hooke's Law given as Eq. (2.2) is not very useful. Indeed, even the use of force and displacement to describe the deformation becomes inappropriate. Consider the atomic arrangements shown schematically in Fig. 2.7. For the four pairs of atoms in (b), one would need to apply four times the force to get the same displacement as a single pair of atoms in (a). For the string of three atoms in (c), one would apply only one-half the force to get the same displacement as the single pair. This implies Hooke's Law would involve body dimension terms as well as k. This is somewhat awkward as it would be preferable to describe the linear elastic response of bulk materials in the simplest way possible. This problem can be overcome by normalizing the force to the cross-sectional area of the body and the displacement to the length of the body.

Consider a body, initial cross-sectional area A_0 and length L_0 being pulled in uniaxial tension by a force F, as shown in Fig. 2.8. The body is assumed to be an isotropic uniform continuum, (i.e., independent of direction with no internal structure). Under the action of the force, the body will increase in length by an

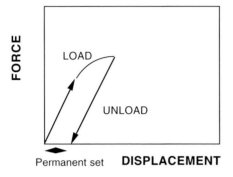

Figure 2.6 Unloading of inelastic materials gives rise to a permanent displacement at zero load.

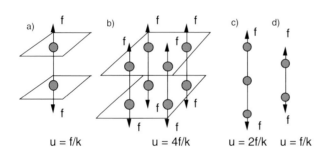

u = f/k   u = 4f/k   u = 2f/k   u = f/k

Figure 2.7 Effect of specimen size on a linear elastic deformation. To obtain the same displacement in b) as in a), one would need four times the applied force. For the same force, one would obtain twice the displacement in c) compared to d).

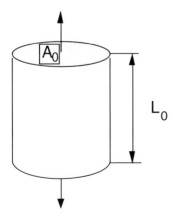

Figure 2.8 Uniaxial tensile deformation of a body.

amount ΔL. Two new parameters, **stress** $\sigma = F/A_0$ and **strain** $\varepsilon = \Delta L/L_0$, can now be defined replacing force and displacement. As a first approximation, let us consider the macroscopic body in Fig. 2.8 as 'strings' of atoms. This ignores interactions between the strings and other longer-range interactions. For one string of atoms, $\varepsilon = u/a_0$ and as the force acting on the string is $f = F/N$, where N is the number of atomic strings, one can rewrite Eq. (2.2) as $F = Nk\varepsilon a_0$ or $\sigma = (N/A)k\varepsilon a_0$. Recognizing that the number of bonds per unit area, $N/A = 1/a_0^2$, one obtains $\sigma = (k/a_0)\varepsilon$. More generally, this is written as

$$\sigma = E\varepsilon \qquad\qquad (2.3)$$

where E is termed the **Young's modulus**, which represents a material constant for a uniaxial tensile deformation. Although the above derivation is approximate and $E = k/a_0$ is only an estimate, this macroscopic version of Hooke's Law is exact for a linear elastic material under a uniaxial tensile or *compressive* stress. It is often forgotten that the simple version of Hooke's Law given in Eq. (2.3) applies only for one type of loading geometry. The application of a more complex stress state will change the form of this equation. Generalized versions of Hooke's Law will be discussed further in Section 2.10. If a uniaxial tension test was performed on a linear elastic material, a response similar to Fig. 2.9 would be obtained, with the slope of the stress-strain curve given by E. In dense polycrystalline ceramics, the value of E is generally in the range of 100–800 GPa and for silicate glasses from 60–80 GPa. The structural factors that influence the magnitude of E will be discussed in Chapter 3. Figure 2.9 also demonstrates the type of behavior obtained if the material remains linear elastic to failure. Such behavior is common in covalently bonded ceramics (e.g., carbides, borides, oxides and nitrides) below ~1000 °C. The points on the two axes corresponding to the failure point are then termed the **fracture stress** (σ_f) and **fracture strain** (ε_f). The fracture stresses of dense ceramics and silicate glasses are often in the range 70–700 MPa but polycrystalline ceramics have been developed recently with fracture stresses (**tensile strengths**) exceeding 2 GPa. Moreover, in the form of single-crystal

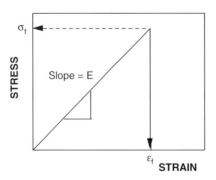

Figure 2.9 Schematic of stress–strain curve for a linear elastic material.

whiskers, the strengths can be even higher (up to ~25 GPa). Using typical numbers for the fracture stress and Young's modulus in Eq. (2.3), fracture strains for polycrystalline ceramics are typically <0.001 (0.1%).

In the uniaxial tension test (Fig. 2.8), there is usually a transverse strain, i.e., a strain perpendicular to the applied stress. This can be used to define a second elastic constant, **Poisson's ratio** (v), as the negative ratio of the transverse strain (ε_T) to the longitudinal strain (ε_L), i.e., $v = -\varepsilon_T/\varepsilon_L$. For isotropic materials, it can be shown from thermodynamic arguments, that $-1 \le v \le 0.5$. For many ceramics and glasses, v is usually in the range 0.18–0.30.

The definitions of stress and strain, given above, do pose a problem in terms of their generality, especially for large deformations. Consider the series of axial tensile deformations shown in Fig. 2.10. We have defined strain by normalizing the longitudinal displacement to L_0; this is termed the **engineering strain** (ε_E). If, however, the initial specimen size was L_3 and the specimens were under uniaxial compression, one would presumably normalize the axial displacements to L_3. This process would imply that the absolute magnitude of the strain in going from L_0 to L_3 would be different than in going from L_3 to L_0. Clearly, one would prefer the two strains to be the same magnitude but with opposite signs. To overcome this problem, one can define strain in terms of the changes in length normalized by the instantaneous length and this is termed the **true strain** (ε_t). With respect to Fig. 2.10, if the true strain is visualized as increments of engineering strain, one can write

$$\varepsilon_t = \left(\frac{L_1 - L_0}{L_0}\right) + \left(\frac{L_2 - L_1}{L_1}\right) + \left(\frac{L_3 - L_2}{L_2}\right) \tag{2.4}$$

By making the increments in the displacements small in such a summation (i.e., $\Delta L = dL$), one obtains

$$\varepsilon_t = \int_{L_0}^{L_F} \frac{dL}{L} = \ln\left(\frac{L_F}{L_0}\right) \tag{2.5}$$

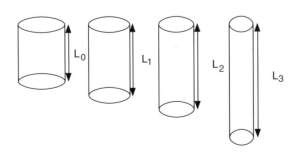

Figure 2.10 A series of deformations on a material to show that engineering strain depends on the stress direction.

where L_0 and L_F are the initial and final lengths of the specimen. From the definition of engineering strain, it is simple to show that $\varepsilon_t = \ln(1 + \varepsilon_E)$. Figure 2.11 shows the two strains in terms of $(\Delta L/L)$ and, for values $<10\%$, the difference between true and engineering strains is found to be very small. Thus, for linear elasticity, the difference is often ignored but in inelastic processes, such as plastic deformation, the difference can become important. There is a similar problem in the definition of stress; our previous definition, normalized by A_0, is termed the **engineering stress**, whereas **true stress** is defined by normalizing the force to the *instantaneous* cross-sectional area. In a similar fashion to strain, the difference between true stress and engineering stress is generally small for linear elastic materials.

In scientific processes one is often concerned with energy changes and thus it is worthwhile considering the energy involved in a linear elastic deformation. Consider the uniaxial tension test shown in Fig. 2.8. The force F, in causing displacements u in the body, is doing work W on the body and it was shown earlier that this work is stored as an increase in the interatomic potential. It is useful to obtain expressions for this **stored elastic strain energy**, which is equivalent to W. In uniaxial tension, the total work done on the body is given by

$$W = \int_0^u F \mathrm{d}u = \int_0^u \sigma A \mathrm{d}u \tag{2.6}$$

For a linear elastic body, using Eq. 2.3, one can write

$$W = \int_0^u EA\varepsilon \mathrm{d}u = \int_0^u EA\left(\frac{u}{L}\right)\mathrm{d}u = \frac{EAu^2}{2L} = \frac{VE\varepsilon^2}{2} \tag{2.7}$$

where V is the volume of the body. Another common energy parameter is the **elastic strain energy density** U, the elastic energy per unit volume, i.e.,

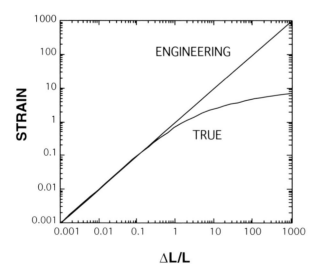

Figure 2.11 Difference between true and engineering strain for various values of fractional length change.

$$U = \frac{W}{V} = \frac{E\varepsilon^2}{2} = \frac{\sigma^2}{2E} = \frac{\sigma\varepsilon}{2} \tag{2.8}$$

In terms of the stress-strain curve shown in Fig. 2.9, the elastic strain energy density for a given stress and strain is equivalent to the area beneath the curve integrated up to those values. It is instructive to consider the magnitude of U that is present in a linear elastic body at fracture. Consider a situation in which $\sigma_f = 800$ MPa and $E = 400$ GPa, one finds from Eq. (2.8) that $U = 800$ kJ/m^3 at fracture. For a body that has a volume, $10 \times 10 \times 10$ mm, $W = 0.8$ J. This value can be better appreciated by considering the change in body energy. Assuming the interatomic spacing is 0.4 nm, the volume associated with the atomic bond is approximately $0.4 \times 0.4 \times 0.4$ nm and thus $W = 5.12 \times 10^{-23}$ J or 3×10^{-4} eV. The energy associated with the primary bonding in ceramic crystals is usually >5 eV and thus one concludes that the change in bond energy is very small, at least on average. The energy is, however, sufficient to have caused fracture. Brittle fracture is the subject of Chapter 8 and it will be shown there why such a small change in average bond energy can still cause failure.

To this point, only one deformation geometry for a linear elastic body has been considered, i.e., uniaxial tension. This led to the definition of two elastic constants, E and v. Engineers and scientists also consider two other primary modes of deformation and these are shown in Fig. 2.12. The first of these, Fig. 2.12(a), is called simple shear and involves a **shape change** in the body. The stress τ is still defined as the force per unit area but the strain is now defined by the change in the angle γ, expressed in radians, which is associated with one of the faces of the body. The shear strain is defined as positive if the angle associated with cartesian axes is reduced, as shown in Fig. 2.12(a). The third elastic constant is the **shear modulus** μ which is defined as the ratio of the stress to the strain, i.e., $\mu = \tau/\gamma$. The final mode of deformation, shown in Fig. 2.12(b), is hydrostatic compression, in which a body is being compressed by a pressure P. The stress associated with this deformation $(-P)$ gives rise to a **volume change** and the strain can be defined as the fractional change in volume $(\Delta V/V)$. To be consistent with our earlier definitions of stress and strain, pressure is denoted as a negative stress and $\Delta V/V$ is negative for a decrease in volume. The ratio of the stress to the strain in hydrostatic compression defines the fourth elastic constant, the

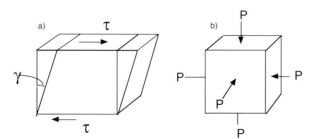

Figure 2.12 Basic deformation mode geometries of a) simple shear and b) hydrostatic compression.

bulk modulus, $B = -P/(\Delta V/V)$. The various elastic moduli for the simple loading geometries have been defined so that the modulus is always a positive quantity. One can easily determine the elastic strain energy density for each of these deformation modes and analogous expressions to Eq. (2.8) are obtained, i.e., $U = \tau^2/2\mu$ and $U = P^2/2B$ for simple shear and hydrostatic compression, respectively. The principle of superposition can also be applied to these elastic strain energy densities. Table 2.1 gives values of the four engineering elastic constants for various ceramic materials. Various trends can be observed in this table and these will be discussed in the next chapter.

2.4   Strain at a point

The definitions of stress and strain developed earlier were for uniform stress states but, often, one has to deal with situations in which the stress and strain are non-uniform and vary from point to point in a body. In these cases, one must consider stresses and strains in a more general way. When a body is loaded in a complex way, different 'particles' of the body will be displaced relative to one another. It is important, therefore, to define both the coordinate of a point and its displacement. The position of a particle P can be defined by its coordinates x_1, x_2, x_3 in a set of cartesian axes X_1, X_2, X_3, which are fixed and independent of the body. As a shorthand, the coordinates can be written as x_i, where $i = 1, 2$ and 3. Suppose, as shown in Fig. 2.13, that the deformation and movement of the body displaces the particle at P to P', such that the new coordinates are $x_1 + u_1$, $x_2 + u_2$ and $x_3 + u_3$. The vector u_i (same subscript notation) is then the displacement of P. There must, however, be a relationship between the vectors u_i and x_i because if u_i was a constant for all the particles in the body, this would only represent a rigid translation of the body and not a deformation. It is the relationship between the two vectors that leads to the concept and precise definition of strain. An essential part of this definition is that u_i varies from one particle to another in a body, that is $u_i = f(x_i)$.

Consider the one-dimensional string shown in Fig. 2.14 with the deformation such that A moves to A', B to B' and the length of the element $AB = dx_1$. If one assumes a continuous variation in the displacement from point to point and ignores the translation of A to A', the displacements of A and B are 0 and $(du_1/dx_1) \, dx_1$ respectively, thus $du_1 = (du_1/dx_1)dx_1$. The relative change in the size of the element (change in length/original length) becomes our definition of **strain** and one can write the strain as $e_{11} = [(du_1/dx_1) \, dx_1]/dx_1 = du_1/dx_1$. That is, strain is a measure of the displacement gradient with respect to position at a point such that $du_1 = e_{11}dx_1$. In this example, no assumptions were made about the functional form of the relationship between displacement and position. If u_1 were a linear function of x_1, the strain would be homogeneous along the string, i.e., e_{11}

Table 2.1 *Elastic constants of selected polycrystalline ceramics (20°C)*

Material	Crystal type	μ (GPa)	B (GPa)	v	E (GPa)
C	cubic	468	417	0.092	1022
SiC	cubic	170	210	0.181	402
TaC	cubic	118	217	0.270	300
TiC	cubic	182	242	0.199	437
ZrC	cubic	170	223	0.196	407
Al_2O_3	trigonal	163	251	0.233	402
$Al_2O_3.MgO$	cubic	107	195	0.268	271
$BaO.TiO_2$	tetragonal	67	177	0.332	178
BeO	tetragonal	165	224	0.204	397
CoO	cubic	70	185	0.332	186
$FeO.Fe_2O_3$	cubic	91	162	0.263	230
Fe_2O_3	trigonal	93	98	0.140	212
MgO	cubic	128	154	0.175	300
$2MgO.SiO_2$	orthorhombic	81	128	0.239	201
MnO	cubic	66	154	0.313	173
SrO	cubic	59	82	0.210	143
$SrO.TiO_2$	cubic	266	183	0.010	538
TiO_2	tetragonal	113	206	0.268	287
UO_2	cubic	87	212	0.319	230
ZnO	hexagonal	45	143	0.358	122
ZrO_2–$12Y_2O_3$	cubic	89	204	0.310	233
SiO_2	trigonal	44	38	0.082	95
CdS	hexagonal	15	59	0.38	42
PbS	cubic	33	62	0.27	84
ZnS	cubic	33	78	0.31	87
PbTe	cubic	22	41	0.27	56
BaF_2	cubic	25	57	0.31	65
CaF_2	cubic	42	88	0.29	108
SrF_2	cubic	35	70	0.29	90
CsBr	cubic	8.8	16	0.26	23
CsCl	cubic	10	18	0.27	25
CsI	cubic	7.1	13	0.27	18
KCl	cubic	10	18	0.27	25
LiF	cubic	48	67	0.21	116
NaBr	cubic	11	19	0.26	28
NaCl	cubic	15	25	0.25	38
NaF	cubic	31	49	0.24	77
NaI	cubic	8.5	15	0.27	20
RbCl	cubic	7.5	16	0.29	21

Note:
All values were calculated from single-crystal data (see Section 3.5).

would be constant. In more general cases, e_{11} would vary from point to point and the strain would be non-uniform.

Consider now a two-dimensional case, in which du_1 is a function of x_1 only and du_2 of x_2 only. The element changes area but not shape (i.e., the angles subtended at the corners remain right angles). This is illustrated in Fig. 2.15 and, using the same approach as above, one obtains $du_1=(\partial u_1/\partial x_1)dx_1$ and $du_2=(\partial u_2/\partial x_2)dx_2$. Thus, the strains are defined as $e_{11}=(\partial u_1/\partial x_1)$ and $e_{22}=(\partial u_2/\partial x_2)$. Partial differentials have now been introduced because displacement changes are being considered with respect to some specific axis. These types of strain are called **normal or extensional strains**; they are positive for tension and negative for compression.

Figure 2.16 shows the case in which the area of the element remains constant and there is only a change in shape. In this example, du_1 is only a function of x_2 and du_2 of x_1. Defining strains as the angles of distortion, one obtains, for small angles, $du_1=e_{12}dx_2$ and $du_2=e_{21}dx_1$, where $e_{12}=(\partial u_1/\partial x_2)$ and $e_{21}=(\partial u_2/\partial x_1)$. These types of strains are called **shear strains** and are considered positive when the deformation rotates one axis towards the other. For a case where there are both shear and normal strains, the displacement equations become $du_1=e_{11}dx_1+e_{12}dx_2$ and $du_2=e_{21}dx_1+e_{22}dx_2$. Extending the argument to three dimen-

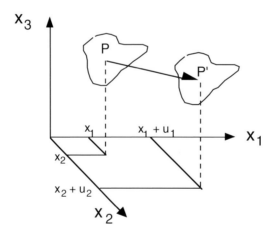

Figure 2.13 Displacement of a point in a body from P to P′.

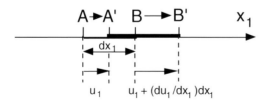

Figure 2.14 Continuum definition of strain for a small element of length dx_1 in a one-dimensional string. In the deformation A moves to A′ and B to B′.

sions is straightforward and, for the case in which a volume element changes both its volume and shape, one obtains

$$du_1 = e_{11} dx_1 + e_{12} dx_2 + e_{13} dx_3 \qquad\qquad (2.9)$$

$$du_2 = e_{21} dx_1 + e_{22} dx_2 + e_{23} dx_3 \qquad\qquad (2.10)$$

$$du_3 = e_{31} dx_1 + e_{32} dx_2 + e_{33} dx_3 \qquad\qquad (2.11)$$

where $e_{ij} = (\partial u_i / \partial x_j)$. For simplicity, a shorthand can be introduced so that Eqs. (2.9)–(2.11) can be written as $du_i = e_{ij} dx_j$, where i or j again have the values 1, 2 and 3 but in addition, a **repeated subscript notation** is being used. In this notation, a summation must be performed for each subscript that is repeated in a term. The values 1, 2 and 3 are used for the repeated subscript. Thus, the term $du_i = e_{ij} x_j$ can be expanded to $du_i = e_{i1} x_1 + e_{i2} x_2 + e_{i3} x_3$ because the j was repeated. After this summation for j, no subscript is repeated in a term. This expanded version must represent three *separate* equations when the three values of $i = 1$, 2 and 3 are inserted and one obtains Eqs. (2.9)–(2.11).

It is worth noting that strain at a point has been considered in terms of the

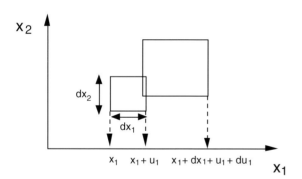

Figure 2.15 Continuum definition of strain for a small two-dimensional element of sides, dx_1 and dx_2 undergoing an area increase.

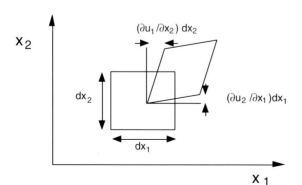

Figure 2.16 Continuum definition of strain for a small two-dimensional element of sides dx_1 and dx_2 undergoing an angular distortion but no change in area.

displacements that occur for a small volume element around that point. In the limit, the gradients of the displacement components with respect to a particular direction become the definition of strain. With this process, nine separate components of strain have been obtained and this is simply a result of relating two vectors (displacement and location), each containing three components. Of the nine e_{ij} components, six are shear strains ($i \neq j$) that measure distortion and three are normal strains ($i=j$) that measure volume change.

There is, however, a problem with our definition of the shear strain, which is illustrated in Fig. 2.17. In particular, e_{ij} does not differentiate between rotations and distortions. For example, if $e_{12}=-e_{21}$, as shown in Fig. 2.17(b), the values of e_{ij} are non-zero even though this only represents a rotation. In order to overcome this problem, it is possible to separate any general shear distortion into a combination of **pure shear** and **rotation** components by writing

$$e_{12}=\varepsilon_{12}+\omega_{12} \text{ and } e_{21}=\varepsilon_{21}+\omega_{21} \tag{2.12}$$

where

$$\varepsilon_{12}=(e_{12}+e_{21})/2, \ \omega_{12}=(e_{12}-e_{21})/2$$
$$\varepsilon_{21}=(e_{21}+e_{12})/2, \ \omega_{21}=(e_{21}-e_{12})/2 \tag{2.13}$$

For pure shear, Fig. 2.17(a), $e_{12}=e_{21}=\varepsilon_{12}=\varepsilon_{21}$ and there is no rotation. For Fig. 2.17(b), $\omega_{12}=e_{12}=-\omega_{21}=-e_{21}$ and, using Eq. (2.13), one finds $\varepsilon_{12}=\varepsilon_{21}=0$, i.e., there is a rotation without any shear strain. Finally, for **simple shear**, shown in Fig. 2.17(c), there is an equal rotation and pure shear, i.e., Eq. (2.13) gives $\varepsilon_{12}=\varepsilon_{21}=\omega_{12}=-\omega_{21}=\gamma/2$. That is, a simple shear is obtained by distorting both the angles equally by $\gamma/2$ and then rotating the element clockwise by $\gamma/2$, as shown in Fig. 2.18. It is important to notice that the earlier definition of shear modulus used γ as the shear strain, whereas in this section distortion is described by a pure shear, $\gamma/2$. For an arbitrary set of strains, Eq. (2.13) allows a general

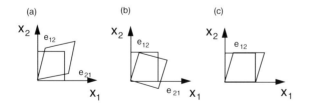

Figure 2.17 Various types of deformation displacements: a) pure shear ($e_{12}=e_{21}$); b) rotation ($e_{12}=-e_{21}$); and c) simple shear ($e_{12}=\gamma$, $e_{21}=0$).

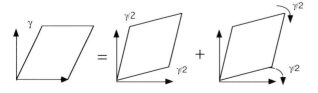

Figure 2.18 A simple shear γ is equivalent to a pure shear of $\gamma/2$ plus a rotation of $\gamma/2$.

shear strain to be split into a pure shear part by averaging the angle changes (half their sum), and a rotation part.

Generalizing Eq. (2.13), using $e_{ij}=(\partial u_i/\partial x_j)$, the nine strain and nine rotation components can be expressed as

$$\varepsilon_{ij}=\frac{1}{2}\left(\frac{\partial u_i}{\partial x_j}+\frac{\partial u_j}{\partial x_i}\right) \tag{2.14}$$

$$\omega_{ij}=\frac{1}{2}\left(\frac{\partial u_i}{\partial x_j}-\frac{\partial u_j}{\partial x_i}\right) \tag{2.15}$$

From these definitions, ε_{ij} represents normal strains when $i=j$ and pure shear strains when $i\neq j$. Moreover, $\varepsilon_{ij}=\varepsilon_{ji}$, so that only six of the components are **independent** and the array of components is said to be **symmetric**. The components are sometimes written down in a matrix type of notation, i.e.,

$$\varepsilon_{ij}=\begin{bmatrix} \varepsilon_{11} & \varepsilon_{12} & \varepsilon_{13} \\ \varepsilon_{21} & \varepsilon_{22} & \varepsilon_{23} \\ \varepsilon_{31} & \varepsilon_{32} & \varepsilon_{33} \end{bmatrix} \tag{2.16}$$

For the rotation components, $\omega_{ij}=0$ when $i=j$ and $\omega_{ij}=-\omega_{ji}$ when $i\neq j$, i.e., there are only three independent components (one for each axis). For the case in which the off-diagonal components are the negative of each other, the set of components is termed **antisymmetric**. In some cases, there are only normal strains at a point. These are termed **principal strains** and the axes that describe these principal values are called the **principal axes**. Consider a situation in which there are only principal strains, as illustrated in Fig. 2.19. The small element that is undergoing these strains is pictured as a cube with sides of unit length. After straining, the element has sides of length $(1+\varepsilon_{11})$, $(1+\varepsilon_{22})$ and $(1+\varepsilon_{33})$. The fractional change in the volume of the cube $\Delta V/V=[(1+\varepsilon_{11})(1+\varepsilon_{22})(1+\varepsilon_{33})-1]/1$. For small strains, any cross terms in strain can be ignored and $\Delta V/V=\varepsilon_{11}+\varepsilon_{22}+\varepsilon_{33}$, i.e., the

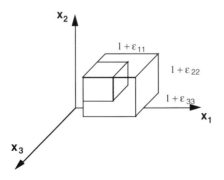

Figure 2.19 Volume element (unit cube) undergoing a volume dilatation with no angular distortion.

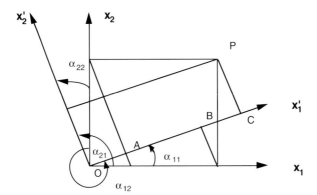

Figure 2.20 Two-dimensional rotation of axes around x_3, and associated angles α_{ij}.

sum of the principal strains. Using the repeated index notation, one can write this sum as ε_{kk}.

A final point to consider in our definition of strain is that the normal and shear components were summed to obtain a displacement in a given direction (Eqs. (2.9)–(2.11)). For large or **finite strains**, one would expect the displacements produced by one component to influence the others. For example, if a body were given a simple shear e_{12} and then, with du_1 held constant, one superimposed a large normal strain e_{22}, the angle of shear γ would change. If, however, both these strains were small (**infinitesimal strain**), the change in γ would be negligible. For this case, the strain components can be superimposed without difficulty.

2.5   **Transformation of strains**

It is often important to be able to determine what happens to a set of strain components when the cartesian axes are rotated. We will now denote these axes by lower case x_1, x_2, x_3. For simplicity, only a rotation in the x_1, x_2 plane will be considered, i.e., around the x_3 axis, as shown in Fig. 2.20. If a set of strains ε_{ij} with respect to the x_i axes is given, one needs to determine the strains ε'_{ij} with respect to the x'_i axes. The first step is to determine the displacements u_i from the strains ε_{ij}. To simplify the mathematics, a state of homogeneous strain will be assumed. From Eqs. (2.9)–(2.11), ignoring the rotation components ω_{ij} that do not contribute to ε_{ij} or ε'_{ij}, one obtains $u_i = \varepsilon_{ij} x_j$. The second step is to define the axes of rotation and this is done by defining a set of nine components, a_{ij}, the direction cosines of the angles associated with the rotation. As shown in Fig. 2.20, the angles of the new axes are defined by an anti-clockwise rotation from the original axes, x_i, to the new axes, x'_i. The angle α_{12} is that between the x_2 and x'_1 giving $a_{12} = \cos\alpha_{12}$. The other direction cosines are defined in a similar fashion so that $a_{ij} = \cos\alpha_{ij}$.[†] The third step in the procedure is to find the displacements u' with

[†] The rotation of the axes is from the old to the new but the subscripts are written in the order new-old.

respect to the new axes, from the original displacements. This is done by constructing various perpendiculars which meet the x_1' axis at A, B and C. One finds that $x_1' = OC = OB + BC = OB + OA = x_1 \cos \alpha_{11} + x_2 \cos \alpha_{12} = a_{11}x_1 + a_{12}x_2$. Continuing this approach to the other coordinates, one obtains

$$x_i' = a_{ij}x_j \text{ or } x_i = a_{ji}x_j' \tag{2.17}$$

Considering the displacement u_1' with respect to x_1', one finds there are three contributions, $a_{11}u_1$, $a_{12}u_2$ and $a_{13}u_3$, which with $u_i = \varepsilon_{ij}x_j$ gives

$$u_1' = a_{1i}u_i = a_{1i}\varepsilon_{ij}x_j \tag{2.18}$$

Repeating this procedure for all the new displacement components one obtains

$$u_k' = a_{ki}u_i = a_{ki}\varepsilon_{ij}x_j \tag{2.19}$$

Using Eq. (2.17), this can be rewritten as

$$u_k' = a_{ki}a_{lj}\varepsilon_{ij}x_l' \tag{2.20}$$

From the definition of strain, however, one can write

$$u_k' = \varepsilon_{kl}'x_l' \tag{2.21}$$

Equating Eqs. (2.20) and (2.21), one obtains the **transformation equation** for the strain components, i.e.,

$$\varepsilon_{kl}' = a_{ki}a_{lj}\varepsilon_{ij} \tag{2.22}$$

With some further manipulation, it can be shown that

$$e_{kl} = a_{ik}a_{jl}\varepsilon_{ij}' \tag{2.23}$$

These transformation equations show that as the axes are rotated, the magnitude of the strain components will change, i.e., the state of strain at a point depends on the reference axes. Consider the example of a transformation, depicted in Fig. 2.21, in which a square (bold) is elongated by a strain ε along the x_1 axis and is contracted by ε along the x_2 axis. The problem is to determine the deformation of the (bold) diamond shape that is inscribed at 45° in the square. From the rotation of the axes, one finds that $a_{11} = a_{22} = a_{12} = -a_{21} = 1/\sqrt{2}$. Using Eq. (2.22) for the case in which the only strains are ε_{11} and ε_{22}, one obtains

$$e'_{kl} = a_{k1}a_{l1}\varepsilon_{11} + a_{k2}a_{l2}\varepsilon_{22} \qquad (2.24)$$

Substituting in the direction cosines and the values of ε_{11} and ε_{22}, one obtains $\varepsilon'_{12} = \varepsilon'_{21} = -\varepsilon$, with the remaining components of e'_{kl} being zero. Thus, although the deformation involved only normal strains with respect to the original axes, it is a pure shear with respect to the new axes. Thus, if the square and the diamond were being considered as small elements around a point at their centers, the magnitude of the strain components and the apparent type of deformation is found to depend on the orientation of the axes. An important theorem concerning this variation of the strain components is that there is *always* one position of the axes (principal axes) in which all the components except ε_{11}, ε_{22} and ε_{33} vanish, i.e., only the principal strains remain. These strains, which involve only a volume change, are usually the easiest for most people to visualize.

Let us consider a two-dimensional state of strain and a rotation about x_3, as shown in Fig. 2.22. A unit square OABC is deformed into OPQR by the strains ε_{11}, ε_{22} and $\varepsilon_{12} (= \varepsilon_{21})$. These strains will be referred to a new set of axes x'_1 and x'_2, that are formed by rotating the original axes by an angle θ. The direction cosines of the transformed axes are $a_{11} = a_{22} = \cos\theta$ and $a_{12} = -a_{21} = \sin\theta$. The shear strain component is then found from Eq. (2.22), i.e.,

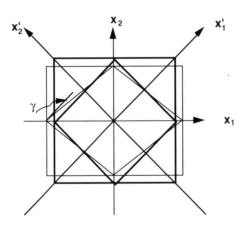

Figure 2.21 Shear of diamond shape inscribed in a square undergoing equal and opposite extensions along the x_1 and x_2 axes.

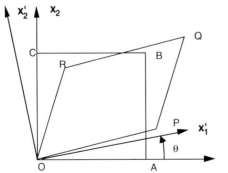

Figure 2.22. Two-dimensional strain in which a unit square OABC deforms to OPQR.

$$\varepsilon'_{12}=a_{11}a_{21}\varepsilon_{11}+a_{11}a_{22}\varepsilon_{12}+a_{12}a_{21}\varepsilon_{21}+a_{12}a_{22}\varepsilon_{22}$$
$$=[(\varepsilon_{22}-\varepsilon_{11})\sin 2\theta]/2+\varepsilon_{12}\cos 2\theta \qquad (2.25)$$

This shear component vanishes when

$$\tan 2\theta=\frac{2\varepsilon_{12}}{(\varepsilon_{11}-\varepsilon_{22})}=q \qquad (2.26)$$

Thus, the angle required to rotate the original axes to the principal axes is when the angle θ is given by $(\tan^{-1}q)/2$ or $(\pi/2)+(\tan^{-1}q)/2$. If one uses Eq. (2.22) to determine the other strain components, it is found that $(\varepsilon_{11}+\varepsilon_{22})=(\varepsilon'_{11}+\varepsilon'_{22})$ and $(\varepsilon_{11}\,\varepsilon_{22}-\varepsilon_{12}\,\varepsilon_{12})=(\varepsilon'_{11}\,\varepsilon'_{22}-\varepsilon'_{12}\,\varepsilon'_{12})$. This implies no matter how the axes are rotated, these combinations of components do not change. For this reason $(\varepsilon_{11}+\varepsilon_{22})$ and $(\varepsilon_{11}\,\varepsilon_{22}-\varepsilon_{12}\,\varepsilon_{12})$ are called the **first and second invariants of strain**, I_{ε} and II_{ε}, respectively. Later, it will be shown that these invariants become more complicated for a general three-dimensional state of strain and that there is an additional invariant, III_{ε}. For the case in which Eq. (2.25) is set to zero, one can obtain a simple quadratic equation that allows the invariants to be used to solve for the principal strains (ε^*), i.e.,

$$\varepsilon^{*2}-I_{\varepsilon}\,\varepsilon^*+II_{\varepsilon}=0 \qquad (2.27)$$

Alternatively, if Eq. (2.27) is expanded, one obtains

$$\varepsilon^*_{11} \text{ or } \varepsilon^*_{22}=\frac{(\varepsilon_{11}+\varepsilon_{22})\pm\sqrt{(\varepsilon_{11}+\varepsilon_{22})^2-4(\varepsilon_{11}\varepsilon_{22}-\varepsilon^2_{12})}}{2} \qquad (2.28)$$

It is sometimes useful to visualize the changes in a two-dimensional set of strain components by using a construction called **Mohr's circle**. Consider the rotation of the axes pictured in Fig. 2.20. If $\alpha_{11}=\theta$, then $a_{11}=a_{22}=\cos\theta$, $a_{12}=-a_{21}=\sin\theta$, $a_{33}=1$ and the remaining direction cosines are zero. If x_1 and x_2 are principal axes, the new strain components can be found from Eq. (2.22), i.e.,

$$\varepsilon'_{11}=\varepsilon^*_{11}\cos^2\theta+\varepsilon^*_{22}\sin^2\theta=(\varepsilon^*_{11}+\varepsilon^*_{22})/2-[(\varepsilon^*_{22}-\varepsilon^*_{11})\cos 2\theta]/2$$
$$\varepsilon'_{22}=\varepsilon^*_{11}\sin^2\theta+\varepsilon^*_{22}\cos^2\theta=(\varepsilon^*_{11}+\varepsilon^*_{22})/2+[(\varepsilon^*_{22}-\varepsilon^*_{11})\cos 2\theta]/2$$
$$\varepsilon'_{12}=-\varepsilon^*_{11}\sin\theta\cos\theta+\varepsilon^*_{22}\sin\theta\cos\theta=[(\varepsilon^*_{22}-\varepsilon^*_{11})\sin 2\theta]/2 \qquad (2.29)$$

These relationships can be represented in a construction (Fig. 2.23), in which the normal strains, ε'_{11} and ε'_{22}, are measured along the horizontal axis and the shear strain ε'_{12} along the vertical axis. Continuing the case under consideration, in which one starts from the principal strains and wishes to determine the new components for some arbitrary rotation, one first plots ε^*_{11} and ε^*_{22} on the horizontal axis as points A and B. The same sign convention for strains is used, so that negative values are plotted on the negative part of the axis and conversely for the positive values. The next step is to draw a circle with AB as the diameter. A line DCE is then drawn at an angle 2θ, measured anti-clockwise about C, where θ is the rotation angle of interest. Perpendicular lines are then drawn from D and E to meet the horizontal axis at F and G. From the geometry of this construction, it is easy to show that $OC=(\varepsilon_{11}^*+\varepsilon_{22}^*)/2$, $CG=FC=\cos 2\theta(\varepsilon_{22}^*-\varepsilon_{11}^*)/2$ and $FD=EG=\sin 2\theta(\varepsilon_{22}^*-\varepsilon_{11}^*)/2$. Comparing these values with Eq. (2.29), one finds $\varepsilon_{11}'=OC-CF=OF$, $\varepsilon_{22}'=OC+CG=OG$ and $\varepsilon_{12}'=EG$. It is important to note that one did not have to start from the principal axes. The points D, E, F and G can be plotted and a circle can then be drawn that passes through E and has its center at the mid-point between F and G. Mohr's circle allows the strains to be determined for some arbitrary rotation or it can be used to determine the principal strains, i.e., from the intersections of the circle with the normal strain axis. One can see in Fig. 2.23 that the maximum shear strain is given by $(\varepsilon_{22}^*-\varepsilon_{11}^*)/2$ and this occurs when the angle of rotation from the principal axes is 45°. There is a three-dimensional version of the Mohr's circle that can be found in more advanced texts. It is, however, useful to realize that the two-dimensional version can be used if x_3 is a principal axis and one is only concerned with rotations about this axis. For this case $(\varepsilon_{22}^*-\varepsilon_{11}^*)/2$ only represents the maximum shear strain in the x_1-x_2 plane. For the three-dimensional case, the maximum shear strain could involve ε_{33}^*.

2.6   Dilatational and deviatoric strains

It has been shown that the magnitude of the normal and shear strain components depends on the choice of axes. Thus, even if there are only principal components, one cannot say there is no shear distortion in the material. Another way is needed to determine how a particular state of strain can be separated into the parts associated with distortion and volume change, respectively. This is accomplished by splitting the strain components into two groups, a dilatational set and a deviatoric set. It was shown previously that $\Delta V/V=\varepsilon_{kk}$ and thus the mean

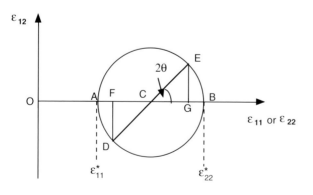

Figure 2.23 Two-dimensional Mohr's circle construction.

normal strain associated with each axis is $\varepsilon_{kk}/3$. Thus, if $\varepsilon_{kk}/3$ is subtracted from each of the normal strains one will be left with the deviatoric strain components (the shear components remain unchanged). This can be written as

$$\varepsilon_{ij} = \left(\frac{\delta_{ij}\varepsilon_{kk}}{3}\right) + \left(\varepsilon_{ij} - \frac{\delta_{ij}\varepsilon_{kk}}{3}\right) = \varepsilon_{ij}^{S} + \varepsilon_{ij}^{D} \qquad (2.30)$$

in which the first term on the right-hand side of the equation represents the **dilatational** components ε_{ij}^{S} and the second term, the **deviatoric** components (ε_{ij}^{D}). Equation (2.30) introduces a new symbol δ_{ij}, which is called the **Kronecker delta**. This parameter represents nine components for the various combinations of i and j, but has the property that $\delta_{ij}=1$ when $i=j$ and $\delta_{ij}=0$ when $i \neq j$. For example, from Eq. (2.30), one would obtain $\varepsilon_{11} = (\varepsilon_{kk}/3) + (\varepsilon_{11} - \varepsilon_{kk}/3)$, $\varepsilon_{12} = (0) + (\varepsilon_{12})$, etc., and the original set of components splits into two new sets of nine components.

2.7   **Strain compatibility**

There is one final concept that is important when understanding strains in a body. If a body is free from strain, the body can be 'cut' into small elements and if necessary, one could 'glue' these elements back together, to perfectly form the body, i.e., the elements are said to be mutually **compatible**. This compatibility must also exist for a body under strain so that the displacements vary smoothly throughout the body, i.e., u_i must be continuous, single-valued functions of x_i. This places restrictions on the strain components and these are called the **compatibility equations**. Another way to think of these equations is that the definition of ε_{ij} represents *six* independent differential equations that are used to determine the *three* components of displacement. This means the system is 'over-determined' and will not, in general, produce a solution for an arbitrary set of strains. There are six distinct compatibility equations that give the necessary and sufficient conditions for these restrictions. Two of these restrictions are given by

$$\left(\frac{\partial^2 \varepsilon_{11}}{\partial x_2^2}\right) + \left(\frac{\partial^2 \varepsilon_{22}}{\partial x_1^2}\right) = \left(\frac{2\partial^2 \varepsilon_{12}}{\partial x_1 \partial x_2}\right) \qquad (2.31)$$

$$\left(\frac{\partial^2 \varepsilon_{11}}{\partial x_2 \partial x_3}\right) = \frac{\partial}{\partial x_1}\left(\frac{-\partial \varepsilon_{23}}{\partial x_1} + \frac{\partial \varepsilon_{31}}{\partial x_2} + \frac{\partial \varepsilon_{12}}{\partial x_3}\right) \qquad (2.32)$$

The other four are obtained by permuting the subscripts 1, 2, 3. The compatibility equations are important in obtaining elastic solutions and will be discussed further in Chapter 4.

2.8   **Tensors**

In describing the properties of materials one is concerned with studying the response of a material (e.g., strain, current flow, heat flow, etc.) to a stimulus (e.g., stress, electric field or thermal gradients). This scientific approach often leads to a description of a material property, i.e., response=property×stimulus. The complexity of these relationships will depend on the ease of describing the stimuli and responses. In the simplest case, stimuli and responses are non-directional (e.g., volume, temperature, etc.) and these are called **scalars**. There is, however, an alternative name, **zero-order tensors**. The next level of complexity are **vectors or first-order tensors** (e.g., force, displacement, momentum, electric field, thermal gradients), and one requires three numbers (e.g., the components along the coordinate axes) to specify their value at a point. The next level of complexity involves **second-order tensors** and in general, one needs nine components for their description, though symmetry can reduce the necessary number of components to six. Nine components are needed because second-order tensors involve the interrelationship between two vectors each with three components ($3\times3=9$). Thus, the analysis already outlined for the description of strain led to the need for a second order tensor, i.e., one that relates the displacement vector (u_i) to the position vector (x_j). Clearly, one could continue the increase in complexity and this gives rise to higher-order tensors. For example, if one wished to relate two second-order tensors as responses and stimuli, one would need a **fourth-order tensor** that would contain, in general, 81 components (9×9). The repeated index notation, introduced earlier, is very useful in allowing the order of a tensor to be identified. This is accomplished by counting the number of **free subscripts** in a term and this number is equivalent to the order of the tensor. A free subscript is simply one that is *not repeated*. For example, the two terms S_{ii} and $T_{ij}T_{ij}$, must be scalars, V_i, V_iT_{ij} vectors, and T_{ij}, v_ix_j, second-order tensors. Tensors may be physical quantities, such as strain, but in some cases they may represent material properties and, hence, are of particular interest to the materials scientist. A simple example of a tensor that represents a material property is α_{ij}, the **linear coefficients of thermal expansion**.

2.9   **Coefficients of thermal expansion**

Suppose a body is uniformly heated to increase its temperature by ΔT. The stimulus is a scalar (ΔT), the response (strain) is a second-order tensor (ε_{ij}). The connecting property must therefore be a second-order tensor. This tensor represents a material property and the components are called the **linear coefficients of thermal expansion,** α_{ij}, i.e.,

$$\varepsilon_{ij} = \alpha_{ij} \Delta T \qquad\qquad\qquad (2.33)$$

Note that both sides of the equation have the same two free subscripts, indicating each side is a second-order tensor. Multiplying a second-order tensor by a scalar gives us a second-order tensor; each expansion coefficient is simply multiplied by ΔT to obtain the corresponding strain component. The tensor ε_{ij} has already been shown to be symmetric, thus α_{ij} must also be symmetric. In addition, it was shown that ε_{ij} can be referred always to its principal axes by an appropriate rotation of the axes. For this particular case, one would only need three expansion coefficients, the diagonal components, α_{11}, α_{22}, α_{33}. For single crystals, the principal axes for the expansion coefficients coincide with the ones used in crystallography.[†] Thus, if one imagines a spherical region in a crystal and the temperature is increased by ΔT, the sphere becomes an ellipsoid, with axes, $(1 + \alpha_{11}\Delta T)$, $(1 + \alpha_{22}\Delta T)$, $(1 + \alpha_{33}\Delta T)$. The number of thermal expansion coefficients needed for a particular crystal type depends on its symmetry and may be less than three. For example, in *cubic* single crystals, the thermal expansion along the principal axes $<100>$ must be identical, as the arrangement of atoms along these directions is identical. For this case, $\alpha_{11} = \alpha_{22} = \alpha_{33}$ and a spherical region expands or contracts uniformly (remaining spherical). Thus, cubic crystals possess **isotropic thermal expansion**, whereas *non-cubic* crystals will have **anisotropic thermal expansion**. This leads to an important effect in materials, especially in ceramic science, in which one often deals with polycrystalline materials. A polycrystalline ceramic often consists of a *random* array of single crystals. If these crystals are non-cubic and the temperature is changed, each crystal will attempt to strain differently than its neighbors. Thus, adjacent crystals will 'push' or 'pull' on each other, creating **residual strains** in the material. These strains are termed residual because they exist even in the absence of *applied* stresses. In brittle materials, these **residual stresses** or **'self-stresses'** can be very high. In some cases, this effect can give rise to cracking and the material may even spontaneously disintegrate. Indeed, this may happen as the material is first cooled down after its (high-temperature) fabrication process. For example, during a sintering process, the material 'flows' as densification occurs and stresses can relax fairly easily at that stage. As the material cools down, however, the residual strains will increase once the material is below the temperature at which stress relaxation cannot occur rapidly.

Table 2.2 gives the thermal expansion coefficients for various ceramic crystals. For example, consider the strain (ε_m) that would arise on cooling $Al_2O_3.TiO_2$ to room temperature, i.e., $\varepsilon_m = \alpha_d \Delta T$, where α_d is the maximum deviatoric component of the thermal expansion. Assuming these strains arise over a temperature

---

[†] Symmetry is also important in composite materials. In orthotropic materials, for example, the principal axes for thermal expansion coincide with the axes of symmetry for the material.

Table 2.2 *Thermal expansion coefficients of selected single crystals*

Crystal	System	α_{11}	α_{22}	α_{33}	Average
20–800 °C					
C	hexagonal	0.0	0.0	28.5	9.5
C	cubic	3.2	3.2	3.2	3.2
Si	cubic	3.8	3.8	3.8	3.8
Ge	cubic	7.1	7.1	7.1	7.1
α-SiC	hexagonal	5.2	5.2	4.8	5.1
β-SiC	cubic	5.2	5.2	5.2	5.2
TiC	cubic	7.4	7.4	7.4	7.4
ZrC	cubic	6.4	6.4	6.4	6.4
HfC	cubic	6.2	6.2	6.2	6.2
NbC	cubic	6.9	6.9	6.9	6.9
TaC	cubic	6.5	6.5	6.5	6.5
WC	hexagonal	4.9	4.9	4.2	4.7
Cr_3Si	cubic	11.5	11.5	11.5	11.5
Mo_3Si	cubic	6.6	6.6	6.6	6.6
$MoSi_2$	tetragonal	7.8	7.8	9.7	8.4
WSi_2	tetragonal	7.5	7.5	9.7	8.2
BeO	hexagonal	8.5	8.5	7.9	8.3
MgO	cubic	13.5	13.5	13.5	13.5
CaO	cubic	13.3	13.3	13.3	13.3
$MgO.Al_2O_3$	cubic	8.6	8.6	8.6	8.6
$FeO.Fe_2O_3$	cubic	13.8	13.8	13.8	13.8
ThO_2	cubic	9.2	9.2	9.2	9.2
Al_2O_3	trigonal	8.1	8.1	8.8	8.3
ZrO_2	monoclinic	7.5	1.1	12.5	7.0
HfO_2	monoclinic	6.8	1.2	11.1	6.4
TiO_2	tetragonal	8.1	8.1	10.1	8.8
$ZrO_2 (2\% Y_2O_3)$	tetragonal	9.0	9.0	13.0	10.3
$ZrO_2 (3\% Y_2O_3)$	tetragonal	10.1	10.1	11.6	10.6
20–1020 °C					
$MgO.2TiO_2$	orthorhombic	2.3	10.8	15.9	9.7
$Al_2O_3.TiO_2$	orthorhombic	−3.0	11.8	21.8	10.2
$Fe_2O_3.TiO_2$	orthorhombic	0.6	10.1	16.3	9.0
300–900 °C					
$Li_2O.Al_2O_3.2SiO_2$	hexagonal	8.2	8.2	−17.6	−0.4
$Al_2O_3.SiO_2$	orthorhombic	2.3	7.6	4.8	4.9
$3Al_2O_3.2SiO_2$	orthorhombic	3.9	7.0	5.8	5.5
$2Al_2O_3.SiO_2$	orthorhombic	4.1	5.6	6.1	5.3
$ZrO_2.SiO_2$	tetragonal	3.6	3.6	6.2	4.5
ZnS	hexagonal	7.8	7.8	7.2	7.6

Table 2.2 (*cont.*)

Crystal	System	α_{11}	α_{22}	α_{33}	Average
ZnS	cubic	7.1	7.1	7.1	7.1
ZnSe	cubic	7.0	7.0	7.0	7.0
PbS	cubic	20.4	20.4	20.4	20.4
PbSe	cubic	20.6	20.6	20.6	20.6
NaCl	cubic	42.2	42.2	42.2	42.2
KCl	cubic	36.9	36.9	36.9	36.9
KBr	cubic	42.0	42.0	42.0	42.0
KI	cubic	37.8	37.8	37.8	37.8
RbCl	cubic	38.5	38.5	38.5	38.5
RbBr	cubic	40.7	40.7	40.7	40.7

interval of 1000 °C, one finds $\varepsilon_m = -13.2 \times 10^{-6}(1000) = -0.0132$ (1.32%). This is a very high strain for a ceramic material when compared to a typical macroscopic fracture strain (see Section 2.3). It is therefore no surprise that this could lead to failure, and this is termed **spontaneous microcracking**. There is, however, an unusual effect in ceramic materials in that the onset of microcracking depends on the grain (crystal) size. If the grain size is below a critical value, no spontaneous microcracking occurs. Thus, microcrack-free ceramics can be fabricated from materials with large thermal expansion anisotropy, provided the grain size is kept below the **critical grain size** for microcracking.

For (random) polycrystals, the thermal expansion coefficient is often estimated by averaging the single-crystal values and these values are included in Table 2.2. In some cases, microcracking caused by thermal expansion anisotropy can influence the overall expansion behavior (see Section 3.7). Although not discussed here, it is also important to note that thermal expansion coefficients often vary with temperature in many ceramic systems.

At this point, it is useful to review the above discussion of second-order tensors, especially in relationship to the ideas put forward on strain and thermal expansion. A second-order tensor (T_{ij}) can be defined as a physical quantity with nine components that transform, such that $T'_{ij} = a_{ik}a_{jl}T_{kl}$ or perhaps more simply, as an operator that relates two vectors. There are two special types of second-order tensors: **symmetric**, in which $T_{ij} = T_{ji}$; and **antisymmetric**, in which $T_{ij} = -T_{ji}$ for $i \neq j$. There are some properties of these tensors that are important to this text and these are given here without proof.

1. Second-order tensors can be split into symmetric and antisymmetric parts, e.g., in Section 2.4, e_{ij} was split into ε_{ij} and ω_{ij}.
2. The symmetry does not depend on the choice of axes.

3. Any symmetric second-order tensor can be transformed by a rotation of axes to a form in which the tensor only contains the three diagonal components, the off-diagonal components being zero. For this state, the diagonal components are called the principal values and the particular set of axes are called the principal axes.

4. Any symmetric second-order tensor (T_{ij}) has three scalar invariants (i.e., they do not change with axis rotation) and they are given by

$$I_T = T_{ii}; \quad II_T = (T_{ii}T_{jj} - T_{ij}T_{ij})/2; \quad III_T = E_{ijk}T_{1i}T_{2j}T_{3k} \tag{2.34}$$

where E_{ijk} is the called the **permutation symbol**. E_{ijk} has a value of 1 when the subscripts are in the order 123, 231, 312, a value of -1 for any other order and is zero if a subscript is repeated.[†] The invariants can be used to determine the principal values (T^*) of a symmetric second-order tensor using

$$T^{*3} - I_T T^{*2} + II_T T^* - III_T = 0 \tag{2.35}$$

Solving this cubic equation gives three values for T^*, i.e., the three principal values. It can be shown that if the second-order tensor is symmetric, the roots of the cubic equation must be real and the axes must be orthogonal. An example to show how Eq. (2.35) is derived will be given in the next section.

2.10   Definition of stress

Stress is the state of force produced throughout a body by the mutual interactions of the particles in their displaced positions. To define stress at a point, consider a small cube of material around that point, as shown in Fig. 2.24. To obtain the stress components, the forces on each face are resolved into components parallel to the coordinate axes. These forces are then divided by the area of the face on which they are acting. This gives nine components, and collectively these are called the stress tensor, σ_{ij}, i.e.,

$$\sigma_{ij} = \begin{bmatrix} \sigma_{11} & \sigma_{12} & \sigma_{13} \\ \sigma_{21} & \sigma_{22} & \sigma_{23} \\ \sigma_{31} & \sigma_{32} & \sigma_{33} \end{bmatrix} \tag{2.36}$$

[†] For Eq. (2.34), it can be shown the third invariant is equivalent to the determinant of the components of the tensor.

The first subscript represents the direction of the face normal on which the stress is acting and the second subscript represents the direction of the stress. The diagonal components are the normal stresses and they are positive in tension and negative in compression. The off-diagonal components are the shear stresses and these are positive if they are directed along a positive axis whose outward normal also points along a positive axis. In Fig. 2.24, the shear stresses are shown as positive.

Although there are nine stress components, one can show that only six are independent because stress, like strain, is a symmetric second-order tensor. Consider a cube with sides of length δL, in which there are two opposing couples, one due to σ_{12} and one due to σ_{21}, as shown in Fig. 2.25. The couples are given by $(\sigma_{12})(\delta L)^3$ and $(\sigma_{21})(\delta L)^3$, respectively. In equilibrium, the couples must balance and hence $\sigma_{12}=\sigma_{21}$. The repetition of this procedure for each shear stress gives $\sigma_{ij}=\sigma_{ji}$.

Just as it is useful to break down strain components into dilatational and deviatoric parts, one can perform a similar procedure with stress. The **hydrostatic stress components** are $\sigma_{ij}^{H}=(\delta_{ij}\sigma_{kk}/3)$ and the **deviatoric stress components**, $\sigma_{ij}^{D}=\sigma_{ij}-(\delta_{ij}\sigma_{kk}/3)$ so that $\sigma_{ij}=\sigma_{ij}^{H}+\sigma_{ij}^{D}$.

When stresses are applied to a body, it is often important to determine the stresses acting normal or parallel to a particular plane at a point. Using Fig. 2.26, consider the tetrahedron OABC with an arbitrary face ABC on which a force F is found. From the stress tensor, one can find the surface forces on planes OBC, OCA and OAB and, using the equilibrium of forces, determine the force

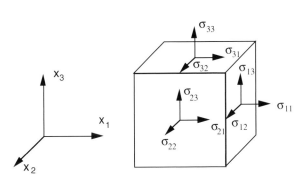

Figure 2.24 Stress components acting on a small element.

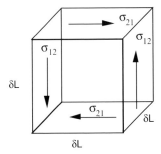

Figure 2.25 Equilibrium leads to symmetry in off-diagonal stress components.

on ABC. The normal to the plane ABC is denoted by the line ON, which is inclined at angles α_1, α_2 and α_3 to the coordinate axes. If the direction cosines are a_i, i.e., $a_1 = \cos \alpha_1$, etc., and if the plane ABC is considered to have unit area, the areas of OBC, OCA and OAB are a_1, a_2 and a_3, respectively. It is now possible to determine the three components of the force F with respect to the cartesian axes in terms of the stress components. For the x_1 direction, three stresses are acting in the negative x_1 direction, σ_{11}, σ_{21} and σ_{31} and these are acting on planes OBC, OCA and OAB, respectively. Using the area of these planes and the equilibrium of forces one finds the component of F in the x_1 direction is

$$F_1 = \sigma_{11} a_1 + \sigma_{21} a_2 + \sigma_{31} a_3 \tag{2.37}$$

Generalizing this result and using the symmetry of the tensor, one can write

$$F_i = \sigma_{ij} a_j \tag{2.38}$$

If one wishes to find the magnitude of the normal stress F_N acting on the plane ABC, F_i is resolved along ON, i.e.,

$$F_N = F_1 a_1 + F_2 a_2 + F_3 a_3 \tag{2.39}$$

and hence, from Eq. (2.38),

$$F_N = \sigma_{ij} a_i a_j \tag{2.40}$$

The shear stress acting on the plane is determined from the parallelogram of forces and its magnitude (F_S) is given by

$$F_S^2 = F_1^2 + F_2^2 + F_3^2 - F_N^2 \tag{2.41}$$

As an example, consider a uniaxial tension test and the resolution of the stress onto a plane with its normal inclined at an angle α to the x_1 (loading) direction.

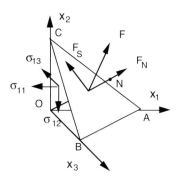

Figure 2.26 Force acting on a plane of unit area leading to shearing and normal forces on and across the plane.

For this case, the only stress component is $\sigma_{11} = \sigma$, the others are zero. In addition, one can write $\alpha_1 = \alpha$, $\alpha_2 = (\pi/2) - \alpha$, and $\alpha_3 = \pi/2$. Using Eqs. (2.38), (2.40) and (2.41), one finds

$$F_N = \sigma_{11} a_1^2 = \sigma \cos^2 \alpha \tag{2.42}$$

$$F_1^2 + F_2^2 + F_3^2 = \sigma^2 \cos^2 \alpha \tag{2.43}$$

$$F_S = \sqrt{\sigma^2 \cos^2 \alpha - \sigma^2 \cos^4 \alpha} = \sigma \cos \alpha \sin \alpha \tag{2.44}$$

The shear stress has a maximum value of $\sigma/2$ on planes $45°$ to the applied stress axis.

Like strain, a stress tensor can be rotated to its principal values. In isotropic materials, the principal stress and principal strain axes coincide. Often, the geometry of a body or the arrangement of applied forces enable the principal axes to be found by simple inspection. For example, a free surface cannot possess shear stresses in its plane and, thus, one of the principal axes must be normal to the free surface (the other two must lie in the surface). Returning to Fig. 2.26 one can show how the principal stresses and principal axes can be obtained, i.e., Eq. (2.35) will be derived for a stress tensor. In Fig. 2.26, the set of angles that makes the normal to the ABC plane a principal axis needs to be determined. If the principal stress acting along the normal is σ^*, the components of the load acting along the cartesian axes are $F_1 = \sigma^* a_1$, $F_2 = \sigma^* a_2$ and $F_3 = \sigma^* a_3$. Equating these components with those in Eq. (2.38), the following set of simultaneous equations are obtained.

$$(\sigma_{11} - \sigma^*) a_1 + \sigma_{12} a_2 + \sigma_{13} a_3 = 0$$

$$\sigma_{21} a_1 + (\sigma_{22} - \sigma^*) a_2 + \sigma_{23} a_3 = 0$$

$$\sigma_{31} a_1 + \sigma_{32} a_2 + (\sigma_{33} - \sigma^*) a_3 = 0 \tag{2.45}$$

The solution to this set of equations for σ^* is obtained by setting the determinant of the coefficients, $|\sigma_{ij} - \delta_{ij} \sigma^*|$, equal to zero. On expansion this gives

$$\sigma^{*3} - I_\Sigma \sigma^{*2} + II_\Sigma \sigma^* - III_\Sigma = 0 \tag{2.46}$$

where $I_\Sigma = \sigma_{ii}$, $II_\Sigma = (\sigma_{ii}\sigma_{jj} - \sigma_{ij}\sigma_{ij})/2$ and $III_\Sigma = E_{ijk}\sigma_{1i}\sigma_{2j}\sigma_{3k}$. This equation will give three real roots for a stress tensor (symmetric) and for each root, one can use Eq. (2.45) to determine a_i for the direction each principal axis makes with the original axes. In solving for a_i, it is often important to remember that $a_i a_i = 1$ for a set of orthogonal axes.

A set of important conditions, the compatibility relations, was imposed on the strains and similar restrictions are needed for stress. In this case, however, one is concerned with the equilibrium conditions and the variation of stress from point to point. Consider the prism in Fig. 2.27 in which the sides δx_i are just large enough to give a significant small variation of stress across the prism. The condition of equilibrium along x_1 is

$$A_1+A_2+B_1+B_2+C_1+C_2+D=0 \tag{2.47}$$

where A_1, A_2, B_1, B_2, C_1 and C_2 are the six forces acting on the faces in this direction and D is the body force in this direction. If H_1, H_2 and H_3 are the resolved components of the total body force per unit mass (H_j) and ρ is the density of the material, one finds $D=H_1\rho\delta x_1\delta x_2\delta x_3$. If the force acting on the face at $x_1=0$ is $-\sigma_{11}\delta x_2\delta x_3$, the force acting at $x_1=\delta x_1$ is $(\sigma_{11}+[\partial\sigma_{11}/\partial x_1]\delta x_1)\delta x_2\delta x_3$. Thus $A_1+A_2=(\partial\sigma_{11}/\partial x_1)\delta x_1\delta x_2\delta x_3$. Repeating this procedure for the other forces, Eq. (2.47) becomes

$$\frac{\partial\sigma_{11}}{\partial x_1}+\frac{\partial\sigma_{21}}{\partial x_2}+\frac{\partial\sigma_{31}}{\partial x_3}+H_1\rho=0 \tag{2.48}$$

Continuing this process for the other two axes and using the symmetry relationships gives the three **stress equilibrium equations**, i.e.,

$$\frac{\partial\sigma_{ij}}{\partial x_j}+H_i\rho=0 \tag{2.49}$$

2.11   General version of Hooke's Law

In Sections 2.4 and 2.10, stress and strain at a point have been discussed as independent entities and these are valid for any body no matter how it deforms, i.e.,

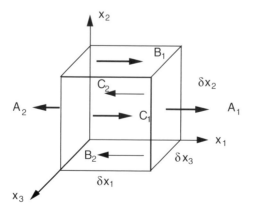

Figure 2.27 Forces acting in single direction must be in equilibrium.

it need not be linear elastic. If, however, one needs to relate stress and strain, the properties of the body must then be considered. The type of deformation behavior observed in a body can vary substantially, e.g., plasticity, viscosity, elasticity, etc. For this chapter, however, linear elastic bodies, in which the strains are linearly related to the stresses, are being considered. In the most general case, each stress component will have a linear relationship with each of the strain components. For example,

$$\sigma_{11} = c_{1111}\varepsilon_{11} + c_{1112}\varepsilon_{12} + c_{1113}\varepsilon_{13} + c_{1121}\varepsilon_{21} + c_{1122}\varepsilon_{22} + c_{1123}\varepsilon_{23} +$$
$$c_{1131}\varepsilon_{31} + c_{1132}\varepsilon_{32} + c_{1133}\varepsilon_{33} \tag{2.50}$$

and this would need to be repeated for each of the nine stress components. The coefficients, c_{ijkl}, are elastic constants and one needs 81 components to completely write down the nine equations for linear elastic behavior. In line with the discussion in Section 2.7, the interrelationship of two second-order tensors would require a fourth-order tensor, i.e., c_{ijkl}. It has already been shown that the strain and stress tensors are symmetric, so that only six components of each are really needed, reducing the number of required elastic constants to 36 (=6×6). At this point, it is useful to introduce a new subscript notation, which simplifies equations such as Eq. (2.50), i.e.,

$$\sigma_{11}\rightarrow\sigma_1,\ \sigma_{22}\rightarrow\sigma_2,\ \sigma_{33}\rightarrow\sigma_3,\ \sigma_{23}\rightarrow\sigma_4,\ \sigma_{13}\rightarrow\sigma_5,\ \sigma_{12}\rightarrow\sigma_6,$$
$$\varepsilon_{11}\rightarrow\varepsilon_1,\ \varepsilon_{22}\rightarrow\varepsilon_2,\ \varepsilon_{33}\rightarrow\varepsilon_3,\ 2\varepsilon_{23}\rightarrow\varepsilon_4,\ 2\varepsilon_{13}\rightarrow\varepsilon_5,\ 2\varepsilon_{12}\rightarrow\varepsilon_6 \tag{2.51}$$

The numbers 1 to 3 are associated with normal stresses and strains, 4 to 6 with the shear components. It is useful to note that the numbers run down the diagonal of the stress and strain tensors and circle back up the third column and along the top row to the starting position. This new notation also removes the difference between simple and pure shear strains discussed earlier. With the new notation, Eq. (2.50) becomes

$$\sigma_1 = c_{11}\varepsilon_1 + c_{12}\varepsilon_2 + c_{13}\varepsilon_3 + c_{14}\varepsilon_4 + c_{15}\varepsilon_5 + c_{16}\varepsilon_6 \tag{2.52}$$

and one would have five more equations like this to complete the linear elastic relations. A more succinct form of the six equations is

$$\sigma_i = c_{ij}\varepsilon_j \tag{2.53}$$

The repeated suffix notation is still being used but now the subscripts take numbers 1 through 6. The components, c_{ij}, are called the **elastic stiffness constants** and they also form a symmetric array, reducing the required number of elastic constants to 21. As these constants are properties of a material, the symmetry of the material may further reduce the number of constants. It is some-times useful to write the general version of Hooke's Law in a different form, i.e.,

$$\varepsilon_i = s_{ij}\sigma_j \tag{2.54}$$

where s_{ij} are the **elastic compliance constants**. This form is particularly useful if strains are required from a set of applied stresses. Either set of elastic constants can be written out in a matrix type of formation, e.g., for c_{ij}

$$c_{ij} = \begin{bmatrix} c_{11} & c_{12} & c_{13} & c_{14} & c_{15} & c_{16} \\ c_{21} & c_{22} & c_{23} & c_{24} & c_{25} & c_{26} \\ c_{31} & c_{32} & c_{33} & c_{34} & c_{35} & c_{36} \\ c_{41} & c_{42} & c_{43} & c_{44} & c_{45} & c_{46} \\ c_{51} & c_{52} & c_{53} & c_{54} & c_{55} & c_{56} \\ c_{61} & c_{62} & c_{63} & c_{64} & c_{65} & c_{66} \end{bmatrix} \tag{2.55}$$

If the elastic constants are divided into four quadrants as shown above, one finds the constants in the upper left quadrant all involve the numbers 1 to 3. This set of constants, therefore, represents the relationships between normal stresses and normal strains. In the lower right quadrant, the only numbers are 4 to 6 and hence, involve only shear stresses and shear strains. In the upper right quadrant, which by symmetry is the same as the lower left, there is the possibility of rela-tionships between normal stresses and shear strains and other such combina-tions. This implies it is possible to apply a normal stress to some bodies and obtain a shear strain or a shear stress to produce a normal strain. For many materials, these upper right quadrant constants may all be zero, but this is not always the case. For example, in some crystal symmetries, such as monoclinic, some of these components are non-zero. A similar effect can be found in angle-ply fiber composites in which, for example, a normal stress applied at an angle to the fibers can give rise to shear strains.

Figure 2.28 shows the orientation conventions for single crystals relative to orthogonal stress and strain axes for description of the elastic constants. Table 2.3 shows the form of the elastic constants for these crystal types. Comparing this table with Eq. (2.55), one can determine which elastic constants are equiva-lent and which are zero. For example, in the triclinic system, one can see that $c_{21} = c_{12}$, etc., indicating the tensor is symmetric. Thus 21 elastic constants are needed to describe the linear elastic behavior of a triclinic crystal. In the hexag-onal system one finds, in addition to symmetry of the tensor, that $c_{11} = c_{22}$, $c_{13} = c_{23}$, $c_{44} = c_{55}$, $c_{66} = (c_{11} - c_{12})/2$ and many other elastic constants are zero. Overall, five elastic constants are needed for hexagonal single crystals. Table 2.3 was written out for the elastic constants c_{ij} but the constants s_{ij} have the same symmetries. In some cases, one may wish to determine the components s_{ij} from

c_{ij} and the appropriate equations are given in Appendix 1. The generalized version of Hooke's Law allows the strain energy density to be expressed as $U = c_{ij}\varepsilon_i\varepsilon_j/2 = \sigma_i\varepsilon_i/2$.

2.12 Elastic behavior of anisotropic materials

If the values of c_{ij} or s_{ij} for any particular single crystal are known, one can use Eqs. (2.53) and (2.54) to determine the stresses required to produce *any* set of strains or vice versa. Although anisotropic elasticity has been discussed primarily with respect to single crystals, there are many other types of materials that are anisotropic and one can use these equations in the same way. Natural materials, such as wood, are macroscopically anisotropic and this is also reflected in the elastic constants. In some cases, the way a material is fabricated can lead to elastic anisotropy. For example, uniaxial hot pressing can often give isotropic properties in the plane perpendicular to the hot pressing direction but different properties parallel to this direction. This is termed **transverse isotropy** and is equivalent to hexagonal symmetry. In the field of fiber composites, **orthotropic elasticity** is often used. This has the same symmetry as orthorhombic crystals provided one can recognize the three principal material directions in the composite. Thus, for an orthotropic material, Hooke's Law for ε_1 is

$$\varepsilon_1 = s_{11}\sigma_1 + s_{12}\sigma_2 + s_{13}\sigma_3 \tag{2.56}$$

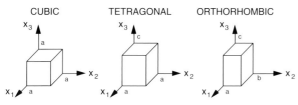

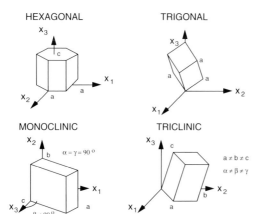

Figure 2.28 Orientation convention of unit cells relative to orthogonal stress and strain axes for description of single–crystal elastic constants.

Table 2.3 *Form of the compliance and stiffness tensors for the crystal classes*[a]

CLASS C_1, S_2;
TRICLINIC SYSTEM (21 CONSTANTS)

$$\begin{pmatrix} c_{11} & c_{12} & c_{13} & c_{14} & c_{15} & c_{16} \\ c_{12} & c_{22} & c_{23} & c_{24} & c_{25} & c_{26} \\ c_{13} & c_{23} & c_{33} & c_{34} & c_{35} & c_{36} \\ c_{14} & c_{24} & c_{34} & c_{44} & c_{45} & c_{46} \\ c_{15} & c_{25} & c_{35} & c_{45} & c_{55} & c_{56} \\ c_{16} & c_{26} & c_{36} & c_{46} & c_{56} & c_{66} \end{pmatrix}$$

CLASS C_{1h}, C_2, C_{2h}
MONOCLINIC (13 CONSTANTS)

$$\begin{pmatrix} c_{11} & c_{12} & c_{13} & 0 & 0 & c_{16} \\ c_{12} & c_{22} & c_{23} & 0 & 0 & c_{26} \\ c_{13} & c_{23} & c_{33} & 0 & 0 & c_{36} \\ 0 & 0 & 0 & c_{44} & c_{45} & 0 \\ 0 & 0 & 0 & c_{45} & c_{55} & 0 \\ c_{16} & c_{26} & c_{36} & 0 & 0 & c_{66} \end{pmatrix}$$

CLASS C_{2v}, V, V_h
ORTHORHOMBIC (9 CONSTANTS)

$$\begin{pmatrix} c_{11} & c_{12} & c_{13} & 0 & 0 & 0 \\ c_{12} & c_{22} & c_{23} & 0 & 0 & 0 \\ c_{13} & c_{23} & c_{33} & 0 & 0 & 0 \\ 0 & 0 & 0 & c_{44} & 0 & 0 \\ 0 & 0 & 0 & 0 & c_{55} & 0 \\ 0 & 0 & 0 & 0 & 0 & c_{66} \end{pmatrix}$$

CLASS C_{3h}, D_{3h}, C_6, C_{6h}, C_{6v}, D_6, D_{6h}
HEXAGONAL (5 CONSTANTS)

$$\begin{pmatrix} c_{11} & c_{12} & c_{13} & 0 & 0 & 0 \\ c_{12} & c_{11} & c_{13} & 0 & 0 & 0 \\ c_{13} & c_{13} & c_{33} & 0 & 0 & 0 \\ 0 & 0 & 0 & c_{44} & 0 & 0 \\ 0 & 0 & 0 & 0 & c_{44} & 0 \\ 0 & 0 & 0 & 0 & 0 & (c_{11}-c_{12})/2 \end{pmatrix}$$

CLASS C_4, S_4, C_{4h}
TETRAGONAL (7 CONSTANTS)

$$\begin{pmatrix} c_{11} & c_{12} & c_{13} & 0 & 0 & c_{16} \\ c_{12} & c_{11} & c_{13} & 0 & 0 & -c_{16} \\ c_{13} & c_{13} & c_{33} & 0 & 0 & 0 \\ 0 & 0 & 0 & c_{44} & 0 & 0 \\ 0 & 0 & 0 & 0 & c_{44} & 0 \\ c_{16} & -c_{16} & 0 & 0 & 0 & c_{66} \end{pmatrix}$$

CLASS C_{4v}, V_d, D_4, D_{4h}
TETRAGONAL (6 CONSTANTS)

$$\begin{pmatrix} c_{11} & c_{12} & c_{13} & 0 & 0 & 0 \\ c_{12} & c_{11} & c_{13} & 0 & 0 & 0 \\ c_{13} & c_{13} & c_{33} & 0 & 0 & 0 \\ 0 & 0 & 0 & c_{44} & 0 & 0 \\ 0 & 0 & 0 & 0 & c_{44} & 0 \\ 0 & 0 & 0 & 0 & 0 & c_{66} \end{pmatrix}$$

CLASS C_3, S_6 TRIGONAL (7 CONSTANTS)[b]

$$\begin{pmatrix}
c_{11} & c_{12} & c_{13} & c_{14} & -c_{25} & 0 \\
c_{12} & c_{11} & c_{13} & -c_{14} & c_{25} & 0 \\
c_{13} & c_{13} & c_{33} & 0 & 0 & 0 \\
c_{14} & -c_{14} & 0 & c_{44} & 0 & c_{25} \\
-c_{25} & c_{25} & 0 & 0 & c_{44} & c_{14} \\
0 & 0 & 0 & c_{25} & c_{14} & (c_{11}-c_{12})/2
\end{pmatrix}$$

CLASS C_{3v}, D_3, D_{3d} TRIGONAL (6 CONSTANTS)

$$\begin{pmatrix}
c_{11} & c_{12} & c_{13} & c_{14} & 0 & 0 \\
c_{12} & c_{11} & c_{13} & -c_{14} & 0 & 0 \\
c_{13} & c_{13} & c_{33} & 0 & 0 & 0 \\
c_{14} & -c_{14} & 0 & c_{44} & 0 & 0 \\
0 & 0 & 0 & 0 & c_{44} & c_{14} \\
0 & 0 & 0 & 0 & c_{14} & (c_{11}-c_{12})/2
\end{pmatrix}$$

CLASS T, T_h, T_d, O, O_h CUBIC (3 CONSTANTS)

$$\begin{pmatrix}
c_{11} & c_{12} & c_{12} & 0 & 0 & 0 \\
c_{12} & c_{11} & c_{12} & 0 & 0 & 0 \\
c_{12} & c_{12} & c_{11} & 0 & 0 & 0 \\
0 & 0 & 0 & c_{44} & 0 & 0 \\
0 & 0 & 0 & 0 & c_{44} & 0 \\
0 & 0 & 0 & 0 & 0 & c_{44}
\end{pmatrix}$$

ISOTROPIC (2 CONSTANTS)

$$\begin{pmatrix}
c_{11} & c_{12} & c_{12} & 0 & 0 & 0 \\
c_{12} & c_{11} & c_{12} & 0 & 0 & 0 \\
c_{12} & c_{12} & c_{11} & 0 & 0 & 0 \\
0 & 0 & 0 & (c_{11}-c_{12})/2 & 0 & 0 \\
0 & 0 & 0 & 0 & (c_{11}-c_{12})/2 & 0 \\
0 & 0 & 0 & 0 & 0 & (c_{11}-c_{12})/2
\end{pmatrix}$$

Notes:

[a] The form of the tensors are the same for s_{ij} but for those cases in which an elastic constant$=(c_{11}-c_{12})/2$ occurs, it is replaced by $2(s_{11}-s_{12})$.

[b] For the trigonal class*, $s_{46}=2s_{25}$ and $s_{56}=2s_{14}$, while for the other trigonal class $s_{56}=2s_{14}$.

as one can deduce from Table 2.3 that $s_{14}=s_{15}=s_{16}=0$. In some texts, the engineering elastic constants are preferred instead of s_{ij}. For this case, the engineering elastic constants must be given subscripts, as they are defined *only* with respect to the principal directions. For example, one can write Eq. (2.56) as

$$\varepsilon_1 = \frac{\sigma_1}{E_1} - \frac{v_{21}\sigma_2}{E_2} - \frac{v_{31}\sigma_3}{E_3} \tag{2.57}$$

where E_1, E_2 and E_3 are the Young's modulus values for the principal direction and v_{ij} represents a transverse strain in the j direction for a uniaxial stress in the i direction. For the shear stresses, three shear moduli must be defined, e.g., μ_4, μ_5 and μ_6. In orthotropic elasticity, nine elastic constants (same as orthorhombic) are needed but, from the above discussion, twelve elastic constants were defined. Clearly, this implies there must be some interrelationships. It can be shown, for example, that

$$\frac{v_{21}}{E_2} = \frac{v_{12}}{E_1} \tag{2.58}$$

If this process is repeated, one finds only three values of Poisson's ratio are needed, not six. For fiber-reinforced materials, the number of elastic constants may be further reduced if other symmetries appear. For example, in some materials short fibers are randomly oriented in a plane and this gives transverse isotropy. That is, there is an elastically isotropic plane but the stiffness and compliance constants will be different normal to this plane (five elastic constants are needed).

The familiarity of the engineering elastic constants, especially Young's modulus, is often used in showing the variation of 'stiffness' with direction in an anisotropic material. For example, a **representational surface** can be produced that shows the variation in the magnitude of Young's modulus with stress direction. The magnitude is represented by the length of a radial vector from the origin of a set of cartesian axes. For example, in cubic crystals it can be shown that

$$\frac{1}{E} = s_{11} - 2(s_{11} - s_{12} - \frac{s_{44}}{2})(a_1^2 a_2^2 + a_2^2 a_3^2 + a_1^2 a_3^2)$$

where a_i represents the direction cosines between the direction of interest and the x_i axes. Appendix 2 gives the complete set of equations for the representational surfaces for all the crystal systems. Figure 2.29 shows the representational Young's modulus surface for potassium fluoride (cubic). It should be realized that the shear modulus and Poisson's ratio will also change with direction. It is, however, not possible to show the variation by a single representational surface. In these cases, the variation with stress direction can only be calculated for a particular plane.

Let us consider cubic crystals in some more detail. From Table 2.3, it is found that three elastic constants are needed for cubic crystals, c_{11}, c_{12} and c_{44}. This is a result of the three independent modes of deformation in this crystal system. The first mode is dilatation by a hydrostatic stress ($\sigma = \sigma_1 = \sigma_2 = \sigma_3$). For this case, using Eq. (2.53) and Table 2.3, Hooke's Law for σ_1 is given by

$$\sigma = \sigma_1 = c_{11}\varepsilon_1 + c_{12}\varepsilon_2 + c_{12}\varepsilon_3 \tag{2.59}$$

From Eq. (2.54) it can also be shown that $\varepsilon_1 = \varepsilon_2 = \varepsilon_3$. As shown in Section 2.4, $\varepsilon_1 + \varepsilon_2 + \varepsilon_3 = \Delta V/V$ and, from the definition of bulk modulus (Section 2.3), one finds

$$B = \frac{-\sigma}{\Delta V/V} = \frac{(c_{11} + 2c_{12})}{3} \tag{2.60}$$

The other two independent modes of linear elastic deformation in cubic single crystals relate to shape changes, in particular to shear on the {100} and {110} planes in the <100> and <110> directions, respectively, as depicted in Fig. 2.30. In cubic crystals, these modes have different shear modulus values because they represent two different ways to deform the material and yet conserve volume. For {100} shear, using Eq. (2.53) and Table 2.3, one can show that the shear modulus μ_0 associated with this deformation is given by

$$\mu_0 = \frac{\sigma_4}{\varepsilon_4} = c_{44} \tag{2.61}$$

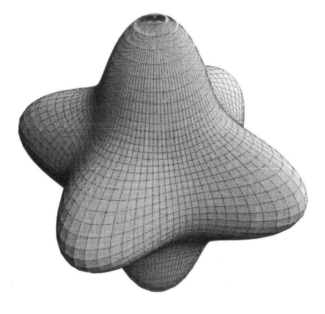

Figure 2.29
Representational surface for the Young's modulus of potassium fluoride. (Courtesy of Ming-Jen Pan.)

The <110> type of shear was demonstrated earlier in Fig. 2.21, when it was shown that a pure shear strain is equivalent to two equal and opposite normal strains along orthogonal directions. From Eq. (2.53), the application of two strains along x_1 and x_2 of ε and $-\varepsilon$ respectively leads to stresses given by

$$\sigma_1 = c_{11}\varepsilon - c_{12}\varepsilon \text{ and } \sigma_2 = c_{12}\varepsilon - c_{11}\varepsilon \tag{2.62}$$

and thus, $\sigma_1 = -\sigma_2 = \sigma$. Using the transformation equation (Eq. (2.22)) to determine resolved shear stresses and strains on the {110} planes, one finds $\sigma_4 = \sigma$ and $\varepsilon_4 = 2\varepsilon$. Thus, the ratio of stress to strain on the {110} planes, i.e., the shear modulus μ_1, is given by

$$\mu_1 = \frac{\sigma}{2\epsilon} = \frac{(c_{11}-c_{12})}{2} \tag{2.63}$$

A measure of the anisotropy in cubic crystals is the ratio of the two shear moduli, which from Eqs. (2.61) and (2.63) is given by

$$Z = \frac{\mu_0}{\mu_1} = \frac{2c_{44}}{c_{11}-c_{12}} \tag{2.64}$$

The parameter Z is known as the **Zener ratio** or the **elastic anisotropy factor**. For the Young's modulus representational surface in cubic crystals, this anisotropy is reflected in different types of surface for $Z<1$ and $Z>1$. This is depicted in Fig. 2.31 in which for $Z<1$, the maximum value of E is along <100> and the minimum is along <111>. For $Z<1$, $\mu_0 < \mu_1$, and one expects that this will translate into a high Young's modulus along <100>. For $Z>1$ the opposite holds true, with Young's modulus being a minimum along <100> and a maximum

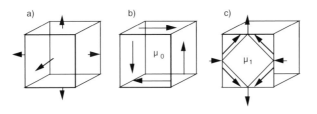

Figure 2.30 Three basic modes of deformation in a cubic single crystal: a) dilatation; b) {100}<100> shear; and c) {110}<110> shear.

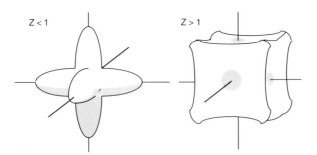

Figure 2.31 The two possible representational surfaces for Young's modulus in cubic crystals. The maximum in the modulus is along <100> if $Z < 1$ and along <111> if $Z > 1$.

along <111>. For the special case $Z=1$, the cubic crystal would also be isotropic and the representational surface would be spherical. Table 2.4 lists the elastic constants and Zener ratios for various cubic crystals and can be seen to form distinctive groups. For example, the alkali metals with a body-centered cubic (bcc) structure have large Z values, whereas alkali halides with the rock-salt structure have the lowest. From Table 2.4, one also notices that anisotropy follows definite trends within particular crystal families. The weakness of bcc metals against μ_1 shear (high Z) can also be noticed. These ideas will be discussed further in the next chapter, in which the elastic constants are related to the atomic structure.

For cases in which central forces exist between atoms in a solid and the atoms are at the crystallographic centers of symmetry, the **Cauchy relation**, $c_{12}=c_{44}$, should be satisfied. In cubic solids, this relation is satisfied only when the Poisson's ratio is exactly 0.25. The Cauchy relationship holds fairly well for many ionic crystals but fails for more covalent materials, such as MgO, Si and Ge. The c_{12}/c_{44} ratio is sometimes used to estimate the degree of covalency in the atomic bonding.

As with other anisotropic materials, Hooke's Law for cubic crystals may be expressed in terms of the engineering elastic constants. Equation (2.57) can be written as

$$\varepsilon_1 = \frac{\sigma_1}{E} - \frac{v\sigma_2}{E} - \frac{v\sigma_3}{E} \tag{2.65}$$

and for the shear stresses ($i=4$, 5 or 6)

$$\varepsilon_i = \frac{\sigma_i}{\mu} \tag{2.66}$$

Comparing this equation and Eq. (2.61) with the general version of Hooke's Law, one can easily show that

$$E = \frac{(c_{11}-c_{12})(c_{11}+2c_{12})}{c_{11}+c_{12}} = \frac{1}{s_{11}} \tag{2.67}$$

$$v = \frac{c_{12}}{c_{11}+c_{12}} = \frac{-s_{12}}{s_{11}} \tag{2.68}$$

$$B = \frac{(c_{11}+2c_{12})}{3} = \frac{1}{3(s_{11}+2s_{12})} \tag{2.69}$$

$$\mu = c_{44} = \frac{1}{s_{44}} \tag{2.70}$$

It is important to remember these elastic constants are defined only for the principal axes and will, in general, be different for other stress directions

Table 2.4 *Elastic stiffness constants of various cubic crystals*

Material	c_{11} (GPa)	c_{12} (GPa)	c_{44} (GPa)	μ_1 (GPa)	B (GPa)	Zener ratio	v
BCC METALS							
Li	13.5	11.4	8.78	1.03	12.1	8.5	0.46
Na	7.39	6.22	4.26	0.59	6.61	7.3	0.46
K	4.14	3.31	2.63	0.42	3.59	6.3	0.44
DIAMOND							
C	932.0	411.2	416.7	260.4	584.8	1.6	0.31
Si	165.7	63.9	79.60	50.90	97.83	1.6	0.28
Ge	131.6	50.9	66.90	40.35	77.80	1.7	0.28
ROCK SALT							
LiF	117.8	43.36	62.81	37.21	68.16	1.7	0.27
LiCl	49.40	22.60	24.90	13.40	31.53	1.9	0.31
LiBr	39.40	18.80	19.10	10.30	25.67	1.9	0.32
LiI	28.50	14.00	13.50	7.25	18.83	1.9	0.33
NaF	97.00	25.60	28.00	35.70	49.40	0.78	0.21
NaCl	46.41	12.66	12.78	16.88	23.91	0.76	0.21
NaBr	38.70	9.70	9.70	14.50	19.37	0.67	0.20
NaI	30.35	9.00	7.20	10.68	16.12	0.67	0.23
KF	65.80	14.90	12.80	25.45	31.87	0.50	0.18
KCl	36.69	1.93	6.41	17.38	13.52	0.37	0.05
KBr	34.50	5.40	5.08	14.55	15.10	0.35	0.14
KI	26.68	4.26	4.20	11.21	11.73	0.37	0.14
RbF	57.00	12.50	9.10	22.25	27.33	0.41	0.18
RbCl	36.45	6.10	4.75	15.18	16.22	0.31	0.14
RbBr	31.85	4.80	3.85	13.53	13.82	0.28	0.13
RbI	25.85	3.75	2.81	11.05	11.12	0.25	0.13
AgCl	60.10	36.20	6.25	11.95	44.17	0.52	0.38
AgBr	56.20	32.80	7.28	11.70	40.60	0.62	0.37
PbS	102.0	38.00	25.00	32.00	59.33	0.78	0.27
PbTe	107.2	7.68	13.00	49.76	40.85	0.26	0.07
MgO	286.0	87.00	148.0	99.50	153.3	1.5	0.23
MnO	223.0	120.0	79.00	51.50	154.3	1.5	0.35
CoO	261.2	147.0	83.00	57.12	185.1	1.5	0.36
SrO	160.1	43.50	59.00	58.30	82.37	1.0	0.21
TiC	500.0	113.0	175.0	193.5	242.0	0.90	0.18
ZrC	472.0	98.70	159.3	186.7	223.1	0.85	0.17
TaC	505.0	73.00	79.00	216.0	217.0	0.37	0.13
UC	320.0	85.00	64.70	117.5	163.3	0.55	0.21
CESIUM CHLORIDE							
CsCl	36.40	9.20	8.00	13.60	18.27	0.59	0.20
CsBr	31.00	8.40	7.50	11.30	15.93	0.66	0.21
CsI	24.50	7.10	6.20	8.70	12.90	0.71	0.22
FLUORITE							
CaF_2	162.8	43.30	33.40	59.75	83.13	0.56	0.21
SrF_2	123.5	43.05	31.28	40.23	69.87	0.78	0.26
BaF_2	90.10	40.30	24.90	24.90	56.90	1.0	0.31
PbF_2	88.80	47.20	24.54	20.80	61.07	1.2	0.35
$ZrO_2\text{--}12Y_2O_3$	405.1	105.3	61.80	149.9	205.2	0.41	0.21
ThO_2	367.0	106.0	79.70	130.5	193.0	0.61	0.22
UO_2	396.0	121.0	64.10	137.5	212.7	0.47	0.23

2.13   Elastic behavior of isotropic materials

Many polycrystalline ceramics consist of a random array of single crystals and, in these cases, the materials are elastically isotropic. As shown in Table 2.3, only two elastic constants are needed to describe a linear elastic deformation, c_{11} and c_{12}. For convenience, the engineering elastic constants can be used, and one obtains Eqs. (2.65)–(2.70). There is, however, an additional relationship between c_{11}, c_{12} and c_{44}, such that Eq. (2.70) can be written as

$$\mu = \frac{(c_{11}-c_{12})}{2} = \frac{1}{2(s_{11}-s_{12})} \tag{2.71}$$

In some texts, the **Lamé constants**, λ and μ, are used.[†] These constants are equal to c_{12} and c_{44}, respectively. In some cases, it is convenient to write Hooke's Law in a form specific for isotropic materials. For example, using the Lamé constants, one can write

$$\sigma_{ij} = \lambda \delta_{ij} \varepsilon_{kk} + 2\mu \varepsilon_{ij} \tag{2.72}$$

In other cases, one may wish to emphasize the volumetric and deviatoric strains and one can write

$$\sigma_{ii} = 3B\varepsilon_{ii} \ \text{ and } \ \sigma_{ij}^{D} = 2\mu \varepsilon_{ij}^{D} \tag{2.73}$$

where σ_{ij}^{D} and ε_{ij}^{D} are the deviatoric stress and strain tensors, respectively.

The 'excess' of elastic constants gives rise to many interrelationships and some of these are given in Appendix 3. For example,

$$E = \frac{3\mu}{[1 + (\mu/3B)]} \tag{2.74}$$

and as usually $3B \gg \mu$, the right-hand side can be expanded to give

$$E \sim 3\mu[1 - (\mu/3B)] \tag{2.75}$$

which shows that E has a strong contribution from the shear modulus compared to the bulk modulus. This is a result of the specimen sides being free to move in a uniaxial tension test and, thus, the resistance to dilatation is not severely tested. As μ/B approaches zero, i.e., as the material becomes 'fluid', E approaches 3μ and, using the relationship,

---

[†] μ is equivalent to the shear modulus.

$$\mu = \frac{E}{2(1+v)} \tag{2.76}$$

one finds v approaches 0.5. Thus, Poisson's ratio primarily measures the relative resistance to shearing. The lateral strain in uniaxial tension is resisted by the shear, so a low value of v usually implies a relatively strong resistance to shear. The elastic constants for a variety of dense polycrystalline materials was shown earlier in Table 2.1. If the single-crystal elastic constants for a material are known, one may wish to predict the elastic constants for a polycrystalline body, which is comprised of a random array of these crystals. Such constitutive relationships exist and these will be discussed in the next chapter.

Engineering design is based primarily on the linear elastic behavior of materials, and materials with high elastic moduli are often the most attractive candidates for an efficient design. In some applications, in which weight is important, the emphasis often switches to the **specific mechanical properties**, i.e., the property is divided by density. The property that needs to be maximized to obtain maximum stiffness for minimum weight depends to some degree on the mode of loading. In uniaxial tension the design parameter is E/ρ, whereas in bending it may be E/ρ^2 or E/ρ^3. As an aid to this process, **material selection maps** have been introduced, in which E is shown as a function of ρ on a log–log plot. A schematic map is shown in Fig. 2.32. The materials with the maximum values of E/ρ, E/ρ^2 or E/ρ^3 can be easily identified using the lines with slopes 1, 2 or 3. As seen in the figure, fiber-reinforced composites and covalently bonded ceramics have high values of E/ρ. It is also clear that foams and natural materials, such as wood, can possess high E/ρ^2 or E/ρ^3 values.

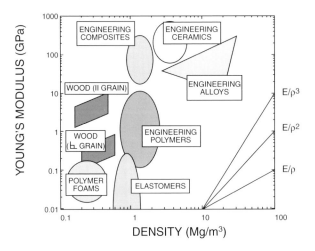

Figure 2.32 Young's modulus as a function of density. The guidelines of constant E/ρ, E/ρ^2 and E/ρ^3 allow selection of materials for minimum weight for a deflection-limited design. (Adapted from Ashby, 1992. Reproduced courtesy of Heineman Publishers, Oxford, UK.)

2.14   Miscellaneous effects on the elastic constants

At the beginning of this chapter, it was indicated that the elastic strains are often very small and one can ignore the higher-order terms associated with displacements. For example, in deriving Eq. (2.2), terms in u^3 or greater were ignored. Clearly, the effect of larger displacements could be introduced, using **higher-order elastic constants** to describe the elastic behavior. This is beyond the scope of this book but it is worthwhile considering, at least qualitatively, some of these effects. For large displacements, non-linearities are expected from the form of the interatomic potential. For example, the application of large pressures to a material should lead to an increase in the elastic constants. This is a result of the increase in interatomic repulsion that occurs as the electron clouds are forced into each other. For high tensile stresses, one would expect the opposite to occur and the elastic constants should decrease at the higher stresses. Temperature changes also affect the elastic constants. Increasing temperature usually decreases the stiffness of the atomic bonds and, hence, decreases the elastic moduli. Figure 2.33 shows Young's modulus of SiC as a function of temperature and the decrease in modulus can be readily seen. These generalizations about the effects of temperature and pressure are useful but, as shown in the next chapter, there can be some subtle structural effects that may give rise to converse effects.

In an ideal elastic solid, a one-to-one relationship between stress and strain is expected. In practice, however, there are often small deviations. These are termed **anelastic effects** and result from **internal friction** in the material. Part of the strain develops over a period of time. One source of anelasticity is **thermoelasticity**, in which the volume of a body can be changed by both temperature and applied stress. The interaction will depend on whether a material has time to equilibrate with the surroundings. For example, if a body is rapidly dilated, the sudden

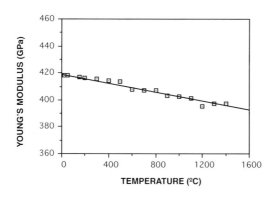

Figure 2.33 Young's modulus of α-SiC as a function of temperature.

increase in the amplitude of atomic vibrations reduces their frequency and the body cools. This is known as the **thermoelastic effect**. If the body is allowed to equilibrate later with the surroundings, a thermal strain, in addition to the elastic strain, will be obtained. Thus, a rapid volume increase gives rise to an elastic strain but also cooling of the body. As the body then equilibrates there will be a further strain and if the thermal expansion coefficient α for the body is positive this will give rise to a further dilatation. Thus, the adiabatic bulk modulus, B_A, is greater than the isothermal modulus, B_I. Using thermodynamics, one can show

$$\frac{B_A - B_I}{B_I} = \frac{\beta^2 T B_A}{\rho C_p} \tag{2.77}$$

where β is the volume expansivity, T is the absolute temperature, ρ is density and C_p is the heat capacity at constant pressure. The left-hand side of Eq. (2.77) is called the **modulus defect**. The difference in the adiabatic and isothermal bulk moduli is often small, $\sim 1\%$. For shear deformation, there is generally no change in volume, so there should be no modulus defect in this case.

The thermoelastic effect can lead to heat flow in a material. For example, in a polycrystalline body composed of anisotropic crystals the application of a fixed strain-rate will give rise to different stress-rates in the constitutive crystals. Thus, it is possible to have heat flow between individual grains during rapid straining. In bend tests, there is a gradient in stress from uniaxial compression at one surface to uniaxial tension at the other. In other words, different parts of the body will be subjected to different stress-rates and the stress may be tensile or compressive. In such situations, cyclic stressing may give rise to heat flow from the compressive to the tensile regions. Anelasticity will be discussed again in Chapter 5 in connection with viscoelasticity.

2.15   Propagation of mechanical disturbances

The forces that act on solids may not always be in static equilibrium and the unbalanced portion of the force will set up motions in the body (dynamics). From Newton's second law of motion, the unbalanced force F can be equated to the rate of momentum change with time, i.e.,

$$F = \frac{d(mv)}{dt} \tag{2.78}$$

where v is velocity, m is mass and t is time.

Consider a long uniform cylindrical rod of material with density ρ and cross-sectional area A. If an unbalanced force is applied to one end of the rod (in the x_1 direction), the material will be set in motion. The motion will then be passed along the rod from one layer to the next by the atomic interactions between the layers. Consider a short element of length δx_1 in the rod. The force $F(x_1)$ acting on one face will be different from that of $F(x_1 + \delta x_1)$ acting on the other. The difference between the forces can be written as $(\partial F/\partial x_1)\delta x_1$ or $A(\partial \sigma/\partial x_1)\delta x_1$. The small element has a mass $\rho A \delta x$. If the displacement in the force direction is u_1, one obtains $mv = \rho A(du_1/dt)\delta x_1$ and when substituted into Eq. (2.78), one obtains

$$\frac{\partial \sigma_1}{\partial x_1} = \frac{\rho \partial^2 u_1}{\partial t^2} \tag{2.79}$$

If a general stress–strain relationship, $\sigma = c\,(\partial u_1/\partial x_1)$, is assumed, Eq. (2.79) can be written as a **wave equation**, i.e.,

$$\frac{\partial^2 u_1}{\partial x_1^2} = \frac{1}{v^2}\frac{\partial^2 u_1}{\partial t^2} \tag{2.80}$$

where $v = \sqrt{(c/\rho)}$. Substituting $y = vt$ into this equation, one obtains

$$\frac{\partial^2 u_1}{\partial x_1^2} = \frac{\partial^2 u_1}{\partial y^2} \tag{2.81}$$

Both x_1 and y have equal status in the equation and, thus, any function of $(x_1 \pm y)$ is a solution, i.e., functions of the form $u = g(x_1 \pm vt)$ are solutions. For example, these solutions would include z^n, z^{-n}, $\log z$, $\exp z$, $\sin z$, etc., where $z = x \pm vt$. These various solutions represent **stress pulses, sound waves or wave packets** running down the rod at a velocity v. Clearly, it is now important to relate the generalized elastic constant c to the elastic constants discussed earlier in the chapter.

If the cylindrical rod discussed above is thin and is being subjected to a uni-axial force, then $c = E$ and the wave velocity is given by

$$v_0 = \sqrt{\frac{E}{\rho}} \tag{2.82}$$

For bodies that are not thin compared to the wavelength of the sound waves, the Poisson's ratio effect sets up lateral stresses and the wave is no longer planar, i.e., a wave with fronts only perpendicular to the propagation direction. Consider next a mechanical disturbance emanating from a point source in an infinite body. At large distances from the source, compared to the wavelength, the spherical wave-front is approximately planar but the wave may now have displacements, u_2 and u_3, perpendicular to the propagation direction. The only non-vanishing strains will be the normal strain $\partial u_1/\partial x_1$ and the shears $\partial u_1/\partial x_2$ and $\partial u_1/\partial x_3$. The normal strain is not equivalent to uniaxial tension as the sides of the a radial cylinder cannot move inwards or outwards. This type of strain, in which all the particles in a layer move equally along x_1, is called **longitudinal** or **irrotational**. The elastic constant for this mode of deformation is c_{11}, and for isotropic materials $c_{11}=B+(4\mu/3)$ and using the above analysis, the **longitudinal wave velocity**, v_1, is given as

$$v_1=\sqrt{\frac{c_{11}}{\rho}} \tag{2.83}$$

The waves associated with the shear strains are known as **transverse** or **equivolu-minal waves**. For these cases, $c=\mu$ and the transverse wave velocity v_2 is

$$v_2=\sqrt{\frac{\mu}{\rho}} \tag{2.84}$$

At the free surface of a solid, the wave velocity is altered because stresses normal to the surface must be zero. Surface waves, called **Rayleigh waves**, propagate with a velocity

$$v_3=f(v)\sqrt{\frac{\mu}{\rho}} \tag{2.85}$$

where $f(v)$ is a function of Poisson's ratio, e.g., for $v=0.25$, $f(v)=0.9553$. Surface waves are important geologically because they spread out in two dimensions and hence do not lose their amplitude as rapidly as spherical waves. In comparing Eqs. (2.83)–(2.85), it is clear that longitudinal waves are the first shocks to arrive after an earthquake. These are followed by the transverse waves and, finally, the devastating surface waves. The later stages are complicated by the arrival of waves resulting from multiple reflections and by other types of surface waves.

    Equations (2.82)–(2.85) can be readily used to determine the magnitude of the wave velocities for ceramics. Covalent ceramics are known to have high elastic moduli and often have low density, giving rise to high sound velocities, compared to other groups of materials. For example, using the data for Al_2O_3 in Table 2.1 and $\rho=3970$ kg/m³, one obtains $v_0=10.1$ km/s, $v_1=10.9$ km/s, $v_2=6.41$ km/s and $v_3\sim6.15$ km/s. These values are approximately twice as high as those found in aluminum and steel. Wave velocities can also be determined for anisotropic materials in a similar way to that described above. In these cases, the velocity of a single wave-type will depend on the propagation direction. For example, if a longitudinal wave is propagating along a principal direction in orthotropic or orthorhombic crystals, the elastic constant in Eq. (2.82) will be c_{11}, c_{22} or c_{33}. Measurement of the various wave velocities are used to determine the elastic constants of a material (see Section 2.17).

2.16   Resonant vibrations

In the last section it was shown that the inertial mass of a material causes elastic waves to travel at finite speeds. These waves will be reflected by the boundaries of the material and, as the incident and reflected waves pass through each other, a pattern of **standing waves** can be formed (principle of superposition). The displacement of an element in the body no longer has the form $u=g(x_1\pm vt)$ and becomes instead $u=u_0 \sin \omega t \sin kx_1$. In Eq. (2.80), x_1 and t occur in separate terms and, thus, one can now look for solutions of the form

$$u=X(x_1)T(t) \tag{2.86}$$

Equation (2.80) becomes

$$\frac{1}{v_0^2 T}\frac{\partial^2 T}{\partial t^2}=\frac{1}{X}\frac{\partial^2 X}{\partial x_1^2} \tag{2.87}$$

The left-hand side of this equation is a function only of t and the right only of x_1 and yet the terms are equal. This implies both terms are constant and this constant can be denoted as $-k^2$. Equation (2.86) can now be split into two separate equations,

$$\frac{\partial^2 X}{\partial x_1^2}+k^2 X=0; \quad \frac{\partial^2 T}{\partial t^2}+k^2 v_0^2 T=0 \tag{2.88}$$

For these equations, solutions of the form

$$X=A \sin kx_1+B \cos kx_1 \tag{2.89}$$

can be used. Suppose that the vibrating rod under discussion is of length L and has fixed ends. This implies that $u=0$ at both $x_1=0$ and $x=L$ for all values of t and, thus, $X=0$ at these points. This requires $B=0$ and $k=n\pi/L$, where $n=1, 2,$ 3, etc., so that nodes occur at the ends of the rod, as shown in Fig. 2.34. The standing waves have wavelength λ which is given by $\lambda=2L/n$. The longest wavelength, **the fundamental mode**, has a wavelength twice the length of the rod. The higher modes, or **harmonics**, have **nodes** within the rod as well as at the ends. The cosine function can be used for rods with free ends and in this case, **antinodes** occur at the ends of the rod. Once k is known, the solution to Eq. (2.88) is straightforward. One obtains $\omega=\pi n v_0/L$ and

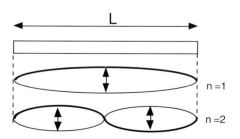

n =1

n =2

Figure 2.34 First two modes of standing waves in a rod with fixed ends.

$$u = u_0 \sin\left(\frac{\pi n v_0 t}{L}\right) \sin\left(\frac{\pi n x}{L}\right) \qquad (2.90)$$

These standing waves are fundamental in vibration problems and are known as **natural vibrations**, or **vibrational modes**. Other types of progressive waves could be analyzed in a similar fashion but this is left to other texts. Resonant vibrations can be used as a technique to measure elastic constants and this will be discussed in the next section.

2.17   Measurement of elastic constants

In order to describe the elastic behavior of a body, the values of the elastic constants are needed. Thus, it is important to understand the experimental techniques that are used to measure these constants. The most obvious approach is to apply a stress and measure the resulting strain field (**static loading**). For such approaches, **strain gages** are often used to measure the strain. These gages are usually electrical resistors that are calibrated such that changes in resistance can be converted to strain. Newly developed optical techniques, such as laser extensometers, are allowing strains to be measured without specimen contact.

Applying a known stress distribution to a material and measuring the resulting strain can be difficult experimentally. For example, in a uniaxial tension test, precise specimen alignment can critically determine the accuracy. Gripping the specimen and shaping the specimen, such that unwanted stress concentrations do not appear, can be troublesome. Moreover, in ceramics the elastic constants are high and, thus, the resulting displacements are often small, making them difficult to measure accurately. Larger deflections can be obtained from bend tests but, again, there can be experimental difficulties, e.g., accurate specimen geometry, friction at the load points, etc. In many cases, it is difficult to obtain experimental accuracies within $\pm 10\%$ for the measurement of elastic constants directly from static loading.

As a result of the above difficulties, alternative techniques were sought and have become established. One approach (**sound velocity**) involves passing ultrasonic waves through the material and determining their velocities. The other approach (**dynamic resonance**) uses the natural vibration of the material. The elastic constants are determined from specimen geometry and the resonant frequency. For these last two techniques, experimental accuracies of $<0.1\%$ are not uncommon. Both of these latter techniques can also be used on anisotropic materials but the current discussion will emphasize isotropic materials.

The velocities v_1 and v_2 of the longitudinal and transverse waves are given in Eqs. (2.83) and (2.84), respectively. Thus, from the measurement of these velocities and density, one can readily calculate B and μ for an isotropic body. The values of E and v are then obtained from the equations in Appendix 3. A schematic of the experimental set-up is shown in Fig. 2.35. In this example, a single transducer is used to send and receive the sound pulse (pulse–echo mode). In an alternative set-up, two separate transducers are used to measure the transit time. In order to apply Eq. (2.83) for longitudinal waves, it is important that the body is large compared to the wavelength. For example, the velocity in a rod will reach the limit $v_0 = \surd(E/\rho)$ as the diameter of the rod approaches zero. To overcome this problem, ultrasonic waves are used such that their wavelength is substantially less than the specimen diameter (usually a factor of ~ 50). For example, in a material in which the velocity is 10 km/s, a sample diameter of 50 mm would be required at a frequency of 10 MHz. Modern techniques allow ultrasonic wave velocities to be measured with accuracies better than 1 part in 10^5. Similar accuracy can be obtained in the density measurement but this can be highly dependent on the technique used. For anisotropic materials, it is necessary to measure these velocities along a variety of directions within the material.

In the dynamic resonance experimental technique, a body is forced to vibrate and the constants are determined from the resonant frequencies. The types of vibration utilized are usually the **longitudinal, flexural or torsional** modes. The first two allow E to be determined and the last gives the shear modulus. It is usually easier to excite flexural waves than longitudinal ones, thus the use of flexural and torsional waves will be emphasized in this discussion. To use the dynamic resonance approach, the solution to the differential equations of motion must be known and this has been accomplished for several specimen shapes. In particular, it is common to use specimens of rectangular or circular cross-section, as solutions are readily available. Vibrations in the **fundamental mode** usually give the largest amplitude and are, therefore, the easiest to detect.

For torsional waves, the relationship between μ and the fundamental resonant frequency f is given by

$$\mu = 4\rho R L^2 f^2 \tag{2.91}$$

where R is a shape factor and L is the specimen length. For cylindrical specimens $R=1$, while for rectangular specimens R depends on the aspect ratio of the cross-

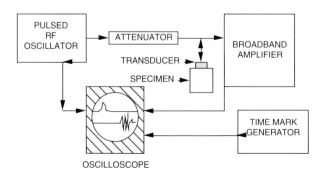

Figure 2.35 Block diagram of apparatus for transit-time measurements of sound velocity in pulse–echo mode. (Adapted from Schreiber *et al.*, 1973, reproduced courtesy of McGraw-Hill, New York.)

section. Various analytical expressions have been suggested for R and these are often tabulated. For a square cross-section, $R=1.185\ 59$.

For flexural vibrations, the equations for the fundamental resonant frequencies can be written for cylindrical and rectangular specimens respectively as

$$E=1.261L^4f^2\rho Q/D^2 \tag{2.92}$$

$$E=0.946L^4f^2\rho S/h^2 \tag{2.93}$$

where Q and S are shape factors, D is the specimen diameter and h is the specimen dimension (rectangular cross-section) parallel to the direction of vibration. Tables for Q and S are available and these parameters approach unity for large values of L/D or L/h. The shape factors actually depend on the value of v, which is unknown initially. Thus, the calculation procedure is to estimate v and then use the experimental values of E and μ to calculate a new value of v. After a few iterations, an accurate value of E is readily obtained.

Figure 2.36 shows a schematic of an experimental set-up proposed by Spinner and Teft (1961). The specimen is suspended from the driver and pick-up, both of which are usually piezoelectric transducers. The oscilloscope and voltmeter (or ammeter) are used to accurately determine the exact resonant frequency and its maximum amplitude (using Lissajous patterns, see Schreiber *et al.* (1973)). Alternatively, electronic audio equipment is now available to induce a wide range of vibration frequencies and the ensuing acoustic spectrum can be obtained by a Fourier transform.

The suspension of the specimen at the nodal points usually gives the most accurate results. For example, in the fundamental flexural vibration, the nodes occur at $0.224L$ and $0.776L$. In order to suspend the specimen, nylon and cotton fibers can be used; for high temperatures, sapphire filaments have been found to be useful. It is important to minimize any mechanical constraint on the specimen, as this will influence the natural vibration frequency. An alternative approach is to allow the specimen to rest on supports at the nodes and to use other techniques to excite and detect the vibrations. For example, gluing metal tabs to the ends of ceramic specimens and using electromagnetic transducers to

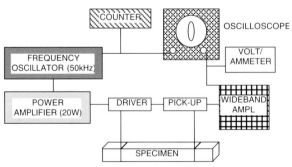

Figure 2.36 Block diagram of apparatus for dynamic bar resonance technique. (Adapted from Schreiber *et al.*, 1973, reproduced courtesy of McGraw-Hill, New York.)

excite and detect the vibrations has been found to be very useful. For flexural vibration, the tabs are located in the mid-point of each end, while for torsion they are in opposite corners. There are many other techniques to induce and detect the vibrations and these have been discussed in the literature. Clearly, it is important to identify the type of resonant frequency and its mode. As the longitudinal vibrations are difficult to excite, it is usually a matter of distinguishing between the torsional and flexural modes. Except for specimens with very low shear moduli, the fundamental frequency for torsional vibration is higher than that for flexure. Determination of the position of the nodes is also useful in detecting the type and mode of vibration.

Problems

2.1   a) Expand and simplify, if possible, the following expressions that involve the indicial (subscript) notation (δ_{ij} is Kronecker delta). Indicate the tensor order.

i) $\delta_{pq} A_{pk}$ ii) $\delta_{ij}\delta_{jk}$ iii) A_{ii} iv) $A_{ij}A_{ij}$ v) $a_i = H_{ijj}$ vi) $\delta_{ij}\delta_{ij}$
vii) $\delta_{ij}T_{kk}/3$ viii) $\delta_{ij}\delta_{ik}\delta_{jk}$ ix) $\delta_{ij}c_{ij}$ x) $\delta_{ij}x_j$ xi) c_{ij} xii)$T_{ij}x_j$
xiii) $\sigma_{ii}=3B\varepsilon_{ii}$ xiv) $\sigma_{ij}=\lambda\delta_{ij}\varepsilon_{nn}+2\mu\varepsilon_{ij}$

   b) Expand $a_i b_i$. What is this product called in vector analysis?
   c) Expand $c_i = E_{ijk}a_j b_k$ (E_{ijk} is the permutation symbol). What is c_i called in vector analysis?
   d) Using indicial notation and the above problems, show $(\mathbf{a}\times\mathbf{b}).\mathbf{a}=0$ (where $\mathbf{a}$ and $\mathbf{b}$ are vectors).
   e) Expand $E_{ijk}A_{1i}A_{2j}A_{3k}$. What is this term called in matrix algebra?
   f) Show $E_{ijk}E_{klm}=(\delta_{il}\delta_{jm}-\delta_{im}\delta_{jl})$.
   g) For the direction cosines a_{ij}, show $a_{pi}a_{qi}=\delta_{pq}$.
   h) Expand and simplify $E_{ijk}E_{kij}$.
   i) Determine the second component (f_2) for the vector $f_i=E_{ijk}T_{jk}$.

2.2   The test data overpage were gathered in a uniaxial tension test on a poly-crystalline alumina specimen. The gage section of the specimen (section of specimen under uniform tension) was cylindrical with a length of 50 mm and a diameter of 3 mm.
Complete the following tasks.
   a) Plot the data in terms of stress and strain. Is the specimen linear elastic?
   b) Determine Young's modulus (use linear regression) and Poisson's ratio.
   c) What is the stress, strain and stored elastic strain energy density at failure?
   d) Calculate the bulk and shear moduli values.

Load (kg)	Change in gage length (µm)
25	4.5
50	9.5
75	13.0
100	17.5
125	22.0
150	26.0
175	30.5
200	35.0
225	39.0
231[a]	40.0

Note:
[a] Specimen failed. Just prior to fracture, the diameter had decreased by 0.55 µm.

2.3 The interionic potential (energy per molecule) in a KCl crystal is given by $\phi=(-25.2/r)+(76{,}000/r^{10})$ eV, where r is the distance between ions in Å. The equation accounts for interactions with all surrounding ions.

a) Plot the interionic potential showing the energy minimum.

b) Determine mathematically the equilibrium ionic spacing (convert to SI units).

c) Determine the spring constant of the ionic bond in KCl (convert to SI units).

d) Calculate the cohesive energy/molecule for KCl (the energy required to separate ions to infinite distance).

2.4 A small strain at a point is defined by the following components of a displacement vector u_i: $u_1=0.04x_1-0.01x_2+0.03x_3$, $u_2=0.01x_1+0.07x_2$, $u_3=-0.03x_1+0.04x_2+0.04x_3$. Determine

a) the deformation tensor, e_{ij}.

b) the Eulerian infinitesimal tensor, ε_{ij}.

c) the linear Eulerian rotation tensor, ω_{ij}.

d) the dilatational and deviatoric components of ε_{ij}.

e) the strain invariants and the principal strains.

f) the angles needed to rotate original axes to the principal axes.

g) the maximum normal and maximum shear strains.

h) the strains if principal axes were rotated 30° anti-clockwise about x_3.

i) the principal values of the deviatoric stress tensor.

2.5   The stresses (in MPa) at a point are $\sigma_{11}=2$, $\sigma_{12}=6$, $\sigma_{13}=-4$, $\sigma_{22}=2$, $\sigma_{23}=-4$, $\sigma_{33}=12$. Determine the principal stresses by solving the cubic equation that involves the three stress invariants (Eq. (2.46)).

2.6   A cylindrical single crystal of MgO, 100 mm long and 10 mm in diameter, is loaded in uniaxial tension along the cylinder axis [001] to a stress of 10 MPa.

      a) Determine the normal and maximum shear stress on the (101) and (111) planes.

      b) Determine the direction [hkl] of the maximum shear stress on (111).

      c) Repeat part a) for the case in which stresses of 10 MPa are applied along both the [001] and [010] directions (equibiaxial tension).

2.7   The elastic constants of cubic ZrO_2 (8 mol.% Y_2O_3) are given by $c_{11}=401$ GPa, $c_{12}=96$ GPa and $c_{44}=56$ GPa.

      a) Determine the strains produced if the following stresses are applied to the {100} faces of a single crystal, $\sigma_1=\sigma_2=\sigma_3=500$ MPa and $\sigma_4=200$ MPa. Determine the normal and shear stresses on the (110) plane.

      b) Calculate the stress components if $\varepsilon_1=\varepsilon_4=0.1\%$.

      c) Determine the Young's modulus values for stressing along [100] and [111].

      d) Determine the Zener ratio for cubic ZrO_2.

      e) Zirconia also exists in a monoclinic and tetragonal form that can undergo a martensitic phase transformation. The lattice parameters for these two crystal structures at room temperature are i) $a=0.5151$ nm, $b=0.5203$ nm, $c=0.5315$ nm and $\beta=99.19°$ and ii) $a=0.5082$ nm and $c=0.5189$ nm. Determine the strain tensor for an unconstrained tetragonal single crystal transforming to monoclinic.

      f) Calculate the maximum and minimum values of Poisson's ratio for uniaxial stressing along [110].

      g) Calculate the shear modulus for {100} as a function of shear stress direction.

2.8   a) By separating stresses and strains into their spherical and deviatoric parts, show the elastic strain density of a solid ($U=\sigma_{ij}\varepsilon_{ij}/2$) can be expressed as a sum of dilatational and distortional energy densities. For an isotropic material, express these two energy densities in terms of the bulk modulus B and the shear modulus μ.

      b) For an isotropic material, determine the relationship between B and the elastic stiffness constants c_{11} and c_{12}. Repeat for s_{11} and s_{12}.

c) Calculate the strain energy density for an isotropic body (B=200 GPa) submerged in 1 km of water.

2.9   a) An isotropic rod is stretched longitudinally in tension but its sides are prevented from moving. Show that the appropriate elastic constant for this deformation is $B+4\mu/3$.

b) Derive the appropriate elastic constant for the equibiaxial stretching of a sheet.

2.10   From first principles, prove that $-1\leq v\leq 0.5$ for an isotropic elastic solid.

2.11   The Lone Ranger, Tonto and a ceramics scientist were riding through the prairie and were trying to impress each other with their talents when they approached a railroad track. Dismounting his horse, Tonto listened to the track and said his scouting skills were so good, he could tell that someone was working on the track in the far distance. The Lone Ranger said his hearing and timing was so good that he could hear that the vibrations came in sets of three pulses and that there were 1.153 s between the first two sets. Not to be outdone, the ceramics scientist, who remembered the Young's modulus of steel was 210 GPa, Poisson's ratio was 0.333 and the density of steel was 7800 kg/m^3, told them he knew the distance to the workman. Needless to say, the Lone Ranger and Tonto were extremely impressed and bought the scientist a drink.

a) How did he calculate the distance and what was his answer?

b) As a bonus the scientist told them the time delay between the second and third pulses. What was this value?

2.12   In your first job, you are asked to work on a new idea, alumina fiber guitar strings! Your job is to design the G string (fundamental frequency, 387.5 Hz). After rummaging through your old notes you find the Young's modulus (400 GPa), the Poisson's ratio (0.22) and the density (3960 kg/m^3) of alumina fibers. In addition, you find the equation for the fundamental frequency of a plucked string in terms of the tension in the string in a physics textbook. In some preliminary tests, you determine the minimum tensile strength of the fibers is 1 GPa. The ceramic fiber is 1 mm in diameter and the guitar string has a vibration length of 600 mm.

a) What tensile load is needed in the fiber and will it withstand this load? Do you foresee any other technical problems in the idea?

b) Suppose the tension in the string is suddenly increased. How long does it take for the stress pulse to reach the other end of the string (treat as a thin body)?

c) What is the position of the first fret to get G$^{\#}$ (410.6 Hz)?

d) What are the speeds of transverse and longitudinal waves in alumina?

References

General reading

M. F. Ashby, *Materials Selection in Mechanical Design*, Pergamon Press, Oxford, UK, 1992.

A. H. Cottrell, *Mechanical Properties of Matter*, Wiley and Sons, New York, 1964.

J. E. Gordon, *The New Science of Strong Materials or Why We Don't Fall Through the Floor*, Penguin Books, 1976.

W. D. Kingery, H. K. Bowen and D. R. Uhlmann, *Introduction to Ceramics,* Wiley and Sons, New York, 1976.

G. E. Mase, *Theory and Problems of*

Continuum Mechanics, Schaum Outline Series, McGraw-Hill, New York, 1970.

J. F. Nye, *Physical Properties of Crystals*, Oxford University Press, Oxford, UK, 1985.

E. Schreiber, O. L. Anderson and N. Soga, *Elastic Constants and Their Measurement,* McGraw-Hill, New York, 1973.

S. Spinner and W. E. Teft, A method for determining mechanical resonance frequencies and for calculating elastic moduli from these frequencies, *Proc. ASTM*, **61** (1961) 1221–8.

Materials data

G. Bayer, Thermal expansion characteristics and stability of pseudobrookite-type compounds Me$_3$O$_5$, *J. Less Common Metals,* **24** (1971) 129–38.

S. P. Clark, Jr, *Handbook of Physical Constants,* Geological Society of America Inc., New York, 1966.

F. H. Gillery and E. A. Bush, Thermal contraction of β-eucryptite (Li$_2$O.Al$_2$O$_3$.2SiO$_2$) by x-ray and dilatometer methods, *J. Am. Ceram. Soc.,* **42** (1959) 175–7.

D. J. Green, R. H. J. Hannink and M. V. Swain, *Transformation Toughening of Ceramics,* CRC Press, Boca Raton, FL, 1989.

Z. Li and R. C. Bradt, Thermal expansion of the hexagonal (6H) polytype of silicon carbide, *J. Am. Ceram. Soc.,* **69** [12] (1986) 863–6.

G. Simmons and H. Wang, *Single Crystal Elastic Constants and Calculated Aggregate Properties*, MIT Press, MA, 1971.

H. Schneider and E. Eberhard, Thermal expansion of mullite, *J. Am. Ceram. Soc.,* **73** [7] (1990) 2073–6.

J. F. Shackleford and W. Alexander, *The CRC Materials Science and Engineering Handbook*, CRC Press, Boca Raton, FL, 1992.

Chapter 3

Effect of structure on elastic behavior

In the last chapter, the formal description of linear elasticity was introduced. It was shown that knowledge of the elastic constants for a particular material allows one to describe the strains produced by any arbitrary state of stress. In materials science one is often interested in 'controlling' a material property and, thus, this chapter is concerned with the influence of structure on the elastic constants. At the most basic level, the elastic constants reflect the ease of deformation of the atomic bonds but it will be shown that other levels of structure can be very important, especially with the use of composite materials.

3.1 Relationship of elastic constants to interatomic potential

In Section 2.1, linear elasticity was considered from the perspective of a single atomic bond. It was shown that the elastic constant was related to the shape of the interatomic potential, notably the curvature of the potential in the vicinity of the equilibrium spacing. For example, consider the interatomic potentials shown in Fig. 3.1 that are typical of the extremes for (a) **ionic** and (b) **covalent ceramics**. The strong directional bonding associated with covalency leads to a deep potential well. For atomic interactions acting over similar distances this would lead to a sharper curvature at the potential minimum, compared to the purely ionic case. Thus, covalent ceramics are expected to have higher melting points (more energy needed to separate the atoms) and higher elastic constants. As shown in Fig. 3.2, there is a rough correlation between melting point and the bulk modulus of ceramic materials. Moreover, the deeper potential well for covalently bonded ceramics is expected to be more symmetric and this leads to lower thermal expansion coefficients. As temperature is increased in a material, the mean inter-atomic position changes and the degree of change is reflected in the symmetry of the potential well. Thus, one concludes that covalent ceramics should have higher melting points, higher elastic moduli and lower thermal expansion coefficients than ionic ceramics. The last two effects can be seen in the data of Tables 2.1 and 2.2. **Metallic bonding** is intermediate in strength between covalent and ionic and, thus, similar comparisons can be made between metals and cova-lent ceramics, though the differences are less extreme.

If an accurate equation for the interatomic potential is known, the elastic con-stants can be calculated from first principles. This analysis is straightforward for cubic ionic crystals. The potential for a pair of positive and negative ions is often written in the form

$$\phi = \frac{-q_c q_a}{4\pi\varepsilon_0 r} + \frac{c}{r^n} \tag{3.1}$$

where the first term on the right-hand side is the electrostatic attraction between the positive and negative ions and the second is an empirical repulsive term, that contains the empirical constants C and n. In Eq. (3.1), r is the distance between

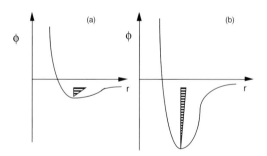

Figure 3.1 Comparison of interatomic potentials for (a) ionic and (b) strongly covalent bonds. The cova-lent potential usually exhibits a deeper, more symmetric energy well and stronger curvature of the potential at the equilib-rium spacing.

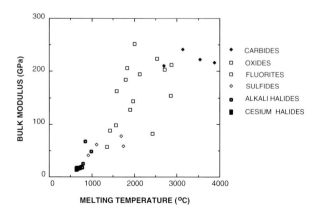

Figure 3.2 Correlation between bulk modulus and melting point for various ceramics.

the centers of the ions, q_c and q_a are the charges on the cation and anion, respectively, and ε_0 is the permittivity of vacuum. A crystal contains many such atomic pairs and one must take into account all the ionic interactions. For the attractive term, a constant called the **Madelung constant** A is used to sum these interactions for a particular crystal structure type. The Madelung constants for various crystal structures are given in Appendix 4. The repulsive term is usually considered to be short range (high n) so only the interactions with the z nearest neighbors are important. Thus, when all the interactions are summed, one obtains

$$\phi_z = \frac{-Aq_cq_a}{4\pi\varepsilon_0 r} + \frac{zC}{r^n} \tag{3.2}$$

If Eq. (3.2) is differentiated and set equal to zero for $r=a_0$ (the interatomic spacing), the parameter zC can be eliminated, i.e.,

$$\frac{d\phi_z}{dr} = \frac{Aq_cq_a}{4\pi\varepsilon_0 r^2} - \frac{znC}{r^{n+1}} = 0 \tag{3.3}$$

which gives

$$zC = \frac{Aq_cq_a a_0^{n-1}}{4\pi\varepsilon_0 n} \tag{3.4}$$

Substituting Eq. (3.4) into Eq. (3.2), one obtains

$$\phi_z = \frac{-Aq_cq_a}{4\pi\varepsilon_0 r}\left(1 - \frac{a_0^{n-1}}{nr^{n-1}}\right) \tag{3.5}$$

To determine the energy ϕ_b between nearest neighbors within a particular crystal structure Eq. (3.5) is divided by z (the coordination number). For rock-salt crystal structures, $z=6$ and so Eq. (3.5) becomes

$$\phi_b = \frac{-Aq_cq_a}{24\pi\varepsilon_0 r}\left(1 - \frac{a_0^{n-1}}{nr^{n-1}}\right) \tag{3.6}$$

The force constant k for the single bond is obtained by differentiating ϕ twice with respect to r and putting $r=a_0$ (Eq. (2.2)). Performing these operations one obtains

$$\frac{d\phi_b}{dr} = \frac{Aq_cq_a}{24\pi\varepsilon_0 r^2}\left(1 - \frac{a_0^{n-1}}{r^{n-1}}\right) \tag{3.7}$$

and

$$\left(\frac{d^2\phi_b}{dr^2}\right)_0 = \frac{Aq_cq_a(n-1)}{24\pi\varepsilon_0 a_0^3} \tag{3.8}$$

If the rock-salt crystal is being subjected to a hydrostatic stress, i.e., σ along each of the $<100>$ directions, the force f acting on a single bond is given by $f=\sigma a_0^2$. If the displacement in a given direction with respect to cartesian axes is u, the fractional change in volume $\Delta V/V = 3u/a_0$. From the definition of bulk modulus (Section 2.3) one has $B = \sigma/(\Delta V/V)$. Using the version of Hooke's Law in Eq. (2.2) and the above information, one obtains $B = (d^2\phi_b/dr^2)_0/3a_0$. Using Eq. (3.8) and re-introducing the nearest-neighbor parameter, one obtains

$$B = \frac{Aq_cq_a(n-1)}{12z\pi\varepsilon_0a_0^4} \tag{3.9}$$

Figure 3.3 shows the relationship between bulk modulus and interionic spacing for materials with the NaCl crystal structure. Excellent agreement is found for the a_0^{-4} relationship predicted by Eq. (3.9). Indeed, this equation seems to fit the behavior of ionic materials extremely well. It is clear from Eq. (3.9) that B is influenced strongly by **molar volume** (through a_0), by **valence** (through q_c and q_a), by the **strength of the repulsive term** (through n) and by **crystal structure** (through A and z). This relationship holds well for higher-valent, rock-salt crystals (e.g., MgO, PbS and PbSe) and other ionic crystal-types (zinc blende, cesium chloride). It is even a good approximation for the bcc alkali metals (i.e., Li, Na, K, Rb and Cs) and materials with the diamond structure (e.g., C, Si, Ge). The relationship does, however, break down for the transition-metal carbides and, presumably, this is a reflection of the assumed ionic potential. From the above discussion, it is often generalized that a decrease in covalency should lead to a decrease in bulk modulus. Thus, the following series: carbides, borides, nitrides, oxides, sulfides, halides, should be one of decreasing bulk modulus. These trends can be identified in Table 2.1. For many materials, it is useful to note that the molar volume, $V_m \sim a_0^3$, so that Eq. (3.9) predicts a $V_m^{-4/3}$ dependence of the bulk modulus.

The above analysis can be applied to other forms of the interatomic potential. For example, in some crystals, the attractive term is of shorter range than a coulombic interaction and is often replaced by a term with an r^{-m} dependence.

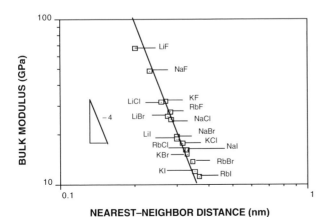

Figure 3.3 Bulk modulus as a function of interionic spacing for alkali halides.

In some approaches, the empirical repulsive term is written in other mathematical forms, such as an $\exp(-r/\lambda)$ dependence, where λ is an empirical constant. For many ceramics, in which the cation to anion ratio is less than 0.414, the structures are considered as close-packed anions with the cations in the interstices. In these structures, the spacing between the anions becomes important and the potential must account for the **anion–anion repulsion**. In many situations the above approach, which links the interatomic potential to the elastic constants, is used in reverse. That is, the elastic constants are used to determine the unknown constants in the interatomic potential, e.g., n in Eq. (3.1). Equation (3.9) is also useful in estimating the elastic constants of newly studied materials using values measured from other materials with the same crystal structure.

Crystal structure has an important role to play in determining the elastic constants. One approach is to consider the number of bonds per unit area available to support a stress. For example, $CaCO_3$ (aragonite) has a slightly higher density than $CaCO_3$ (calcite) and this leads to a slight increase in the corresponding elastic constants. **Bond density** is also very useful in understanding the elastic moduli of silicate glasses ($B \sim 40$ GPa and $\mu \sim 30$ GPa). These more-open structures lead to lower elastic constants than the corresponding crystalline oxides due to decrease in bond density. In silicate glasses, the elastic constants are sometimes considered in terms of the number of **non-bridging oxygens**. The openness of the glass structure is often determined by the **field strength** of the **modifying ions**. Figure 3.4 shows data on the Young's modulus of sodium aluminosilicate glasses, in which the sodium to aluminum ratio was varied to give a change in the number of non-bridging oxygens; increasing the number of non-bridging oxygens decreases the bond density and, hence, leads to a decrease in the modulus.

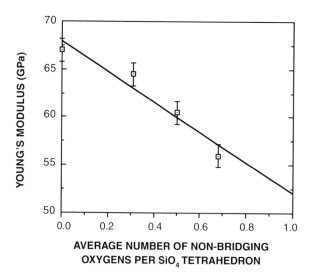

Figure 3.4 Variation of Young's modulus with non-bridging oxygen concentration in sodium aluminosilicate glasses (Adapted from Coon, 1986.)

3.2   Elastic anisotropy and atomic structure

The discussion in the previous section indicated that the elastic constants are related to the stiffness of the atomic bonds. In addition, one also expects the geometry of the bonding will be reflected in the symmetry and magnitude of the elastic constants. Returning to Table 2.4, compare the **Zener ratios** for various cubic crystals. For the alkali metals, one notes that $Z>1$. This is expected, as the primary bonds in a bcc structure are aligned along <111>. For materials with the rock-salt structure, one finds primarily that $Z<1$, reflecting <100> as the primary bond directions. For the rock-salt crystal structures that satisfy these conditions, one also notes that anisotropy tends to increase as one moves down the periodic table, reflecting the increase in the polarizability of the anions. For example, these trends can be seen in the sodium, potassium and rubidium halides as one moves from the fluoride to the iodide. For the lithium halides and the transition-metal oxides in Table 2.4, it is found that $Z>1$, at odds with the above generalizations. If one considers the crystal structure of these materials, one notes that there is a large difference between the size of the anions and cations. It is better, therefore, to consider these structures as close-packed anions, with the cations in the interstices. For example, Fig. 3.5 shows a schematic of the MgO and KCl structures with the anions drawn approximately to scale. It is clear that the MgO crystal will be dominated by anion–anion repulsion, so that uniaxial stressing along <110> becomes much more difficult. The <110> directions are closer to <111> than <100> and, thus, Z increases above unity. The increase in the relative ease of deformation for these groups is also reflected in higher v values for uniaxial stressing along <100>. A similar contrast can also be found between the alkali metals and the cesium halides. Both groups have bcc structures but have opposite Z values (greater and lesser than unity, respectively). In the cesium halides, anion–anion repulsion becomes important, increasing the stiffness along <100>.

One of the difficulties in using Z values as a measure of anisotropy is the dis-

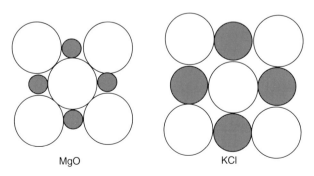

MgO   KCl

Figure 3.5 The relative ion sizes for the MgO and KCl rock-salt structures for {100}. In MgO, the larger radius ratio for the ions allows the anions (not shaded) to be in closer proximity, giving a strong repulsion when uniaxially stressed along [100]. (Adapted from Newnham, 1975.)

parate ranges of Z values below and above unity. For $Z<1$, values range from zero to unity whereas, for $Z>1$, values range from unity to infinity. For this reason, other anisotropy parameters have been suggested. For example, one simple approach would be to form the ratio of the absolute difference between the shear moduli μ_0 and μ_1 (see Section 2.11) and twice their average. This would give a new parameter $Z'=(Z-1)/(Z+1)$ and the Z' values would then range from -1 to $+1$.

Atomic trends in anisotropy also exist for crystals with lower symmetry than cubic but these changes are more difficult to visualize. For hexagonal crystals, Eq. (2.64) can still be used to assess the anisotropy but, clearly, additional anisotropy parameters would be needed to completely describe the behavior. This increase in complexity is beyond the scope of this text except for a couple of simple examples, discussed below.

As a preliminary situation, consider systems where bonds of two different stiffness values (k and K) are placed in series or in parallel. The situation is shown schematically in Fig. 3.6. For the series arrangement the force F applied to the system is given by $F=c_s A_s u_s/L_s$, where u_s is the total displacement, A_s is the area and c_s is the stiffness constant for the series system. For this arrangement, both spring types (length, $L_s/2$) are subjected to the same force, i.e., $F=2n_s k u_1/L_s=2n_s K u_2/L_s$, where n_s is the number of links, and u_1 and u_2 are the displacements of the compliant (k) and stiff (K) springs, respectively. In addition, $u_1+u_2=u_s$ and with some manipulation, one obtains

$$c_s=\frac{n_s L_s}{A_s}\left(\frac{2kK}{k+K}\right)\tag{3.10}$$

For the parallel arrangement, both types of spring are subjected to the same displacement. It is left to the reader to show that the stiffness constant c_p for the parallel system with an equal number of the two spring types is given by

$$c_p=\frac{n_p L_p}{A_p}\left(\frac{k+K}{2}\right)\tag{3.11}$$

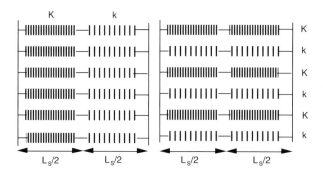

Figure 3.6 Two possible spring models for mixed stiff and compliant bonding. In one case (*left*) the springs are in series, compared to a parallel arrangement in the other case.

Table 3.1 *Comparison of longitudinal elastic stiffness constants for various single crystals (units GPa)*

Material	c_{11}	c_{22}	c_{33}
Diamond	1076	1076	1076
Graphite	1076	1076	47
α-Quartz	87	87	107
Biotite	186	186	54
Muscovite	178	178	55
Phlogopite	179	179	52

where n_p is the total number of springs. To assess the combination of stiff and compliant bonding, it is useful to ignore the geometric terms, i.e., to assume $n_s L_s/A_s = n_p L_p/A_p$. For this case, $c_s/c_p = 4kK/(k+K)^2$ and one finds the ratio is less than unity when $K/k > 1$. Thus, the displacement in the parallel case will be dominated more by the stiffer springs than in the series case, i.e., most of the energy will be stored in the stiffer springs. It is concluded that the parallel system will have the higher stiffness and, hence, in crystals with a mixture of weak and strong bonding, directions in which the bonds are in series will be the most compliant.

Comparison of the elastic constants for graphite and diamond, as given in Table 3.1, shows that although the elastic constants are comparable for uniaxial strains in the x_1 and x_2 directions (strongly covalent), the value of c_{33} for graphite is much lower. This is a reflection of the weak secondary bonding (compliant 'springs') between the layers in graphite. Table 3.1 also includes elastic constant data on some micas. In these crystals, Si_6O_{18} rings are joined to each other to form the covalently bonded tetrahedral layer giving high values of c_{11} and c_{22} but, again, weaker ionic bonding occurs between the layers, giving low c_{33} values.

In Section 2.14 it was argued that the pressure derivatives of the elastic constants are usually positive and the temperature derivatives negative. Table 3.2 shows the pressure derivatives for four oxide minerals and some derivatives are negative. For crystals consisting of open structures, bond rotations can occur under the action of pressure and the shearing rotations lead to negative derivatives. Table 3.3 gives the temperature coefficients for the elastic stiffness constants of α-quartz and it is found that Tc_{66} is positive, converse to the generalization. This crystal has unusual expansion behavior and, in particular, some bonds tend to rotate and 'straighten' as the temperature is increased. Thus, if the bond is subjected to a tensile force at high temperature, i.e., when it is straight, it is not subject to as much bending and the stiffness is higher. This effect is illustrated schematically in Fig. 3.7.

Table 3.2 *Comparison of pressure derivatives for elastic stiffness constants ($\partial c_{ij}/\partial P$) for four oxide minerals*

(After Newnham, 1975.)

ij	Beryl	Quartz	Corundum	Forsterite
11	4.5	3.3	6.2	8.3
22	4.5	3.3	6.2	5.9
33	3.4	10.8	5.0	6.2
44	−0.2	2.7	2.2	2.1
55	−0.2	2.7	2.2	1.7
66	0.3	−2.7	1.5	2.3
12	3.9	8.7	3.3	4.3
13	3.3	6.0	3.7	4.2
23	3.3	6.0	3.7	4.2
14	0	1.9	0.1	0

Table 3.3 *Temperature coefficients of elastic stiffness (Tc_{ij})[a] for α-quartz*

(After Newnham, 1975.)

ij	11	33	44	66
Tc_{ij}	−0.5	−2.1	−1.6	1.6

Note:
[a] Units: 10^{-4} / K

3.3   Elastic behavior of particulate composites

Moving to a larger scale, let us now look at the influence of microstructure on elastic behavior. As indicated in the previous section, the elastic constants are a fundamental property of single crystals through the geometry and stiffness of the atomic bonds. Thus, one may expect elastic behavior to be controlled simply by the choice of material. By using **composite materials**, however, one can control the final set of elastic properties with some precision, i.e., by mixing phases with different elastic constants. Clearly, it is useful to be able to predict the elastic constants of a composite from those of its constituents. This has been accomplished for many types of composite microstructures. For this section, however, the emphasis will be on (elastically) isotropic composites, i.e., composites contain-

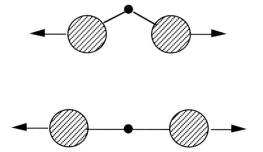

Figure 3.7 In some open structures, bonds may be bent at lower temperatures but straighten at higher temperatures as the structure expands. The increase in bond linearity increases its stiffness, offsetting the decrease normally associated with a temperature increase.

ing a random array of particles (**particulate composites**). Using stress and strain averaging techniques, Voigt and Reuss[†] were able to predict upper and lower bounds for the **constitutive relationships**. For the bulk modulus B and the shear modulus μ of the composites, simple 'Rule of Mixtures' equations were obtained, giving **Voigt–Reuss bounds** in the form

$$M = M_1 V_1 + M_2 V_2 \tag{3.12}$$

$$\frac{1}{M} = \frac{V_1}{M_1} + \frac{V_2}{M_2} \tag{3.13}$$

where M is either the bulk or shear modulus, V is the volume fraction and the subscripts 1 and 2 denote the individual phases. The values of Young's modulus and Poisson's ratio are subsequently calculated from the bulk and shear moduli. Equations (3.12) and (3.13) are very similar to the results of the analysis in the previous section, in which springs were combined either in parallel (equal strain) or in series (equal stress). From the above equations, it is clear that the addition of covalent materials (high modulus) can be useful in increasing substantially the stiffness of low-modulus materials, such as polymers, glasses and some metals. Figure 3.8 shows an example of the microstructure of a particle-reinforced aluminum alloy. Clearly, the above theoretical approach could be extended for materials with more than two phases. Equations (3.12) and (3.13) will simply have additional terms for each new phase.

Hashin and Shtrikman[†] have determined bounds using basic elasticity energy theorems rather than stress and strain averaging and their predictions can be written in the form

$$\frac{M - M_1}{M_2 - M_1} = V_2 \left[1 + \frac{V_1(M_2 - M_1)}{M_1 + H} \right]^{-1} \tag{3.14}$$

† See Watt *et al.*, 1976.

where $H=4\mu_2/3$ or $H=4\mu_1/3$ for the bounds on the *bulk* modulus and

$$H=\frac{\mu_2(9B_2+8\mu_2)}{6(B_2+2\mu_2)}; \ H=\frac{\mu_1(9B_1+8\mu_1)}{6(B_1+2\mu_1)} \tag{3.15}$$

for the bounds on the *shear* modulus if $(\mu_2-\mu_1)(B_2-B_1)\geq0$. For the case when $(\mu_2-\mu_1)(B_2-B_1)<0$, Eq. (3.15) is replaced by

$$H=\frac{\mu_2(9B_1+8\mu_2)}{6(B_1+2\mu_2)}; \ H=\frac{\mu_1(9B_2+8\mu_1)}{6(B_2+2\mu_1)} \tag{3.16}$$

The **Hashin-Shtrikman (HS) bounds** are found to always lie within the Voigt–Reuss (VR) bounds and are thus more useful. The HS bounds on the bulk modulus have been shown to be the best possible limits, given only the volume fraction and moduli of the constituent phases. There does not appear to be a similar proof for the shear modulus. In order to show that the bulk modulus bounds were the best limits, Hashin (1983) demonstrated that the bounds were identical to the exact results for a composite sphere assemblage. This assemblage is shown in Fig. 3.9, in which a sphere of one phase is embedded in a concentric shell of the other phase. If the radii of the inner and outer spheres are a and b, the volume fraction of the inner phase, $V_1=(a/b)^3$. The solid body is then produced by adding additional spheres, decreasing in size, until all the interstices in the structure are filled. The bounds are obtained by transposing the phases in the com-

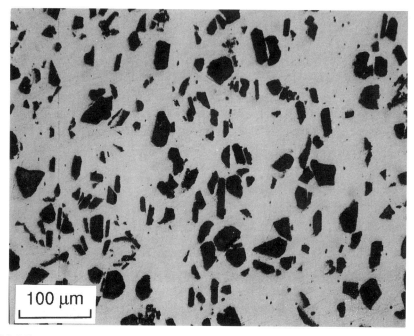

100 μm

Figure 3.8 Microstructure of aluminum alloy reinforced with alumina particles; optical micrograph. (Courtesy of Venkata Vedula.)

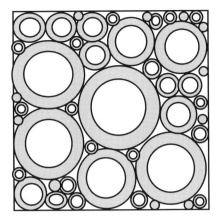

Figure 3.9 Composite sphere assemblage used by Hashin (1983) for exact solutions to the elastic behavior of statistically isotropic composites.

a)

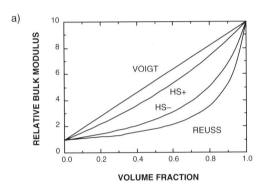

b)

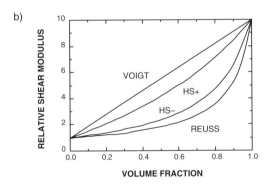

Figure 3.10 Relative value of a) bulk and b) shear modulus for a particulate composite in which the bulk (shear) modulus of one phase is ten times that of the other phase.

posite spheres, i.e., the inner phase becomes the concentric shell and vice versa. This sphere assemblage is useful in giving a physical interpretation to the HS bounds if one notes the 'inner' phase is isolated in a connected matrix of the 'outer' phase. When the stiffer phase is the matrix material, one obtains the HS upper bound and the lower bound when it is the inner core material. It is important to remember the HS bounds do *not* actually assume a geometry for the phases. Figure 3.10 shows a comparison of the VR and HS bounds for the case when the ratio of (a) the bulk and (b) the shear modulus of the two phases is 10. The width

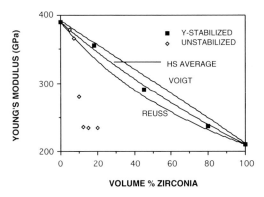

Figure 3.11 Comparison of predicted Young's modulus for alumina–zirconia composites with experimental data. The agreement is good if zirconia is retained in the tetragonal phase (by stabilization with yttria) but is poor for the unstabilized zirconia at higher volume fractions.

of the bounds decreases as the modulus ratio of the constituents decreases. From the above discussion one concludes that the HS bounds are superior to the VR bounds and are extremely useful for composites in which the modulus ratio is less than 10. If the modulus ratio is greater than 10, the bounds become less useful and, to make further progress, information on the geometry of the phases is needed.

For bounding types of solutions, various procedures for **averaging** the upper and lower bounds have been put forward but often the simple arithmetic average is used. For the VR bounds this idea was suggested by Hill[†] and hence this average is often called the **VRH average**. If one accepts the HS bounds are preferable to the VR bounds, then the arithmetic average of the HS bounds should also be preferred. For example, one could quote the HS average with $\pm$ value to indicate the width of the bounds. Another difficulty with the VRH average is that it can lie outside the HS bounds in some composite systems.

Figure 3.11 shows elastic constant data for the **alumina-zirconia** (Al_2O_3–ZrO_2) **composite** system as predicted by Eqs. (3.12)–(3.14). Values for the Young's moduli for the alumina and zirconia were taken as 390 and 210 GPa, respectively. The Poisson's ratio values were assumed to be 0.23 and 0.30 for alumina and zirconia, respectively, though the calculation is not very sensitive to these values. The HS bounds were found to be very close to each other and, thus, only the arithmetic average is shown in the figure. Experimental data for the composite system is also included in the figure for cases in which zirconia is either **unstabilized** (no alloying additive) or is **stabilized** with yttria. All the experimental data were gathered from hot-pressed specimens, which were close to theoretical density (> 98%). No attempt was made to include any effects due to porosity differences. For the stabilized system, the data lie close to the HS average. Elastic-constant data lying below the HS lower bound can be taken as evidence for poor mechanical integrity in the composites. This could be due to poor bonding between the phases, porosity or microcracking effects. Many of the composites containing unstabilized ZrO_2, fall below the lower bound in Fig. 3.11. The source of this discrepancy will be discussed later (Section 3.6).

[†] See Watt *et al.*, 1976.

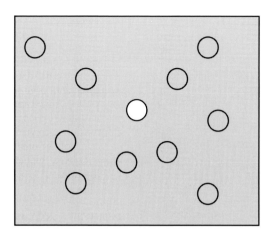

Figure 3.12 The self-consistent scheme considers a particle with a given geometric shape, embedded in an effective medium that possesses the elastic properties of the composite.

3.4   Advanced constitutive relationships for composites

As pointed out in the previous section, if the constituent phases have widely differing moduli, the bounding solutions become less useful. One approach to obtaining more information about such composites is to specify the geometrical structure of the composite. Some exact solutions are available for dilute concentrations of the second phase, notably the results of Eshelby (1957) for ellipsoidal inclusions. These solutions will not be discussed, as they are the limiting case for the self-consistent solutions, to be discussed next.

The **self-consistent (SC) solutions** are an approximate approach, in which a particle of one phase of given shape (e.g., spheres, needles or discs) is surrounded by the *composite* material. This is illustrated in Fig. 3.12, in which a spherical particle is surrounded by an **effective medium** that represents the elastic properties of the composite. For the spherically shaped particles, one can again use Eq. (3.14) but now $H=4\mu/3$ is used for the bulk-modulus calculation and

$$H=\frac{\mu(9B+8\mu)}{6(B+2\mu)} \tag{3.17}$$

for the shear-modulus calculation. The parameter H now uses the elastic properties of the composite, *not* the constituents. The equations for B and μ for the composite are now coupled and, thus, an iteration procedure must be used to obtain their values. The convergence for the solutions is usually very rapid, especially if one starts with a bounding solution as the first approximation, e.g., the average of Eqs. (3.12) and (3.13). The solutions for the disc-shaped particles are the same as the HS bounds. The SC solutions for spherical and disc-shaped particles are shown in Fig. 3.13 (a) and (b) for the bulk and shear moduli of a composite system, in which $B_1/B_2=0.1$ and $\mu_1/\mu_2=0.1$, respectively. The SC solutions lie at or between the HS bounds. The sphere solution is found to be symmetric, i.e., it does not matter which phase is chosen as the sphere. The sphere solutions are also tangential to the upper HS bound when the compliant phase is at low concentration in the stiff matrix, and tangential to the lower HS bound for the opposite case. Thus, if the spheres are the stiffer phase they will be isolated at low volume fractions and will, therefore, be nearer to the lower bound (and vice versa). One can argue that discs are more likely to 'percolate' the structure than spheres. Hence, if they are more compliant, they will reduce the modulus more than spheres for a given volume fraction (and vice versa). The SC solution for needles lies between the sphere and disc estimates.

The SC solutions appear to run into difficulties when there is a large **elastic mismatch** between the constituent phases, for example, at high concentrations of a rigid phase in a compliant matrix or of a porous phase in a stiff matrix. The latter situation will be discussed in Section 3.6. One approach to this problem is known as the **Generalized Self-Consistent Approach**, the concept behind which is illustrated in Fig. 3.14. Instead of a single inclusion in an effective medium, a composite sphere is introduced into the medium. As in the composite sphere assemblage discussed in the last section, the relative size of the spheres reflects the volume fraction, i.e., $V_1 = (a/b)^3$ as before. Interestingly, this approach leads to the HS bounds for the bulk modulus. The solution for the shear modulus is complex but can be written in a closed form.

Two other approaches have gained significant attention: the **Differential**

a)

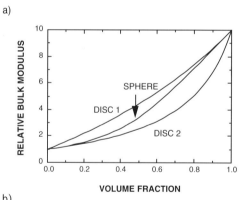

b)

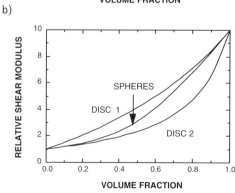

Figure 3.13 Relative values of a) bulk and b) shear modulus for a particulate composite in which the bulk (shear) modulus of one phase is ten times that of the other phase using self-consistent solutions.

Method; and the **Mori–Tanaka Method** (Mori and Tanaka, 1973). In the former technique, the composite is viewed as a sequence of dilute suspensions and, thus, one can use the exact solutions for these cases to determine the effective composite properties. For example, the solution by Eshelby (1957) for ellipsoidal inclusions can be used. The increments of added inclusions are taken to an infinitesimal limit and one obtains differential forms for the bulk and shear moduli, which are then solved. The Mori–Tanaka method involves complex manipulations of the field variables. This approach also builds on the dilute suspension solutions (low V_1) and then forces the correct solution as $V_1 \rightarrow 1$.

The various approaches to the constitutive relationships for the elastic constants of statistically isotropic composites have also been used to describe other material properties, e.g., thermal and electrical conductivity. Of interest to this text is the thermal expansion behavior. The thermal expansion α of an isotropic composite is given by

$$\alpha = \alpha_1 + \frac{B_2(\alpha_2 - \alpha_1)(B_1 - B)}{B(B_1 - B_2)} \tag{3.18}$$

where α_1 and α_2 are the thermal expansion coefficients for the constituents, B_1 and B_2 are the bulk moduli of the constituents and B is the bulk modulus of the composite. In order to progress further, the bulk modulus of the composite is needed and this can be predicted by the methods outlined previously. For example, the HS bounds on the bulk modulus will give bounds on the thermal expansion of the composite.

Many composites are elastically anisotropic and, thus, more than the two elastic constants are needed to describe their elastic behavior. This is a very large topic and, for this text, just some of the basic ideas that apply to unidirectional fiber composites will be discussed. Consider a two-phase material, with the two geometries shown in Fig. 3.15, being subjected to a uniaxial tensile stress. For the structure in Fig. 3.15(a), the two phases are subjected to equal strain whereas,

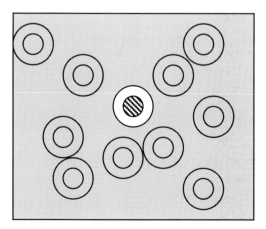

Figure 3.14 The generalized self-consistent approach in which a composite particle is embedded in an effective medium that possesses the elastic properties of the composite. The relative sphere size in the composite particle represents the volume fraction of the phases.

in the other structure, the phases are subjected to equal stress. For Fig 3.15(a), the applied force F must be divided between that on phase 1 (F_1) and that on phase 2 (F_2), i.e., $F=F_1+F_2$. In terms of the stresses, one obtains

$$\sigma_b A_b = \sigma_1 A_1 + \sigma_2 A_2 \tag{3.19}$$

where A_b is the total bulk cross-sectional area and A_1 and A_2 are the areas of the two phases. Using Hooke's Law for uniaxial tension, $\sigma = E\varepsilon$, and, recognizing the bulk strain is the same as that on the individual phases, one obtains

$$E_L = E_1 V_1 + E_2 V_2 \tag{3.20}$$

for the longitudinal geometry, where V_1 and V_2 are the volume fractions of phases 1 and 2, respectively, and are equal to their respective area fractions. This is a simple 'rule-of-mixtures' type of solution and is analogous to the analysis associated with Eq. (3.11). It is left to the reader to show that

$$\frac{1}{E_T} = \frac{V_1}{E_1} + \frac{V_2}{E_2} \tag{3.21}$$

for the transverse geometry in Fig. 3.15(b). These equations are the simplest approximations for the longitudinal and transverse Young's moduli of a unidirectional fiber composite. To completely describe the elastic behavior of such a transversely isotropic material, one needs five elastic constants.

The various techniques for developing the constitutive equations of particulate composites, e.g., the HS and SC approaches, have also been applied to fiber composites. Equation (3.20) for the longitudinal Young's modulus is usually considered to be an excellent approximation. In addition, it has been shown the Poisson's ratio v_L for uniaxial stressing in the fiber direction is given by

$$v_L = v_1 V_1 + v_2 V_2 \tag{3.22}$$

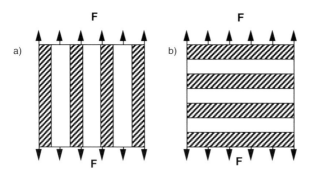

a)

b)

Figure 3.15 Simple slab models for determining (a) longitudinal and (b) transverse Young's modulus of unidirectional fiber composites.

Equation (3.21) for the transverse modulus, however, is usually considered a poor approximation, primarily because of the constraint produced by the fibers in the transverse contraction. It is not appropriate to discuss all the alternative approaches in this text. Instead, only the empirical equations developed by Halpin and Tsai (Halpin, 1984) will be given, as these can be used as simple approximations. For these equations, x_1 is the fiber direction. In addition to Eqs. (3.20) and (3.22), the other elastic constants M for a unidirectional fiber composite can be written as

$$\frac{M}{M_2} = \frac{1 + \xi X V}{1 - X V} \tag{3.23}$$

where M_2 is the corresponding Young's or shear modulus of the matrix, V is the fiber volume fraction, ξ is a constant that depends on the reinforcement geometry,

$$X = \frac{M_1 - M_2}{M_1 + \xi M_2} \tag{3.24}$$

and M_1 is the Young's or shear modulus of the fiber. For the composite shear modulus c_{66}, $\xi = 1$; for the transverse Young's modulus, E_T, $\xi = 2$; and for the other shear modulus, c_{44}, $\xi = 1/(4 - 3v_2)$, where v_2 is the Poisson's ratio of the matrix. In fiber composites, the architecture of the fibers is usually more complex than unidirectional and this can significantly impact the constitutive relationships.

3.5   Constitutive relations for random polycrystals

Many ceramics are used in a random polycrystalline form and thus, it is useful to be able to predict the elastic constants from those of the single crystals. The approaches outlined in the last two sections are used for this procedure by considering the random polycrystal as an infinite number of phases with all possible orientations. For example, Voigt and Reuss[†] used a technique based on averaging the stiffness or compliance constants and obtained upper and lower bounds. The Voigt upper bounds for the bulk (B) and shear (μ) moduli of the composite can be written as

$$B = \frac{C + 2D}{3}; \ \mu = \frac{C - D + 3E}{5} \tag{3.25}$$

where $C = (c_{11} + c_{22} + c_{33})/3$, $D = (c_{12} + c_{23} + c_{31})/3$ and $E = (c_{44} + c_{55} + c_{66})/3$. The Reuss lower bounds are given by

$$B = \frac{1}{3(Q + 2R)}; \ \mu = \frac{5}{(4Q - 4R + 3S)} \tag{3.26}$$

[†] See Watt *et al.*, 1976.

where $Q=(s_{11}+s_{22}+s_{33})/3$, $R=(s_{12}+s_{23}+s_{31})/3$ and $S=(s_{44}+s_{55}+s_{66})/3$. The above equations are simple to use and, again, the arithmetic VRH average is often quoted. Hashin and Shtrikman[†] have improved the VR bounds. For *cubic* poly-crystals, the bulk modulus is unambiguously given by

$$B=\frac{c_{11}+2c_{12}}{3} \tag{3.27}$$

while the shear modulus is bounded by

$$\mu=\mu_1+3\left(\frac{5}{\mu_0-\mu_1}-4\beta_1\right)^{-1} \tag{3.28}$$

and

$$\mu=\mu_0+2\left(\frac{5}{\mu_1-\mu_0}-6\beta_2\right)^{-1} \tag{3.29}$$

where $\mu_0=c_{44}$, $\mu_1=(c_{11}-c_{12})/2$, and

$$\beta_1=\frac{-3(B+2\mu_1)}{5\mu_1(3B+4\mu_1)}; \beta_2=\frac{-3(B+2\mu_0)}{5\mu_0(3B+4\mu_0)} \tag{3.30}$$

Expressions are available for the HS bounds for the other crystal classes but the mathematics increases in complexity and, thus, the VRH averages are often used. The data in Table 2.1 were calculated using this procedure and Table 3.4 compares some experimental values with the theoretical values. It appears from the data that these constitutive equations are extremely successful in estimating the elastic constants of the polycrystalline material.

3.6   Effects of porosity and microcracking on elastic constants

As indicated earlier, the upper and lower bounds for the elastic constants become wider as the modulus ratio of the constituents increases. Thus, the bounding approaches become much less useful. This is particularly true when the second phase is porosity. In this case, the lower bounds are zero and there is, therefore, great difficulty in accurately estimating the elastic constants of porous materials with this approach. The upper bounds in Eqs. (3.12) and (3.14) can still be used by setting the bulk and shear moduli of the second phase to zero. Porosity will always lead to a decrease in elastic modulus as not only is the load-bearing area of a material being reduced by the pores but also the stress becomes 'concentrated' near the pores (see Section 4.9).

[†] See Watt *et al.*, 1976.

Table 3.4 *Comparison of experimental data on the elastic constants of polycrystalline ceramics with the values obtained from single-crystal data*

Material		E (GPa)	μ (GPa)
Al_2O_3	VRH average	401	161
	Experiment	402	163
MgO	HS average	306	131
	Experiment	318	134
$MgO.Al_2O_3$	HS average	272	107
	Experiment	276	107
ThO_2	HS average	250	97.2
	Experiment	256	98.7
α-SiC	HS average	420	179
	Experiment	430	191
TiB_2	HS average	579	262
	Experiment	569	259

Note:
The experimental data are values extrapolated to zero porosity.

Exact solutions exist for bodies containing a dilute suspension of pores. For example, for a body containing a low concentration of spherical pores, the MacKenzie (1950) solution for the Young's modulus E of the porous body is given by

$$E = E_0(1 - AP + BP^2) \tag{3.31}$$

where E_0 is the Young's modulus of the dense material, P is the porosity fraction and A and B are constants of the order of 1.9 and 0.9, respectively. An alternate approach has been put forward by Gibson and Ashby (1988) in which the deformation of a porous unit cell was analyzed. This was formulated primarily for solid foams with very high values of P (i.e., $P > 0.7$). In this work, it was suggested that

$$E = E_0(1 - P)^2 \tag{3.32}$$

and

$$\mu = \mu_0(1 - P)^2 \tag{3.33}$$

where μ and μ_0 are the shear moduli of the porous and dense material, respectively. These equations lie close to the HS upper bound. The similarity between Eq. (3.31) and (3.32) led Gibson and Ashby to the suggestion that Eqs. (3.32) and (3.33) could be used as approximations for all values of P. Experimental data on ceramic foams support the square power relationships in Eqs. (3.32) and (3.33) but the theory overestimates the magnitude of the elastic constants, as shown in Fig. 3.16. The microstructure of these high-porosity ceramics is shown in Fig. 3.17. Equations (3.32) and (3.33) also overestimate the elastic constants for many sintered ceramics, except at low P values.

The next level of refinement in developing constitutive relationships for porous bodies is to assume a pore shape. Thus, the self-consistent solutions can also be

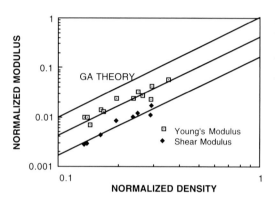

Figure 3.16 Comparison of elastic constants of an open-cell alumina foam with the Gibson and Ashby (GA) theory.

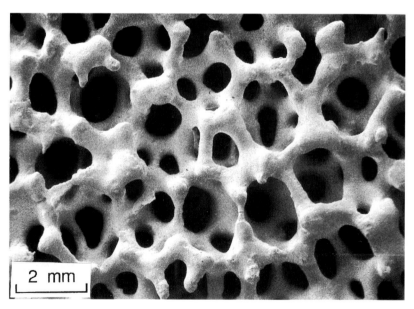

Figure 3.17 Microstructure of an open-cell alumina foam. The structure is usually produced by coating a polymeric foam with a ceramic slip. The polymer is removed in the firing process.

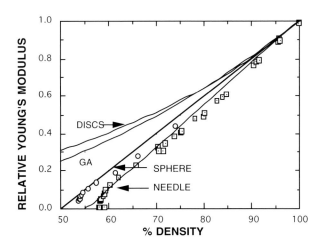

Figure 3.18 Comparison of experimental data on partially sintered alumina (○ and ⊡ represent bodies with two green densities, 52 and 58%, respectively) with the Gibson and Ashby (GA) equation and the self-consistent solutions.

used, e.g., from Eqs. (3.14) and (3.17) for spherical pores. Figure 3.18 shows a comparison of the Gibson and Ashby approach and the self-consistent sphere, disc and needle solutions. One of the difficulties with the self-consistent approach is that the moduli approach zero for the spherical and needle cases for $P \sim 0.4$ to 0.5. This is clearly not very useful for high-porosity materials, such as for the data shown in Fig. 3.16. For sintered ceramic materials, one of the difficulties in using a specific geometry approach is that pore shape is often not specific. Once P exceeds ~ 0.1, the pores are usually interconnected with a complex topology. Figure 3.18 includes data on alumina as a function of P and one can see that there is substantial deviation from the HS upper bound and Eq. (3.32). The data do appear to lie between the self-consistent needle and sphere solutions but one could argue that this is somewhat fortuitous, in that the pores do not have these specific shapes. Moreover, the intercept that gives a modulus close to zero is simply a reflection that the body was sintered from a powder and the green density 'happened' to have a value of P in that range. One expects the modulus to approach zero (in tension) when the body is in the form of a powder compact, i.e., at the green density. As the green body undergoes sintering, however, the solid quickly becomes continuous. There have been some recent attempts to use **percolation theory** to describe this behavior, i.e., sintering allows the solid to 'percolate' the structure.

The idea that the elastic moduli of ceramic powder compacts are close to zero has also led to the idea that the change in the elastic constants is simply given by the **degree of densification**. Thus, it has been suggested that

$$\frac{E}{E_0} = \left(1 - \frac{P}{P_g}\right) \tag{3.34}$$

where P_g is the porosity fraction in the green body. Thus, one obtains $E=0$ when $P=P_g$ and $E=E_0$ when $P=0$. Figure 3.19 compares experimental data for alumina bodies of three different green densities with this theory. Although the theory works well for $1-(P/P_g)>0.1$, there is some discrepancies at lower values. Part of this discrepancy is a result of Young's modulus not being zero even in the green body. Adhesion between the powder particles and the use of binders give finite values for the elastic constants. The other difficulty with the theory is that sintering can occur without densification, e.g., by surface diffusion or evaporation–condensation mechanisms. Thus, necks can grow between particles during sintering, increasing the body stiffness even when there is no densification. The data in Fig. 3.19 have been interpreted using the idea that surface diffusion occurs in alumina at low sintering temperatures ($<$1100 °C), increasing the modulus without densification.

One type of 'pore' that is worthy of further consideration is the extreme case of a microcrack. In Section 2.9 it was shown that thermal expansion anisotropy can lead to residual stresses in ceramics and, in some cases, the formation of localized (spontaneous) microcracking. Microcracks may only represent a small fraction of porosity in a body but their ability to concentrate stress can lead to substantial reductions in the elastic constants. For a random array of circular microcracks, radius a, the SC approach shows the elastic constants μ and B of the microcracked material can be approximated by

$$\frac{\mu}{\mu_0}=1-\frac{32(1-v)(5-v)\Omega}{45(2-v)} \tag{3.35}$$

and

$$\frac{B}{B_0}=1-\frac{16(1-v^2)\Omega}{9(1-2v)} \tag{3.36}$$

where μ_0 and B_0 are the shear and bulk moduli of the non-microcracked materials, Ω is the crack density ($=(1/V)\Sigma\, a^3$) and v is the Poisson's ratio of the microcracked body. These equations can be solved for v and substituted back to obtain B and μ.

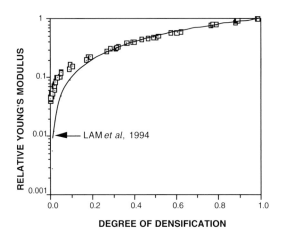

Figure 3.19 Comparison of theory by Lam *et al.* (1994) with experimental data on partially sintered alumina with three different green densities (ranging from 52 to 58% theoretical) combined in a single group.

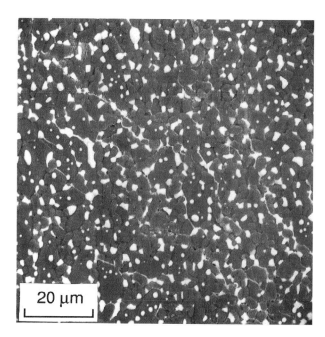

Figure 3.20 Use of silver nitrate to highlight grain boundary microcracks in a zirconia-toughened alumina. The high contrast for the silver nitrate is obtained in the backscattered mode of the SEM.

Figure 3.11, discussed earlier, showed experimental data for alumina-zirconia composites. Materials containing above ~7.5 vol.% unstabilized ZrO_2 have undergone a dilatational phase transformation and the Young's modulus is well below the HS lower bound. The sudden drop in the elastic constant was taken as evidence that the phase transformation had led to microcracking. Indeed, a silver nitrate dye, applied in the vicinity of a hardness indentation, was used to show that the material contained microcracks (Fig. 3.20). The high-contrast dye fills microcracks and these are seen as bright lines. Transmission electron microscopy (TEM) has also shown the presence of microcracks in these materials. Thus, elastic-constant measurements that show anomalously low values are sometimes used as indirect evidence for the presence of microcracks in a ceramic material. In this example, microcracks were formed as a result of a phase transformation. Indeed, if one part of a material undergoes a volume change with respect to the surrounding material, residual stresses arise and in brittle materials, can lead to microcrack formation. Thus, microcracks can also form in composite materials if thermal expansion differences exist between the phases **(thermal expansion mismatch)** or if a chemical reaction occurs that involves a volume change. Figure 3.21 shows a microcrack at the interface of a TiB_2 grain in a SiC–TiB_2 particulate composite. The thermal expansion of TiB_2 is larger than SiC and thus, after fabrication, the TiB_2 contracts more than the SiC, which can give rise to microcrack formation. The microcracking leads to a reduction in the elastic moduli, as shown in Fig. 3.22. As another example, Fig. 3.23 shows elastic-constant data for a SiC platelet–alumina composite for two different platelet sizes. For the larger platelet size, the Young's modulus at high volume fraction is reduced by microcracking. Reducing the platelet size prevents micro-cracking and is another example of the critical size effect mentioned in Section 2.9.

3.7   Thermal expansion behavior of polycrystalline ceramics

Thermal energy input causes atoms to vibrate with greater amplitude and, as indicated earlier, this usually causes expansion. As the atomic spacing increases, the 'spring' constants of the atomic bonds decrease and the atoms vibrate more slowly. The tendency for expansion is opposed by the bulk modulus. The quantitative development of this theory leads to

$$\alpha = \frac{gC_V}{3V_0B_0} \tag{3.37}$$

where g is Gruneisen's constant, C_v is the specific heat per gram, V_0 is the volume per gram at 0 K and B_0 is the bulk modulus at 0 K. The expansion coefficient increases with the vibrational specific heat, as temperature is increased. In strongly bonded solids, this increase can occur quite slowly, so the thermal

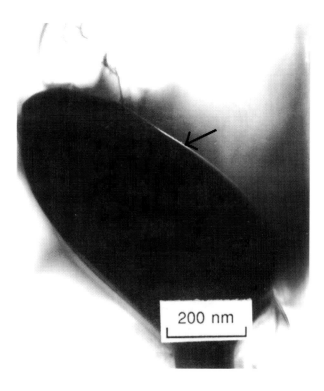

Figure 3.21 Microcrack resulting from thermal expansion mismatch in a SiC–TiB$_2$ particulate composite. (From Hoffman, 1992, reproduced courtesy of The American Ceramic Society, Westerville, OH.)

200 nm

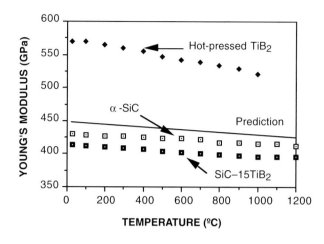

Figure 3.22 Elastic constant data for α-SiC, TiB$_2$ and a SiC material containing 15 vol.% TiB$_2$ particles. The data for the composite lie below the values predicted from the constitutive relationships, indicating the presence of microcracks. The data are all corrected to zero porosity. (Reproduced courtesy of Plenum Press, New York.)

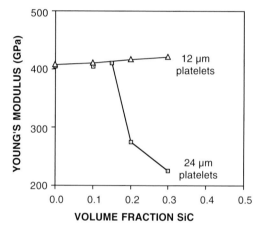

Figure 3.23 Addition of SiC platelets to alumina leads to microcracking at high volume fractions if the platelets are above a critical size.

expansion coefficients increase with temperature even above room temperature. For example, Fig. 3.24 shows the thermal expansion coefficients for single-crystal alumina.

As with the elastic constants, constitutive equations have also been put forward for calculating the thermal expansion coefficient of a polycrystalline aggregate from the single-crystal values. Consider the simple model structure in Fig. 3.25, consisting of three rods (equal cross-section) attached rigidily to upper and lower supports. The rods have different thermal expansion coefficients, α_1, α_2, and α_3 ($\alpha_1 > \alpha_2 > \alpha_3$) and different Young's moduli, E_1, E_2, and E_3. Assume the structure is initially in equilibrium and is then cooled. If the overall strain induced in the structure by cooling is ε, it is clear that the rods will be subjected to **residual stresses**, as rod 1 would prefer to shrink more than rods 2 and 3. Thus, rod 1 will end up in uniaxial tension and rod 3 in uniaxial compression. The value of ε will be determined by a force balance. Assuming the materials are linear elastic, the stresses in the rods are $E_1(\alpha_1\Delta T - \varepsilon)$, $E_2(\alpha_2\Delta T - \varepsilon)$ and $E_3(\alpha_3\Delta T - \varepsilon)$

and using a force balance, one obtains the averaged thermal expansion coefficient, $\varepsilon = \alpha \Delta T$, as

$$\alpha = \frac{E_1 \alpha_1 + E_2 \alpha_2 + E_3 \alpha_3}{E_1 + E_2 + E_3} \qquad (3.38)$$

For simplicity, let us assume the Young's modulus values are the same, so $\alpha = (\alpha_1 + \alpha_2 + \alpha_3)/3$. The average of the three thermal expansion coefficients, $\alpha = (\alpha_1 + \alpha_2 + \alpha_3)/3$ is often used for estimating the thermal expansion behavior of a polycrystalline body from the single-crystal values. Hashin (1984) has shown this average to be the same as one of the exact bounds and close to the self-consistent approximation for hexagonal, tetragonal and trigonal crystals. As with Eq. (3.38), the other bound is dependent on the elastic constants. Table 3.5 gives the thermal expansion coefficients for various polycrystalline ceramics and this can be compared with the single-crystal data in Table 2.2.

The data in Table 3.5 show the trends one expects for an inverse dependence on bulk modulus, i.e., thermal expansion coefficients increase in the order: carbides, silicides, oxides and halides. The main exception appears to be the silicates, which often have quite low thermal expansion coeffiecients. Silicate structures are often open and coupled transverse modes of vibration can lead to a decrease

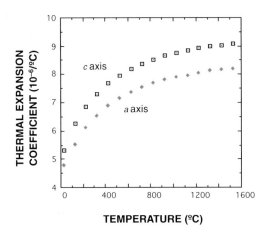

Figure 3.24 Thermal expansion coefficients for single-crystal α-Al$_2$O$_3$.

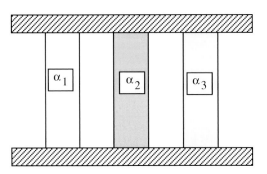

Figure 3.25 A simple model for the expansion of three rods with different thermal expansion coefficients.

Table 3.5 *Thermal expansion coefficients of selected ceramic polycrystals*

Material	Thermal expansion coefficient[a] ($10^{-6}/°C$)
B_4C	5.5
HfC	6.3
α-SiC	4.7
β-SiC	5.1
TaC	6.7
TiC	7.8
WC	4.9
ZrC	6.6
HfB_2	5.5
TiB_2	7.5
ZrB_2	6.0
AlN	5.6
BN	13.3
Si_3N_4	3.7
$MoSi_2$	8.5
WSi_2	8.2
Al_2O_3	8.2
$3Al_2O_3.2SiO_2$	5.1
$Al_2O_3.MgO$	8.4
BeO	9.0
CeO_2	8.9
Cr_2O_3	8.6
MgO	13.6
$2MgO.2Al_2O_3.5SiO_2$	2.8
α-SiO_2	19.4 (25–500 °C)
β-SiO_2	-1.2 (600–1000 °C)
ThO_2	9.4
TiO_2	8.8
ZrO_2 (2 mol.% Y_2O_3 tetragonal)	10.4 (25–800 °C)
ZrO_2 (3 mol.% Y_2O_3 tetragonal)	10.6 (25–800 °C)
ZrO_2 (12 mol.% Y_2O_3 cubic)	11.1 (25–1500 °C)
ZrO_2 (14 mol.% MgO cubic)	11.2 (25–1500 °C)
$ZrO_2.SiO_2$	4.6

Note:
[a] 25–1000 °C unless noted otherwise

in overall thermal expansion. In some cases, the thermal expansion coefficient may be negative for certain crystallographic directions.

The presence of porosity in a material is not expected to influence the thermal expansion behavior. This follows from Eq. (3.18) but, in addition, it can be envisioned that a 'hole' in a body will not constrain the thermal expansion of the surrounding material. Microcracks, on the other hand, can lead to a change in the thermal expansion behavior and **low thermal expansion materials** are sometimes a result of this phenomenon. This effect can occur in non-cubic polycrystalline materials in which there is significant thermal expansion anisotropy. Returning to Fig. 3.25, if the tensile stresses in rod 1 become so high that failure occurs, the stress in this rod is released. The equilibrium is now defined by rods 2 and 3, so $\alpha=(\alpha_2+\alpha_3)/2$, and because α_1 has the highest magnitude, the overall thermal expansion coefficient is reduced. Thus, it is concluded that 'microcracking' can lead to an overall reduction in thermal expansion coefficient, primarily because the large α value was 'removed' from the averaging procedure.

3.8   Elastic behavior of sandwich panels

The structural scale now moves up closer to a macroscopic level and sandwich panels, which consist of two stiff faces separated by a lightweight core (usually porous), will be discussed. This structure is prevalent in natural materials but has also been used in aircraft and sailboat components, as well as skis. An example of a ceramic sandwich panel was shown in Fig. 1.7, but ceramic structures have not yet been exploited in technological applications. Consider the sandwich beam shown in Fig. 3.26. The stresses that arise in such a beam will be quantified in Chapter 4. It should, however, be clear that the greatest tensile and compressive strains are occurring at the beam surfaces, whereas these strains will be low in the center of the beam. For example, one expects the lower face to be in tension as it is stretched. Clearly, having the material with the greatest stiffness at the surfaces and a lightweight material (low stiffness) in the center will be beneficial. By control over the geometry and material properties, structures of a given stiffness with minimum weight can be produced. For example, Gibson and Ashby (1988) have outlined the procedure to determine the optimum core thickness, face thickness and core density for this condition.

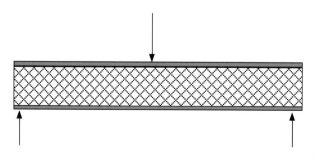

Figure 3.26 Three-point bending of a sandwich beam that consists of stiff faceplates on a lightweight core.

Problems

3.1 Choose a cubic single-crystal material (Table 2.4) and use the elastic stiffness constants to perform the following tasks.

a) Determine the elastic compliance constants.

b) Determine the shear moduli for {100}<100> and {110}<110> deformations. Calculate the elastic anisotropy factor (Zener ratio).

c) If the crystal was given a strain $\varepsilon_1 = 0.1\%$ and the other strains were zero, determine the values of the stress components.

d) If the crystal was stressed with $\sigma_3 = 200$ MPa, determine the strain components.

e) Determine the strain tensor for the stress state: $\sigma_1 = 200$ MPa, $\sigma_2 = 200$ MPa, $\sigma_5 = 300$ MPa.

f) Determine the Young's modulus values for uniaxial extension along <100> and <111>. What is the value of Poisson's ratio for uniaxial tension along <100>?

g) Estimate the engineering elastic constants for a (random) poly-crystalline body for the selected crystal. (Use the arithmetic average of the Voigt and Reuss bounds and the Hashin-Shtrikman bounds.)

3.2 The bulk modulus of NaF is 46.5 GPa and the nearest-neighbor ion spacing is 0.2317 nm.

a) Suggest an equation for the interionic potential for a NaF crystal.

b) If the equilibrium spacings are 0.2820 nm, 0.2989 nm and 0.3237 nm for NaCl, NaBr and NaI, respectively, estimate the bulk modulus of each of these crystals from the data on NaF.

c) If the bulk moduli for CaF_2, SrF_2 and BaF_2 are 82, 67 and 35 GPa, respectively, estimate the equilibrium spacing from the data on NaF.

3.3 Particulate composites with an alumina matrix and zirconia particles are of interest for structural applications. For a 25 vol.% zirconia parti-cle composite:

a) Use Hill's average of Voigt–Reuss bounds to estimate Young's modulus.

b) How would the above answer change, if the composite contained 5 vol.% porosity?

c) How would the answer to part a) change if the composite contained a microcrack density of $10^{16}/m^3$ and their average radius was 10 µm.

3.4 ZrO_2 can form a solid solution with Y_2O_3 and the elastic constants are given overpage for various compositions.

mol.% Y_2O_3	c_{11} (GPa)	c_{12} (GPa)	c_{44} (GPa)
8.13	401	96	55.8
11.09	403.5	102.4	59.9
12.08	405.1	105.3	61.8
15.52	397.6	108.6	65.8
17.88	390.4	110.8	69.1

a) For the 8.13 mol.% composition calculate
   i) Young's modulus for uniaxial stressing along [100] and [111].
   ii) Poisson's ratio with respect to stressing along the principal axes.
   iii) Young's modulus and the shear moduli for a (random) poly-
       crystalline body.
b) Plot Young's moduli as a function of Y_2O_3 content. Suggest an
   explanation for the variation of Young's modulus.
c) How does the Zener ratio change with composition? Does the crystal
   become more anisotropic with the increase in mol.% Y_2O_3? Suggest
   an explanation for the anisotropy trend.
d) Cubic ZrO_2 crystals are very difficult to tell apart from real dia-
   monds. Could the measurement of Young's modulus be used to distinguish them?

3.5   The energy of a single bond in ionic crystals is often described by

$$\phi = C \exp\left(\frac{-r}{\rho}\right) - \frac{A q_a q_c}{4\pi\epsilon_0 z r}$$

where C and ρ are material constants, A is the Madelung constant, q_a
and q_c are the charges on the anion and cation, z is the number of nearest
neighbors and r is the nearest-neighbor distance. For NaCl, the equilibrium spacing is 0.321 nm and the bulk modulus is 24.0 GPa. Calculate
the parameters C and ρ.

3.6   The Young's moduli of polycrystalline alumina and zirconia are 400 and
205 GPa and the Poisson's ratio values are 0.23 and 0.31, respectively.
a) Calculate the Voigt, Reuss and Hashin–Shtrikman bounds for the
   bulk and shear moduli of alumina–zirconia composites as a function
   of volume fraction. Compare the arithmetic average between the two
   types of bounds. Calculate the Young's modulus and Poisson's ratio
   for a 20 vol.% zirconia composite.
b) Determine the microcrack density (radius 1 µm) that would reduce
   the shear modulus of this 20 vol.% composite by a factor of 10.
c) Accurate experimental values for the Young's modulus of a 20 vol.%

ZrO$_2$ composite (fully dense) were found to be above the upper Hashin–Shtrikman bound. Suggest possible explanations. Repeat the discussion for the case in which experimental values were below the lower bound.

3.7   The elastic constants of α-SiC single crystals are: c_{11}=504 GPa, c_{12}=98 GPa, c_{13}=56 GPa, c_{33}=566 GPa and c_{44}=170 GPa. The elastic constants of TiB$_2$ single crystals are: c_{11}=690 GPa, c_{12}=410 GPa, c_{13}=320 GPa, c_{33}=440 GPa and c_{44}=250 GPa.

a) Determine the Young's modulus and Poisson's ratio for a poly-crystalline body of SiC (Hashin–Shtrikman average). Repeat the calculations for TiB$_2$.

b) Calculate the Voigt, Reuss and Hashin–Shtrikman bounds for the bulk and shear moduli of SiC–TiB$_2$ composites as a function of volume fraction. Compare the arithmetic average between the two types of bounds.

c) Use the self-consistent sphere solution to calculate Young's modulus and Poisson's ratio of a 15 vol.% TiB$_2$ composite. Compare the values to the bounding estimates.

d) Briefly discuss the assumptions in the Hashin–Shtrikman and self-consistent solutions for the isotropic elastic behavior of particulate composites.

3.8   The Young's modulus and Poisson's ratio of polycrystalline Al$_2$O$_3$ are 400 GPa and 0.23. Use the self-consistent sphere and disc solutions to calculate the bulk and shear moduli of alumina as a function of the fraction porosity. Suggest simple equations that approximate the self-consistent solutions for the effect of porosity on the elastic constants.

3.9   Calculate the Madelung constant for a line of equally spaced, alternating positive and negative ions.

3.10   The energy (interionic potential) of a single bond in ionic crystals is often described by

$$\phi_b = \frac{-Aq^2}{4\pi z \varepsilon_0 r^m} + \frac{C}{r^n}$$

where C, m and n are material constants, A is the Madelung constant, q is the charge on the anion and cation, z is the number of nearest neighbors and r is the nearest-neighbor distance. Derive a new version of Eq. (3.9) to determine the bulk modulus B for this potential.

3.11 Use the data in Table 2.4 to plot the bulk modulus for crystals with the fluorite and cesium chloride structures as a function of interionic distance. Do the data agree with Eq. (3.9)?

3.12 Using the data in Table 2.2, estimate the change in the thermal expansion coefficient if microcracks form normal to the c axis in polycrystalline $Al_2O_3.TiO_2$. Sketch the variation in the thermal expansion coefficient with temperature for microcracked $Al_2O_3.TiO_2$.

3.13 Use the Halpin–Tsai equations to determine the five elastic constants of a unidirectional fiber composite in which alumina fibers are dispersed in a glass matrix. The Young's modulus and Poisson's ratio of polycrystalline Al_2O_3 are 400 GPa and 0.23 and for the glass, 70 GPa and 0.20.

3.14 Assume the effect of a uniaxial stress σ on an array of atoms can be described by

$$\sigma = \frac{A}{x^n} - \frac{B}{x^m}$$

where x is the distance between atoms ($m > n$).
a) Determine an equation for the equilibrium interionic spacing a_0, in terms of the parameters in the interionic potential.
b) Determine an equation for Young's modulus $E = (x d\sigma/dx)$ in terms of the equilibrium interionic spacing a_0 and the constants A, m and n.
c) Show that the theoretical cleavage strength (maximum stress) can be given by

$$\sigma_0 = \left(\frac{E}{m}\right)\left(\frac{n}{m}\right)^{n/(m-n)}$$

d) Determine the variation in σ_0 for ionic materials as a fraction of Young's modulus ($n=2$ and m is in the range 6–12).

References

General reading

R. C. Bradt, *Introduction to the Mechanical Properties of Ceramics*, Class Notes, The Pennsylvania State University, 1980.

R. Brezny and D. J. Green, The mechanical behavior of cellular ceramics, pp. 463–516 in *Structure and Properties of Ceramics*, edited by M. V. Swain, VCH Publishers, Weinheim, Germany and New York, 1994.

B. Budiansky and R. J. O'Connell, Elastic moduli of a cracked solid, *Int. J. Solid Struct.*, **12** (1976) 81–97.

R. M. Christensen, A critical evaluation for a class of micromechanics models, *J. Mech. Phys. Solids*, **38** (1990) 379–404.

T. W. Clyne and P. J. Withers, *An Introduction to Metal Matrix Composites*, Cambridge University Press, 1993.

A. H. Cottrell, *Mechanical Properties of Matter*, Wiley and Sons, New York, 1964.

J. D. Eshelby, The determination of the stress field of an ellipsoidal inclusion, *Proc. Roy. Soc.*, **A241** (1957) 376–96.

L. J. Gibson and M. F. Ashby, The mechanics of three-dimensional cellular materials, *Proc. R. Soc. London*, **A382** (1982) 43.

L. J. Gibson and M. F. Ashby, *Cellular Solids: Structure and Properties*, Pergamon Press, Oxford, UK, 1988.

D. J. Green, C. Nader and R. Brezny, The elastic behavior of partially sintered alumina, pp. 345–56 in *Ceramic Transactions, Vol. 7*, The American Ceramic Society, Westerville OH, 1990.

J. C. Halpin, *Primer on Composite Materials*, Technomic Publishing Co., Lancaster, PA, 1984.

Z. Hashin, Analysis of composite materials, *J. Appl. Mech.*, **50** (1983) 481–505.

Z. Hashin, Thermal expansion of polycrystalline aggregates, *J. Mech. Phys. Solids*, **32** (1984) 139–57.

D. C. C. Lam, F. F. Lange and A. G. Evans, Mechanical properties of partially dense alumina produced from powder compacts, *J. Am. Ceram. Soc.*, **77** (1994) 2113–17.

J. K. MacKenzie, The elastic constants of a solid containing spherical holes, *Proc. Phys. Soc. London*, **B63** (1950) 2–11.

T. Mori and K. Tanaka, Average stress in matrix and average elastic energy of materials with misfitting inclusions, *Acta Metall.*, **21** (1973) 571–4.

R. E. Newnham, *Structure-Property Relations*, Springer-Verlag, New York, 1975.

J. P. Watt, G. F. Davies and R. J. O'Connell, The elastic properties of composite materials, *Rev. Geophys. Space Phys.*, **14** (1976) 541–63.

Materials Data

S. P. Clark, Jr, *Handbook of Physical Constants,* Geological Society of America Inc., New York, 1966.

D. N. Coon, *Slow Crack Growth in $Na_2O.xAl_2O_3.(3-x)SiO_2$ Glasses,* Ph.D Thesis, The Pennsylvania State University, 1986.

CRC Handbook of Chemistry and Physics, 69th Edition, CRC Press, Boca Raton, FL, 1988.

E. A. Dean and J. A. Lopez, Empirical dependence of elastic moduli on porosity for ceramic materials, *J. Am. Ceram. Soc.,* **66** (1983) 366–70.

D. J. Green, Critical microstructures for microcracking in Al_2O_3-ZrO_2 composites, *J. Am. Ceram. Soc.,* **65** (1982) 610–14.

D. J. Green, R. H. J. Hannink and M. V. Swain, *Transformation Toughening of Ceramics,* CRC Press, Boca Raton, FL, 1989.

P. A. Hoffman, *Thermoelastic Properties of Silicon Carbide-Titanium Diboride Particulate Composites,* MS Thesis, The Pennsylvania State University, 1992.

F. F. Lange, Transformation toughening. Part 4: Fabrication, fracture toughness and strength of Al_2O_3/ZrO_2 composites, *J. Mater. Sci.,* **17** (1982) 247–54.

M-J. Pan, *Microcracking Behavior of Particulate Titanium Diboride-Silicon Carbide Composites and Its Influence on Elastic Properties,* Ph.D. Thesis, The Pennsylvania State University, 1994.

G. Simmons and H. Wang, *Single Crystal Elastic Constants and Calculated Aggregate Properties,* MIT Press, MA, 1971.

J. F. Shackleford and W. Alexander, *The CRC Materials Science and Engineering Handbook,* CRC Press, Boca Raton, FL, 1992.

Chapter 4

Elastic stress distributions

It is important to be able to solve elasticity problems for a variety of different loading geometries. The equations of elasticity do not possess unique solutions unless residual stresses are ignored but fortunately the **principle of superposition** allows one to deal with these as a separate issue. This principle can also be used to build up solutions of complicated problems by superposing a set of simpler problems. In this chapter, some simple elastic stress distributions for linearly elastic isotropic bodies will be shown and discussed. In these solutions there will be simplifying features that allows the general equations of elasticity to be by-passed.

An immediate difficulty in studying loaded bodies is that external forces are often applied to the body in complicated ways, e.g., through bolts, pins, collars, etc. This means that stresses can be complex in the loading region; they may vary sharply, especially if there are changes in the contours of the body. Fortunately, these stresses are generally localized to the region of contact and will not change the stress and strain distribution away from the contact region. This effect is expressed by **St Venant's Principle**, 'If the forces acting on a small part of a body are replaced by a statically equivalent set of forces (i.e., with the same resultant force and couple) acting on the same area, the stress state will be changed negligibly at large distances compared to this area.' As an example, Fig. 4.1 shows a set of forces in which two forces f act symmetrically about a force $2f$. Using St Venant's Principle, one concludes that there is no resultant force or couple acting and, thus, the forces will not give rise to any stresses away from this region. Brittle materials are sensitive to contact damage in a loaded region but such damage will be ignored in this chapter. Instead, there will be an emphasis on the influence of the loading on the overall deformation of the body away from the loading region. Contact damage will be discussed in Chapter 8.

4.1   Statically determinate and indeterminate problems

For some problems, the stresses can sometimes be found simply by using **statics** and the loading is then termed **statically determinate**. For example, Fig. 4.2(a) shows a **pin-jointed frame** in which two rods are loaded by a force F. This frame is sometimes termed a **perfect frame**, as it has just sufficient linkages to prevent any rotations about the joints within the plane of the diagram. For three-dimensional constraint, an additional rod would be needed, e.g., using a tetrahedral configuration of rods. A frame that could not prevent a rotation is termed **imperfect**. In pin-jointed frames, the flexible joints ensure that the rods only transmit tensile and compressive forces. For Fig. 4.2(a), one can use statics to obtain,

$$F = 2f_1 \cos \theta \tag{4.1}$$

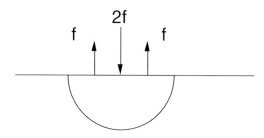

Figure 4.1 A complex loading configuration can be simplified at large distances from the (shaded) contact region by St Venant's Principle.

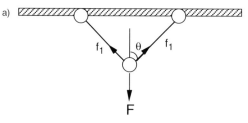

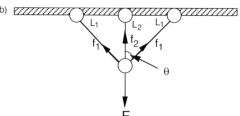

Figure 4.2 Pin-jointed frames in which the stresses are a) statically determinate and b) statically indeterminate.

as the condition for static equilibrium. Contrast the frame shown in Fig. 4.2(b), in which an extra (redundant) rod has been added and the forces become **statically indeterminate**. Assuming there are no self-stresses (residual stresses), there is only one relationship for the static equilibrium,

$$F = 2f_1 \cos\theta + f_2 \qquad\qquad (4.2)$$

and it is not possible to solve for the forces, f_1 and f_2. In order to determine the stresses, one must make an *assumption* about the load–displacement behavior of the rods. For example, the rods could be assumed to be linearly elastic. In order to solve for the stresses, one must realize that the displacement at the point where all three joints meet must be compatible. If the inclined rods (length L_1) increase in length by δL_1 and the vertical rod (length L_2) by δL_2, one can equate the vertical displacements to obtain $\delta L_1 = \delta L_2 \cos\theta$. For simplicity, assume all the rods have unit cross-sectional areas and the same Young's modulus E so that $f_1 = E\delta L_1/L_1$ and $f_2 = E\delta L_2/L_2$. Rearrangement of these relationships allows a second equation between f_1 and f_2 to be obtained, i.e.,

$$f_1 = f_2 \cos^2\theta \qquad\qquad (4.3)$$

The simultaneous solution of Eqs. (4.2) and (4.3) allows the forces and stresses in the various rods to be determined. It should be noted that even though linear elasticity is assumed, the terms for stresses do not involve the elastic constants. This is not true, however, for the strains. In the last chapter, the geometry shown in Fig. 3.25 was statically indeterminate, and to solve the problem the rods were assumed to be linear elastic.

4.2   Thin-walled pressure vessels

A simple example of a statically determinate problem is a thin-walled cylindrical pressure vessel with spherical end-caps, containing a gas or fluid exerting a pressure P. The geometry is shown in Figs. 4.3 and 4.4; the cylindrical section has a radius r, length L and thickness t, such that $L \gg r \gg t$. The cylinder is being expanded by the internal pressure and the principal axes are easily

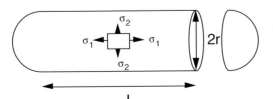

Figure 4.3 Thin-walled cylindrical pressure vessel, with cut to show cross-section.

identified. The tensile stress σ_1 along the tube length is termed the **longitudinal stress** and the circumferential stress σ_2 the **hoop stress**. As the wall thickness is assumed to be small, one can assume σ_1 and σ_2 are constant. The third principal stress σ_3 varies from $-P$ inside the vessel to zero on the outside surface. The hoop stress can be determined from the force balance shown in Fig. 4.4(a). The pressure produces an outward force $2rLP$ and this is 'opposed' by force produced by the hoop stress, $2tL\sigma_2$, so that $\sigma_2 = Pr/t$. The equilibrium in the end-caps is shown in Fig. 4.4(b), showing the outward force is $\pi r^2 P$ and this is opposed by $2\pi rt\sigma_1$, giving $\sigma_1 = Pr/2t$. Thus, one finds the hoop stress to be twice the longitudinal stress. The above approach can also be used for thin-walled spherical pressure vessels, by considering only the end-caps and obtaining $\sigma_1 = \sigma_2 = Pr/2t$.

4.3   Bending of beams

Another set of problems that can be solved (approximately) using simplifying geometric assumptions is the bending of beams. Consider, first, the shearing forces and bending moments that arise during bending. The bending geometry shown in Fig. 4.5 is termed **three-point bending**. Statics can be used to determine the shearing force V and bending moment M at any point along the beam. This

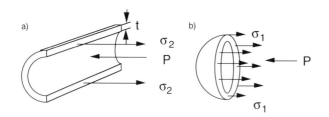

Figure 4.4 Cross-sections of pressure vessel showing the force balances: a) longitudinal section; b) transverse section.

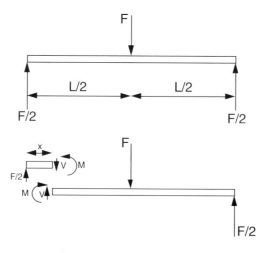

Figure 4.5 Three-point loading geometry. Making a cut in beam allows the bending moments and shearing forces to be determined.

is accomplished by making an imaginary cut in the beam at a point, distance x along the beam (see Fig. 4.5). To retain equilibrium, a shearing force V and a bending moment M must be applied to the 'cut' ends. Assuming static equilibrium, one obtains $M = Px/2$ and $V = P/2$. Continuing this procedure, one can construct **bending-moment** and **shearing-force diagrams**, as shown in Fig. 4.6. For a given load, the bending moment increases linearly to a maximum at the central section of the beam. The shearing force is constant along the beam but changes sign from one side of the beam to the other. Three-point bending is often used to load and fracture ceramic materials. Another common loading geometry is **four-point bending**, which is shown in Fig. 4.7, along with its associated bending-moment and shearing-force diagrams. For this geometry, the central section of the beam is not subjected to any shearing forces but only to a constant bending moment. This latter type of loading is termed **pure bending**.

Consider the stresses that arise in a regular beam, of thickness h, subjected to bending such that the beam has a curvature R, as shown in Fig. 4.8. The central

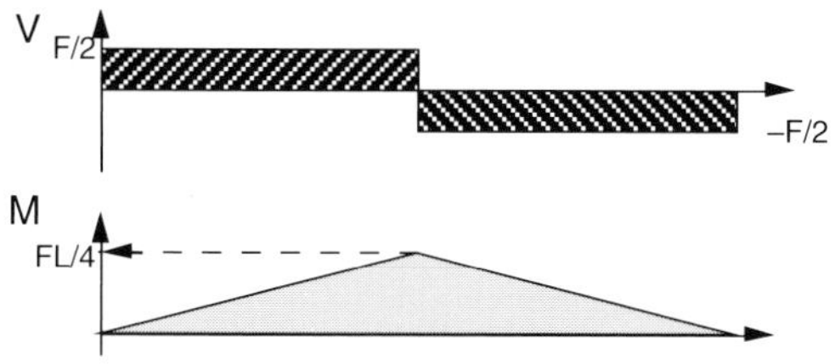

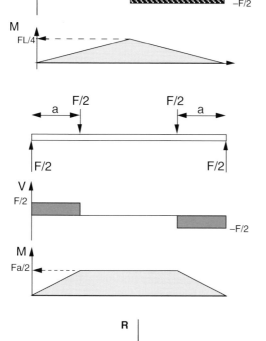

Figure 4.6 Shearing force V and bending moment M diagrams for three-point bend geometry.

Figure 4.7 Four-point bend geometry and associated shearing force and bending moment diagrams.

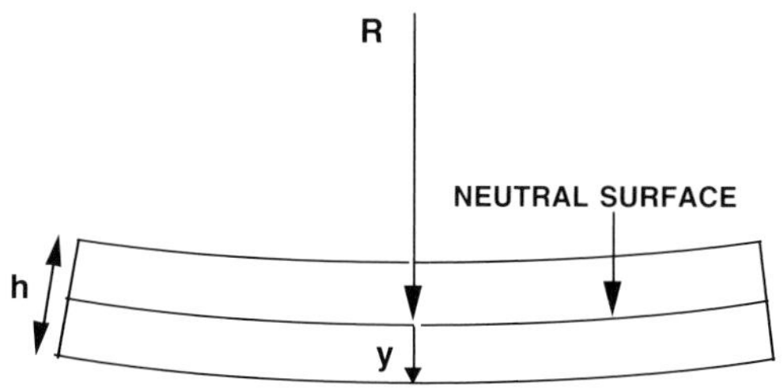

Figure 4.8 Beam of thickness h bent to radius of curvature R. The beam will have a (neutral) surface that has undergone no strain.

surface in the beam is called the **neutral surface** because it does not change its length. In contrast, the upper and lower surfaces will decrease and increase their lengths, respectively. This change in length will occur in a continuous fashion as one moves away from the neutral surface and, thus, the upper half of the beam is in compression, while the lower half is in tension. From simple geometrical considerations, the surface at a distance y from the neutral surface will alter its length by the ratio $(R+y)/R$, so the strain is y/R. The bending geometry is statically *indeterminate* but if one assumes the material to be linear elastic, the longitudinal stresses are Ey/R. If the width of the beam is $b(y)$, the force on a layer of width dy is $Eyb(y)dy/R$. In order to determine the radius of curvature, consider pure bending, so that the longitudinal force over any cross-section of the beam is zero, i.e.,

$$\frac{E}{R}\int yb(y)dy=0 \tag{4.4}$$

Each layer also exerts a force df and, hence, a moment ydf about an axis in the neutral surface. The moments from successive layers cooperate in rotating the cross-sections about this axis. The total bending moment M is given by

$$M=\frac{E}{R}\int_{-h/2}^{h/2} y^2b(y)dy=\frac{EI}{R} \tag{4.5}$$

where I is given by the above integral and is termed the **second moment of area** of the cross-section. Using $\sigma=Ey/R$, derived above, one obtains the following expression for the normal stresses in the beam

$$\sigma=\frac{My}{I} \tag{4.6}$$

For a rectangular beam, $b(y)$ is a constant ($=b$) and the integral in Eq. (4.5) can be solved to give $I=bh^3/12$. It is useful to note that Young's modulus can be determined from Eq. (4.5) provided a technique is available to measure the radius of curvature (e.g., by optical interference or with a profilometer).

The stress state in a bent beam is usually more complex than that encountered in pure bending, with **non-uniform bending moments** and shearing forces. In these cases, the normal stresses produced by the bending moment are still given by Eq. (4.6) but, in addition, shearing stresses τ may be present such that

$$\tau=\frac{VQ}{Ib} \tag{4.7}$$

where Q is the **first moment of area**, i.e.,

$$Q=\int_{y}^{h/2} yb(y)dy \tag{4.8}$$

Consider the **cantilever beam** shown in Fig. 4.9 in which it can be shown that $M=F(L-x)$ and V is constant along the beam length. Using Eqs. (4.7) and (4.8) for a beam of rectangular cross-section, one finds the shear stresses are parabolic, being zero at the free surfaces of the beam and rising to a maximum of $3F/2bh$ at the neutral surface.

It is often important to determine the **beam deflection** Y and several approaches are available to derive the appropriate equations. Deflections are expected to arise from both the normal and shear stresses but these latter deflections are usually small and are often neglected. In the **double integration approach**, one can substitute $1/R=d^2 Y/dx^2$ into Eq. (4.5), i.e.,

$$\frac{d^2 Y}{dx^2} = \frac{M}{EI} \tag{4.9}$$

For the cantilever beam, $M=F(L-x)$ and, performing the first integration with the boundary condition $dY/dx=0$ at $x=0$, one obtains

$$EI\frac{dY}{dx} = FLx - \frac{Fx^2}{2} \tag{4.10}$$

After the next integration, with $Y=0$ at $x=0$

$$EIY = \frac{FLx^2}{2} - \frac{Fx^3}{6} \tag{4.11}$$

The maximum deflection Y_0 occurs at $x=L$, i.e., $Y_0=FL^3/3\ EI$ and this equation could be used to determine E from the beam deflection. If one inspects Fig. 4.5, the three-point geometry can be viewed as two 'attached' cantilever beams. Replacing F by $F/2$ and L by $L/2$ in the cantilever beam deflection formula, the maximum deflection in three-point bending can be determined, i.e., $Y_0=FL^3/48EI$. The resistance of a beam to bending depends on EI (Eq. (4.9)), which is termed the **flexural rigidity**.

Structures are often designed to give maximum stiffness for minimum weight. For beams, the cross-sectional shape is often modified to meet this goal. For example, material near the neutral surface is relatively unstressed and is not really 'needed'. This leads to the use of **I-beams**, in which the central area is reduced to a thin web. This shape change allows the second moment of area to

Figure 4.9 Cantilever beam geometry with 'built-in' end, and a vertical load F on its free end.

be reduced, decreasing the flexural rigidity and increasing the stresses. The decrease in stiffness is, however, more than compensated by the reduction in beam weight. For materials weak in tension, the beam width can be made larger on the tensile side to reduce the stresses in this location.

As mentioned in Section 3.8, use can be made of **composite beams**. In a sandwich beam, such as that shown in Fig. 3.26, the core usually has lower values of Young's and shear modulus than the thin faces. The Young's modulus of the faces and core will be denoted by E_F and E_C, respectively. One approach to stress analysis in these sandwich beams, is to 'transform' the cross-section into a geometry with an equivalent flexural rigidity but consisting of a single material. This transformation is shown in Fig. 4.10 in which the core is replaced by the same material as the faceplates but with a width (bE_C/E_F). In sandwich beams, the faces are usually much thinner than the cores and the **equivalent flexural rigidity** can be written approximately as

$$(EI)_{eq}=\frac{E_F bth^2}{2} \tag{4.12}$$

The normal stresses σ_F and σ_C in the face and core, respectively, are given by

$$\sigma_F=\frac{MyE_F}{(EI)_{eq}} \tag{4.13}$$

and

$$\sigma_C=\frac{MyE_C}{(EI)_{eq}} \tag{4.14}$$

The shear stresses have been ignored in this analysis but these can be significant, especially in the low-modulus core. From Eqs. (4.13) and (4.14), the ratio of the maximum stress in the face to that in the core is given by E_F/E_C. As $E_F \gg E_C$, the faces are therefore found to carry a much higher normal stress than the core. In most composite structures, the stress is carried by the high-modulus materials. If the high-modulus material also has the higher strength, this is clearly advantageous. In a sense, the high-modulus material is 'protecting' the low-modulus material and can give rise to **modulus strengthening**. This effect is often utilized in **fiber composites**, in which the high-modulus fibers carry the majority of the load. For example, in unidirectional fiber composites, the ratio of the lon-

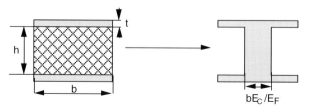

Figure 4.10 Composite beams can be analyzed by transforming beam cross-section into a beam made only of the face material but with an equivalent flexural rigidity.

gitudinal stress in the fiber to that in the matrix is also given by the Young's modulus ratio.

4.4   Elastic stability and buckling

To this point, the possibility that elastic structures may be unstable has not been considered. Clearly, it is very important to determine the **elastic stability** of a structure. For example, if a stressed structure is given a small additional deflection, the structure may not produce the restoring forces necessary to return the structure to equilibrium and this is termed **elastic instability**. There are many forms of elastic instability but, in this text, only the **buckling** of struts under compressive loading will be considered. Consider an initially straight strut of length L being slightly bent, as shown in Fig. 4.11. The curvature of the beam at a general point P is given by

$$\frac{d^2Y}{dx^2} = -\frac{1}{R} = -\frac{M}{EI} = -\frac{FY}{EI} \tag{4.15}$$

If one writes $C^2 = F/EI$, the following differential equation is obtained,

$$\frac{d^2Y}{dx^2} + C^2Y = 0 \tag{4.16}$$

The solution for the equation has the form $Y = m\sin Cx + n\cos Cx$. The strut is symmetric about $x=0$ and this is consistent with the cosine term ($m=0$). Denoting the maximum deflection as Y_0 and noting that $Y=0$ at $x=\pm L/2$, one obtains $Y_0\cos(CL/2)=0$. For $Y_0 \neq 0$, this is satisfied if $CL=\pi$ to give

$$F_c = C^2EI = \frac{\pi^2EI}{L^2} \tag{4.17}$$

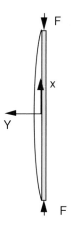

Figure 4.11 Buckling of struts under axial compressive loads.

and this is **Euler's formula** for the critical buckling force. As $Y_0 \cos(CL/2)=0$, the value of Y_0 is *indeterminate*, so that the strut is in neutral equilibrium at $F=F_c$. At this point, the buckling and restoring forces are balanced for any value of Y_0. For $F<F_c$, the restoring forces dominate and allow the strut to be restored to its original (straight) shape, giving stable equilibrium. For $F>F_c$, the slightest disturbance from linearity will cause the strut to buckle. In the limit, this is expected to lead to complete folding of the strut. In many cases, the strut will pass its elastic limit and the strut will break or yield during the collapse. The strut shown in Fig. 4.11 must have pin-jointed ends to allow the necessary rotation. Other strut geometries can often be solved by recognizing which part of a strut will behave like that shown in Fig. 4.11. This allows a general form of Euler's buckling formula to be stated as

$$F_c = \frac{n\pi^2 EI}{L^2} \tag{4.18}$$

where n is determined by the particular loading geometry. The tendency for elastic collapse increases as L increases, while at lower L values the strut or column will undergo compressive failure rather than collapse.

4.5   Plane stress and plane strain

To this point, elastic solutions have been found by making some simplifying assumption and the more general elasticity equations have been by-passed. Such general solutions can be complex, but problems in which the geometric simplification of **plane stress** or **plane strain** can be made allow a relatively straightforward scientific approach and solution.

Plane stress is a loading configuration in which the only stress components

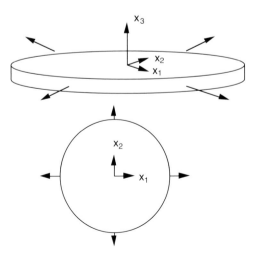

Figure 4.12 In-plane loading of a thin sheet, as an example of plane stress.

that need to be considered are those that lie within a plane. In terms of stress, the problem becomes two-dimensional. As an example, consider a thin sheet loaded by forces in its plane, as shown in Fig. 4.12. The upper and lower surfaces are stress-free and, thus, the stress components acting outwards from these planes must all be zero, i.e., $\sigma_{13}=\sigma_{23}=\sigma_{33}=0$. In other words, x_3 is a principal axis on the outer surfaces and the other two principal axes must lie within the surfaces. If the sheet is very thin, the stresses σ_{13}, σ_{23} and σ_{33}, can be assumed to be zero throughout the thickness. One, therefore, only needs three stress components, σ_{11}, σ_{12} and σ_{22}, and these will be constant through the thickness. It is important to note that although the stresses in the x_3 direction are assumed to be zero, this does not imply the strains are zero. As with the Poisson effect in a uniaxial tension test, there is no transverse stress but there is often a significant transverse strain.

In plane strain problems, the displacements that exist in a particular direction are assumed to be zero. If this direction is x_3, it follows from the definition of strain (Eq. (2.14)) that $\varepsilon_{13}=\varepsilon_{23}=\varepsilon_{33}=0$, i.e., the strains are two-dimensional. As an example, consider the problem shown in Fig. 4.13; a knife edge indenting a thick block of material. Most of the displacements are occurring in the x_1 and x_2 directions, i.e., the material is being pushed downwards or sideways. The only exceptions are in the vicinity of the front and back surfaces, where displacements in the x_3 direction are possible. Overall, the components of the displacement vector at any point can be assumed to be independent of x_3. From Hooke's Law, the assumption that $\varepsilon_{13}=\varepsilon_{23}=\varepsilon_{33}=0$, implies that $\sigma_{13}=\sigma_{23}=0$. As with plane stress, only the stress components σ_{11}, σ_{22} and σ_{12} are needed to define the problem in plane strain. This does not, however, imply that $\sigma_{33}=0$, rather one finds from Hooke's Law that $\sigma_{33}=v(\sigma_{11}+\sigma_{22})$.

In Chapter 2, stress and strain were defined, the compatibility and equilibrium equations were introduced and the relationship between stress and strain was defined. Thus, any solution that satisfies all these equations and the appropriate boundary conditions will be the solution that gives the stress and strain distribution for a particular loading geometry. For the most general problems, the scientific process can be difficult but for plane stress and plane strain problems in elastically isotropic bodies the solution involves a single differential equation.

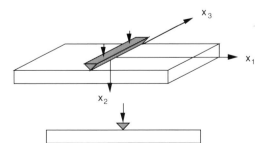

Figure 4.13 Loading of a large plate by a line force, as an example of plane strain.

As a result of the vanishing stresses and strains for plane stress and strain, the compatibility equations (Eqs. (2.31) and (2.32)) simplify to a single equation

$$2\frac{\partial^2\varepsilon_{12}}{\partial x_1 \partial x_2} = \frac{\partial^2\varepsilon_{11}}{\partial x_2^2} + \frac{\partial^2\varepsilon_{22}}{\partial x_1^2} \tag{4.19}$$

In the absence of body forces, the stress equilibrium equations (Eq. (2.49)) become

$$\frac{\partial\sigma_{11}}{\partial x_1} + \frac{\partial\sigma_{12}}{\partial x_2} = \frac{\partial\sigma_{22}}{\partial x_2} + \frac{\partial\sigma_{12}}{\partial x_1} = 0 \tag{4.20}$$

and Hooke's Law for isotropic materials (Eqs. (2.65) and (2.66)) can be written

$$\varepsilon_{11} = \frac{1}{E}(\alpha\sigma_{11} - \beta\upsilon\sigma_{22}); \quad \varepsilon_{22} = \frac{1}{E}(\alpha\sigma_{22} - \beta\upsilon\sigma_{11}); \quad 2\varepsilon_{12} = \frac{2(1+\upsilon)\sigma_{12}}{E} \tag{4.21}$$

where $\alpha = \beta = 1$ for plane stress and $\alpha = (1-\upsilon^2)$ and $\beta = (1+\upsilon)$ for plane strain. Equations (4.20) and (4.21) allow Eq. (4.19) to be written as

$$\frac{\partial^2(\sigma_{11} + \sigma_{22})}{\partial x_1^2} + \frac{\partial^2(\sigma_{11} + \sigma_{22})}{\partial x_2^2} = 0 \tag{4.22}$$

If Eqs. (4.20) and (4.22) are solved, the three stresses σ_{11}, σ_{22} and σ_{12} can be found. Indeed, if one defines a function χ such that

$$\sigma_{11} = \frac{\partial^2\chi}{\partial x_2^2}; \quad \sigma_{22} = \frac{\partial^2\chi}{\partial x_1^2}; \quad \sigma_{12} = -\frac{\partial^2\chi}{\partial x_1 \partial x_2} \tag{4.23}$$

χ is found to solve Eq. (4.20) automatically. Equation (4.22) can then be written as

$$\frac{\partial^4\chi}{\partial x_1^4} + \frac{2\partial^4\chi}{\partial x_1^2 \partial x_2^2} + \frac{\partial^4\chi}{\partial x_2^4} = 0 \tag{4.24}$$

Thus, any solution to Eq. (4.24) that also fits the boundary conditions will be the elastic solution to the problem being sought. Conversely, mathematical functions that satisfy Eq. (4.23) are often studied to find the associated elastic problem. The function χ is called the **Airy stress function** and Eq. (4.24) is called the **Biharmonic equation** This latter equation does not require any elastic constants for its solution, indicating the stress distributions are independent of the elastic properties. If, however, the strains are needed, the elastic constants appear once Hooke's Law is introduced.

The simplest elastic solutions are in the form of polynomials. For example, $\chi = ax_1^2 + bx_1x_2 + cx_2^2$ represents a state of uniform stress, where a, b and c are constants. This stress function satisfies Eq. (4.24) and, using Eq. (4.23), the stresses are found to be $\sigma_{11} = 2c$, $\sigma_{12} = -b$ and $\sigma_{22} = 2a$, i.e., the loading geometry involves two constant normal stresses and a constant shear stress. The Airy stress function must be at least second-order in x_i for the stresses not to be zero. The simplest non-uniform stress is for $\chi=ax_2^3$ which, from Eq. (4.23), gives $\sigma_{11}=6ax_2$, $\sigma_{12}=\sigma_{22}=0$ and represents the pure bending of a beam (Section 4.3) if $6a=E/R$. It is often useful to note that stress functions can be superposed so that complicated solutions can be built up from a series of simpler solutions.

4.6   Cylindrical polar coordinates

Although cartesian coordinates are useful for setting up the basic elastic equations, they are not always appropriate for solving a particular problem. The symmetry of the problem can be such that a different coordinate system makes the mathematics much easier. In the next four sections, three elastic problems involving regions with circular symmetry will be solved. For these problems, it becomes useful to change the cartesian coordinate system to cylindrical polar coordinates. This procedure changes the form of the biharmonic equation (Eq. (4.24)) and the relationship between the stresses and the stress function (Eq. (4.23)). Figure 4.14 shows this coordinate system, in which a point is defined by the coordinates r, θ and z, rather than x_i. The relationships between the two coordinate systems are

$$r^2=x_1^2+x_2^2; \quad x_1=r\cos\theta; \quad x_2=r\sin\theta; \quad x_3=z \tag{4.25}$$

The change in the coordinate system affects the strain components and, after some manipulation, one obtains

$$e_{rr}=\frac{\partial u_r}{\partial r}; \quad \varepsilon_{\theta\theta}=\frac{\partial u_\theta}{r\partial\theta}+\frac{u_r}{r}; \quad \varepsilon_{zz}=\frac{\partial u_z}{\partial z};$$

$$2\varepsilon_{r\theta}=\frac{\partial u_\theta}{\partial r}-\frac{u_r}{r}+\frac{\partial u_r}{r\partial\theta}; \quad 2\varepsilon_{\theta z}=\frac{\partial u_z}{r\partial\theta}+\frac{\partial u_\theta}{\partial z}; \quad 2\varepsilon_{rz}=\frac{\partial u_r}{\partial z}+\frac{\partial u_z}{\partial r} \tag{4.26}$$

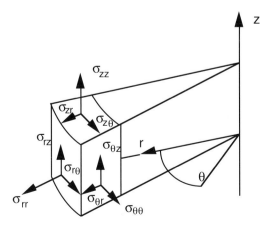

Figure 4.14 Cylindrical polar coordinates and associated stress components.

The various stress components are shown in Fig. 4.14 and the shear components can be shown to be symmetric. The equilibrium equations have a different form in the new coordinate system. This leads to new relationships between the stresses and the Airy stress function. Using the same approach as that outlined in the last section, the following revised versions of Eqs. (4.23) and (4.24) are obtained

$$\sigma_{rr} = \frac{1}{r}\frac{\partial \chi}{\partial r} + \frac{1}{r^2}\frac{\partial^2 \chi}{\partial \theta^2}$$

$$\sigma_{r\theta} = \frac{1}{r^2}\frac{\partial \chi}{\partial \theta} - \frac{1}{r}\frac{\partial^2 \chi}{\partial r \partial \theta}$$

$$\sigma_{\theta\theta} = \frac{\partial^2 \chi}{\partial r^2} \qquad (4.27)$$

and

$$\left(\frac{\partial^2 \chi}{\partial r^2} + \frac{1}{r}\frac{\partial \chi}{\partial r} + \frac{1}{r^2}\frac{\partial^2 \chi}{\partial \theta^2}\right)^2 = 0 \qquad (4.28)$$

4.7 Pressurized thick-walled cylinders

There are many important elastic problems that involve circular symmetry. Of interest here is the solution for a thick-walled cylinder under the action of internal and external pressures P_a and P_b respectively, as shown in Fig. 4.15. For this problem, the symmetry is such that the stresses will not depend on θ. Hence $\sigma_{r\theta}$ and all $\partial \chi/\partial \theta$ terms vanish, which allows Eq. (4.28) to be reduced to the ordinary differential equation,

$$\left(\frac{d^4\chi}{dr^4} + \frac{2}{r}\frac{d^3\chi}{dr^3} - \frac{1}{r^2}\frac{d^2\chi}{dr^2} + \frac{1}{r^3}\frac{d\chi}{dr}\right) = 0 \qquad (4.29)$$

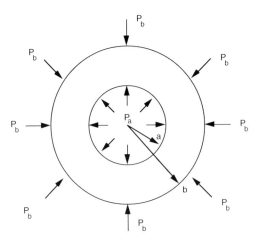

Figure 4.15 Thick-walled cylindrical vessel under the action of internal and external pressures.

This equation has the general solution

$$\chi = A\ln r + Br^2\ln r + Cr^2 + D \tag{4.30}$$

where A, B, C and D are integration constants. For the case being considered, it can be shown from the displacements that $B=0$. Using Eqs. (4.27) and (4.30), the stresses can be determined in terms of the constants, i.e.,

$$\sigma_{rr} = \frac{A}{r^2} + 2C; \quad \sigma_{\theta\theta} = -\frac{A}{r^2} + 2C; \quad \sigma_{r\theta} = 0 \tag{4.31}$$

The constants A and C are found from the boundary condition, $\sigma_{rr} = -P_a$ at $r=a$ and $\sigma_{rr} = -P_b$ at $r=b$. The final solutions for the stresses are

$$\sigma_{rr} = \frac{a^2b^2(P_b - P_a)}{r^2(b^2 - a^2)} + \frac{P_a a^2 - P_b b^2}{(b^2 - a^2)} \tag{4.32}$$

$$\sigma_{\theta\theta} = -\frac{a^2b^2(P_b - P_a)}{r^2(b^2 - a^2)} + \frac{P_a a^2 - P_b b^2}{(b^2 - a^2)} \tag{4.33}$$

For the case where there is only an internal pressure ($P_b = 0$), Eqs. (4.32) and (4.33) reduce to

$$\sigma_{rr} = \frac{P_a a^2(r^2 - b^2)}{r^2(b^2 - a^2)} \tag{4.34}$$

and

$$\sigma_{\theta\theta} = -\frac{P_a a^2(r^2 + b^2)}{r^2(b^2 - a^2)} \tag{4.35}$$

The radial stress is found to be compressive and the tangential stress is tensile. The maximum value of the tensile stress is usually of concern, as failure processes usually initiate in such locations. For the above case, this occurs when $r=a$, i.e.,

$$\sigma_{\theta\theta} = \frac{P_a(a^2 + b^2)}{(b^2 - a^2)} \tag{4.36}$$

The maximum tensile stress is always greater than P_a but approaches this value as b increases. Clearly, the stresses rise as the wall thickness decreases and, in the limit, the same solution as that discussed in Section 4.2 is obtained. If $b=3a$, $\sigma_{\theta\theta} = 1.25P_a$ and there is only a small gain in reducing the maximum stress by further increases in b ($\sigma_{\theta\theta} = P_a$ is the limit). The localized nature of the stress field can be recognized from the dependence on r and, in this case, the stresses fall off fairly rapidly, i.e., as $1/r^2$. If a thick-walled cylinder is being used as a pressure vessel, the above elastic solution is clearly important in the engineering design process. If the designer was concerned about the maximum stress, one way to protect the tube is to 'shrink-fit' a second cylinder on the outside of the first. This introduces residual stresses into the tubes and will produce a more uniform circumferential stress in the inner tube. Residual stresses also arise in composites from thermal expansion mismatch and these will be considered in the next section. Another use of the above solutions is in strength testing of tubes, which can be broken by internal pressurization (burst tests).

4.8   Residual stresses in composites

Residual stresses arise in a body if one part has a different thermal expansion coefficient than another. As an illustration, consider the problem of the shrink-fit of a hollow cylinder onto a solid cylinder. This geometry can be used as an analog for a fiber composite, with the inner cylinder representing a fiber and the outer cylinder the matrix. As shown in Fig. 4.16, the fiber, radius r_f, is being subjected to an external pressure, while the fiber is itself exerting a pressure on the inside surface of the outer cylinder (the matrix). In equilibrium, these two pressures must be equal (P). Using Eqs. (4.32) and (4.33) for the pressure on the fiber with $P_a=0$, $P_b=P$, $a=0$ and $b=r_f$, one finds $\sigma_{rr}=\sigma_{\theta\theta}=-P$. The matrix is being subjected to an internal pressure and the stresses are given by Eqs. 4.34 and 4.35, with $P_a=P$ and $a=r_f$. For the case in which the matrix is considered infinite (low fiber fraction), $r_f^2/b^2=0$ and the stress field in the matrix is given by $\sigma_{\theta\theta}=Pr_f^2/r^2=-\sigma_{rr}$. In order to relate P to the thermal expansion misfit, the displacements that occur at the interface are needed.

Figure 4.17 shows a small portion of the interface and the position of the interfaces if the cylinders were allowed to contract freely. The fiber is under compression. Thus, although its strain-free diameter is on the right, it will be displaced to the left by an amount $-u_r$ to attain its final equilibrium position. By similar reasoning, the 'hole' in the matrix has a strain-free position on the left but moves to the final position by a displacement u_r' because of the pressure exerted by the fiber. The total displacement $u_r'-u_r$ represents the misfit in diameter between the fiber and the matrix, i.e.,

$$(u_r'-u_r)=-r_f(\alpha_f-\alpha_m)(T_0-T_A) \tag{4.37}$$

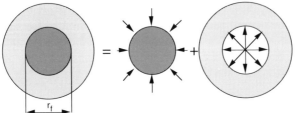

Figure 4.16 Shrink-fit of hollow cylinder onto a solid cylinder leads to residual stresses in the cylinders. The inner cylinder is under an external pressure while the outer cylinder experiences an internal pressure.

where α_f and α_m are the thermal expansion coefficients of the inner and outer cylinders, respectively, T_0 is the temperature at which the misfit stresses start to arise and T_A is the final (ambient) temperature. To relate $(u'_r - u_r)$ to the pressure P, one needs to return to the elastic solutions and consider the strains. From Hooke's Law, $E\varepsilon_{rr} = (\sigma_{rr} - v\sigma_{\theta\theta})$ and, using Eq. (4.26), the radial displacements can be obtained,

$$(u'_r - u_r) = \frac{P(1-v_f)r_f}{E_f} + \frac{P(1-v_m)r_f}{E_m} \tag{4.38}$$

where E and v are Young's modulus and Poisson's ratio, with the subscripts f and m referring to the fiber and matrix, respectively. Using Eqs. (4.37) and (4.38), P can be determined, i.e.,

$$P = \frac{E_f E_m (\alpha_m - \alpha_f)(T_0 - T_A)}{E_m(1-v_f) + E_f(1+v_m)} \tag{4.39}$$

When fiber composites are fabricated, they are often densified at a high temperature. At these temperatures the materials tend to 'flow' and stresses can relax. As the material is cooled, however, if there is a difference between the contraction behavior of the fiber and its surrounding matrix, a state of residual stress is obtained. For example, if the thermal expansion of the fiber is less than the matrix, the matrix contracts more in cooling than the fiber, placing the fiber in compression. For the opposite case, the fiber is under radial and circumferential tension. For this latter case, a perfect interfacial bond must be assumed.

For cases in which the fiber is considered to be bonded to the matrix, it is necessary to consider the longitudinal stresses σ_{zz}. In the z direction, there must be a force balance between the forces on the matrix and fiber, i.e., $\sigma_{zzf}V_f + \sigma_{zzm}(1-V_f) = 0$, where V_f is the fiber volume fraction. Thus, for the case of an infinite matrix $\sigma_{zzm} \sim 0$. Using the same scientific approach as outlined before and converting to plane strain, the equations for the tangential and radial stresses remain the same but P is now replaced by

$$P = \frac{E_f E_m (\alpha_m - \alpha_f)(T_0 - T_A)(1+v_f)}{(1+v_f)E_m(1-2v_f) + E_f(1+v_m)} \tag{4.40}$$

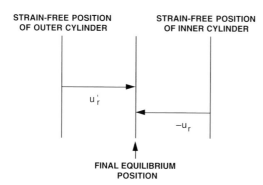

STRAIN-FREE POSITION
OF OUTER CYLINDER

STRAIN-FREE POSITION
OF INNER CYLINDER

u'_r

$-u_r$

FINAL EQUILIBRIUM
POSITION

Figure 4.17
Displacements of inner and outer cylinders so that interface ends in the equilibrium position.

In the fiber, the longitudinal stress is given by

$$\sigma_{zz} = -2v_f P + E_f(\alpha_f - \alpha_m)(T_0 - T_A) \tag{4.41}$$

The residual stress field around a misfitting fiber can now be estimated. These stresses can have important consequences in terms of the fracture behavior of fiber composites. If these stresses are high, failure may initiate near the fiber. For the case in which the fiber is in tension ($\alpha_f > \alpha_m$), the interface or the fiber may fail as σ_{rr}, $\sigma_{\theta\theta}$ and σ_{zz} are tensile in the fiber. For the case in which the fiber is in compression ($\alpha_f < \alpha_m$), the circumferential and longitudinal stresses in the matrix are tensile and this can lead to matrix cracking. In order to account for changes in the residual stresses with the volume fraction of fibers, several different approaches have been taken. For example, Budiansky *et al.* (1986) have derived exact solutions for a composite cylinder. The approach is the same as that already outlined, except that the outer cylinder is assumed not to be infinite. Instead, the outer radius (r_m) is determined by the relative volume fractions, using $V_f = r_f^2/r_m^2$. In fiber composites, there is usually a variability in fiber spacing and, at high volume fractions of fibers, there will be a need to account for fiber–fiber interactions.

The approach outlined above for a fiber in an infinite matrix can also be used for a spherical particle in an infinite matrix, using the elastic solution for a thick-walled spherical shell under the action of internal and external pressures. For this case, the stresses are given by $\sigma_{rr} = \sigma_{\theta\theta} = -P$ in the particle and

$$\sigma_{rr} = -2\sigma_{\theta\theta} = \frac{-r_p^3 P}{r^3} \tag{4.42}$$

in the matrix. The pressure at the interface is given by

$$P = \frac{2E_m E_p(\alpha_m - \alpha_p)(T_0 - T_A)}{2E_m(1 - 2v_p) + E_p(1 + v_m)} \tag{4.43}$$

where the subscript p refers to the properties of the particle. The stress field for the misfitting sphere is more localized than the misfitting fiber, with the stress falling off as $1/r^3$. Assuming that the residual stresses arise during cooling after composite fabrication, if $\alpha_p > \alpha_m$ the radial stresses are tensile and failure would be expected to occur by particle fracture or debonding. For $\alpha_p < \alpha_m$, the tangential stresses are tensile and failure would be expected to occur by radial cracking. These possible failure modes are shown schematically in Fig. 4.18.

As with fiber composites, theoretical approaches have been advanced to account for changes in the residual stress with particle volume fraction and for particle–particle interactions. For example, Pan (1994) used the generalized self-consistent approach to show that the stress in the particle is given by

$$\frac{12B_pB_m\mu_m(\alpha_p-\alpha_m)(T_0-T_A)(1-V_p)}{3B_pB_m+4\mu_mB_m+4\mu_m(B_p-B_m)V_p} \tag{4.44}$$

where B is the bulk modulus, μ is the shear modulus and V_p is the volume fraction of particles. It is straightforward to show that Eq. (4.44) is equivalent to Eq. (4.43) as V_p approaches zero.

Residual stresses can also play an important role in layered structures. For example, consider the sandwich structure shown in Fig. 4.19. If the outside faces (subscript F) have a lower thermal expansion coefficient than the core (subscript C), they will be placed in biaxial compression as the structure cools after fabrication. Conversely, a higher expansion coefficient for the faces will place them in biaxial tension. As with fiber composites, the forces between the layers must balance, so $\sigma_c h+2\sigma_F t=0$. Using Hooke's Law, the force balance can be written as

$$\frac{2\varepsilon_F E_F t}{(1-v_F)}+\frac{\varepsilon_C E_C h}{(1-v_C)}=0 \tag{4.45}$$

The misfit strain ε_0 between unconstrained layers is given by $\varepsilon_0=-\varepsilon_F+\varepsilon_C=(\alpha_C-\alpha_F)(T_0-T_A)$ which, with Eq. (4.45), gives

$$\varepsilon_F=\frac{-\varepsilon_0 E_C h(1-v_F)}{2E_F t(1-v_C)+E_C h(1-v_F)}; \quad \varepsilon_C=\frac{2\varepsilon_0 E_F t(1-v_C)}{2E_F t(1-v_C)+E_C h(1-v_F)} \tag{4.46}$$

For the case where the faces are very thin ($t\sim0$), the stress in the faces becomes equivalent to $-\varepsilon_0$ and in the core it is approximately zero.

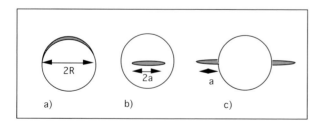

Figure 4.18 Schematic representations of idealized microcracking geometries: a) interfacial failure; b) inclusion failure; and c) radial matrix cracking. For the first two geometries the inclusion is in hydrostatic tension and for the last, hydrostatic compression.

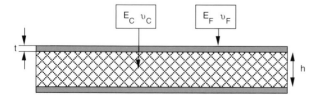

Figure 4.19 Residual stresses can arise in a sandwich beam, if the thermal expansion coefficient of the faces is different from that of the core.

4.9   Stress concentrations due to pores and inclusions

Another important problem in elastic stress distributions, that relates to micro-structures, is the disturbance of a stress field by an inclusion with different elastic properties, i.e., there is an **elastic mismatch**. Extreme examples occur when the inclusion is a pore or a rigid particle. Consider a plate under uniaxial tension that contains a circular hole, Fig. 4.20. As the hole surface is free from applied stresses, $\sigma_{rr} = \sigma_{r\theta} = 0$ at $r = r_0$. At large distances from the hole, the disturbance in the stress field vanishes, so $\sigma_{11} = \sigma$, $\sigma_{12} = \sigma_{22} = 0$ as $r \to \infty$. From Eqs. (2.42) and (2.44), the applied stresses can be resolved into

$$\sigma_{rr} = \sigma\cos^2\theta = \sigma(1 + \cos 2\theta)/2$$
$$\sigma_{r\theta} = -\sigma\sin\theta\cos\theta = -\sigma\sin 2\theta/2$$
$$\sigma_\theta = \sigma\sin^2\theta = \sigma(1 - \cos 2\theta)/2 \tag{4.47}$$

This can be considered as the superposition of two stress fields, σ' and σ'', where

$$\text{and} \quad \sigma'_{rr} = \sigma/2; \quad \sigma'_{r\theta} = 0; \quad \sigma'_{\theta\theta} = \sigma/2$$
$$\sigma''_{rr} = \sigma\cos 2\theta/2; \quad \sigma''_{r\theta} = -\sigma\sin 2\theta/2; \quad \sigma''_{\theta\theta} = -\sigma\cos 2\theta/2 \tag{4.48}$$

Each field must satisfy the boundary conditions at $r = r_0$, i.e.,

$$\sigma'_{rr} = \sigma'_{r\theta} = 0 = \sigma''_{rr} = \sigma''_{r\theta} = 0 \text{ at } r = r_0 \tag{4.49}$$

The stress field σ' does not depend on θ and, thus, Eq. (4.31) can be used for the solution. Incorporating Eqs. (4.48) and (4.49) into Eq. (4.31), one obtains

$$\sigma'_{rr} = \frac{\sigma}{2}\left(1 - \frac{r_0^2}{r^2}\right); \quad \sigma'_{\theta\theta} = \frac{\sigma}{2}\left(1 + \frac{r_0^2}{r^2}\right) \tag{4.50}$$

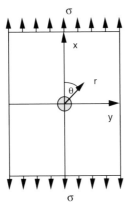

Figure 4.20 Small circular hole, radius r_0, in a plate subjected to uniaxial tension.

For the field σ'', try $\chi''=f(r)\cos2\theta$, where $f(r)$ is a function that depends only on r. The biharmonic equation (Eq. (4.28)) reduces to

$$\left(\frac{d^2f(r)}{dr^2}+\frac{1}{r}\frac{df(r)}{dr}-\frac{4f(r)}{r^2}\right)^2=0 \tag{4.51}$$

and has the general solution

$$f(r)=Cr^2+Fr^4+\frac{G}{r^2}+H \tag{4.52}$$

Expressing the stresses in terms of this equation, using Eqs. (4.27) and $\chi''=f(r)\cos2\theta$, one finds $C=-\sigma/4$, $F=0$, $G=-\sigma r_0^4/4$ and $H=\sigma r_0^2/2$. Adding the solutions for σ' and σ'', one obtains

$$\sigma_{rr}=\frac{\sigma}{2}\left[1-\frac{r_0^2}{r^2}+\left(1+\frac{3r_0^4}{r^4}-\frac{4r_0^2}{r^2}\right)\cos2\theta\right]$$

$$\sigma_{\theta\theta}=\frac{\sigma}{2}\left[1+\frac{r_0^2}{r^2}-\left(1+\frac{3r_0^4}{r^4}\right)\cos2\theta\right] \tag{4.53}$$

$$\sigma_{r\theta}=-\frac{\sigma}{2}\left(1-\frac{3r_0^4}{r^4}+\frac{2r_0^2}{r^2}\right)\sin2\theta$$

The only stress at the surface of the hole is $\sigma_{\theta\theta}=\sigma(1-2\cos2\theta)$, which has a maximum value of 3σ at $\theta=\pm\pi/2$ and a minimum value of $-\sigma$ at $\theta=0$ and π. In engineering design, **stress concentrations** in which applied stresses are amplified are very important, as failure is expected to initiate in such locations. It is important to clearly distinguish between residual stresses and stress concentrations, though both may be important in terms of failure processes. The stress field produced by the circular hole falls off fairly rapidly, primarily as $1/r^2$ and, thus, at $r=3r_0$ the stresses are approaching the applied values.

The above solution can also be used if σ is an applied uniaxial compressive stress. In this case, the maximum tensile stress is $-\sigma$ and the minimum is 3σ. Thus, plates can fail in a tensile mode, even if the plate is in compression. This idea is used extensively in rock mechanics, as most rocks are in a state of compression but can undergo a tensile failure in the vicinity of holes and pores. In general, pores in brittle materials are not expected to be bounded by a smooth surface and the stress concentrations can be much larger. Consider what would happen to the stresses if another circular hole with a much smaller radius was placed at $\theta=\pm\pi/2$. This process could be repeated to approximate surface roughness or holes with sharp radii of curvature. It is clear from such a process that the stress concentration would quickly exceed 10σ. As will be shown in Chapter 8, the ability of cracks to concentrate stress is a key aspect in understanding the fracture process.

As an example of the effect of hole shape on stress concentration, consider the problem of a large plate containing an elliptical hole. The geometry is shown in Fig. 4.21. The maximum normal stress occurs at the ends of the major axis, in the same direction as the applied stress, and is given by

$$\sigma_{11}^{max}=\sigma\left(1+2\sqrt{\frac{c}{\rho}}\right)$$
(4.54)

where ρ is the radius of curvature at the end of the major axis. Clearly, if ρ approached atomic dimensions the stress concentration would become very large.

In many ceramics, pore shapes can be complex and although elastic solutions are available for ellipsoidal pores, these solutions are complex. For a spherical pore, the maximum tensile stress occurs at the pore surface for $\theta=\pm\pi/2$ and is given by

$$\sigma_{\theta\theta}^{max}=\sigma\left(\frac{27-15v}{2(7-5v)}\right)$$
(4.55)

At $\theta=0$ or π, the circumferential stress is given by

$$\sigma_{\theta\theta}=-\sigma\left(\frac{3+15v}{2(7-5v)}\right)$$
(4.56)

For typical values of v, $\sigma_{11}^{max}\sim2\sigma$ and, thus, the **stress concentration factor** $(\sigma_{\theta\theta}^{max}/\sigma)$ is less for a spherical pore than a circular hole.

Holes and pores are extreme examples of a mismatch in elastic behavior. In many cases one is concerned with inclusions, and solutions for ellipsoidal inclusions are available. For the case of spherical inclusions, it is useful to classify inclusions into those that are *more rigid* or *more compliant* than the matrix. For more-compliant spherical inclusions, the stress distribution is similar to that of a spherical pore but the maximum stress is less extreme. For more-rigid inclusions, the maximum tensile stress moves to the poles of the inclusion ($\theta=0$ or π).

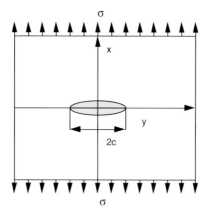

Figure 4.21 Small elliptical hole in a plate subjected to uniaxial tension.

4.10   **Contact forces**

Ceramics are susceptible to contact damage and, thus, it is important to under-stand the stress fields that arise around contacts between two bodies. Consider the simple problem of a body being contacted by a linear force per unit thickness F_t as illustrated in Fig. 4.22. The stress function for this problem can be expressed as $\chi = Cr\theta\sin\theta$, where C is a constant. From Eq. (4.27), the stress components are $\sigma_{rr} = 2C\cos\theta/r$, $\sigma_{\theta\theta} = \sigma_{r\theta} = 0$. The constant C can be determined using statics for a force balance in the vertical direction, i.e.,

$$F_t = -2\int_0^{\pi/2} \sigma_{rr}\cos\theta r d\theta = -2\int_0^{\pi/2} 2C\cos^2\theta d\theta = -C\pi \tag{4.57}$$

giving

$$\sigma_{rr} = \frac{-2F_t}{\pi}\frac{\cos\theta}{r} \tag{4.58}$$

The stresses fall off somewhat more slowly than the other solutions considered in this chapter. In addition, one finds a singularity occurs in the stress field at $r=0$. Hooke's Law would be expected, therefore, to fail in some region near the contact and the material would be expected to yield, crush or fracture. At larger distances from the contact, the material is expected to be elastic. If the material yields in the contact area, the permanent deformation gives rise to a residual stress. This is illustrated in Fig. 4.23, in which the elastic material outside the yield zone can no longer return to zero strain on unloading because of the pres-ence of the permanently deformed material. The elastic region 'presses' on the yield zone and, in some ways, the situation is similar to that already discussed for a matrix contracting onto a low-expansion inclusion.

   This chapter has only touched on the elegant field of elasticity and a myriad of exact elastic solutions have been obtained in the literature. Later in the book,

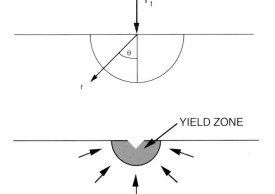

Figure 4.22 A free surface being subjected to a line force F_t.

Figure 4.23 Permanent deformation at an inden-tation site leads to a state of residual stress, if stresses cannot relax.

other important solutions will be discussed, viz., cracks, dislocations and other types of contact. Complex loading geometries can now be solved easily using computers, often with a technique called **finite-element analysis**. In this technique, a mesh is generated for the chosen geometry and the appropriate boundary conditions are applied. Figure 4.24 shows a mesh chosen to study the problem of a circular hole in a plate under a uniaxial tensile stress. Unlike the problem studied in Section 4.9, the plate can no longer be considered large compared to the hole size. Figure 4.25 shows an example of an output from the finite-element analysis. In this example, the stresses (in MPa) in the x direction are shown as **stress contours**, very much akin to a topographical map.

It is also useful to be able to study stress distributions experimentally. Some materials, especially glasses and polymers, are **photoelastic**, i.e., they undergo a change in refractive index under the application of mechanical stress. When viewed under polarized light, mechanical models made from these materials give rise to an optical pattern that can be related quantitatively to the principal stress distribution in the loaded body. Figure 4.26 shows a phototelastic fringe pattern

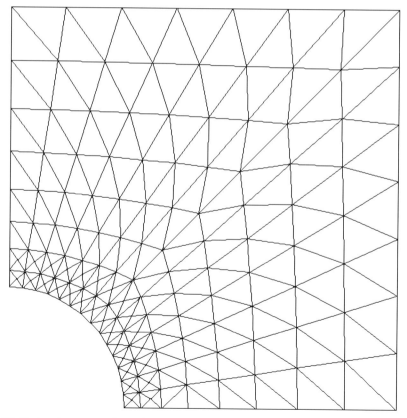

Figure 4.24 Finite-element mesh used to solve problem of a circular hole in a thin plate under uniaxial tension.

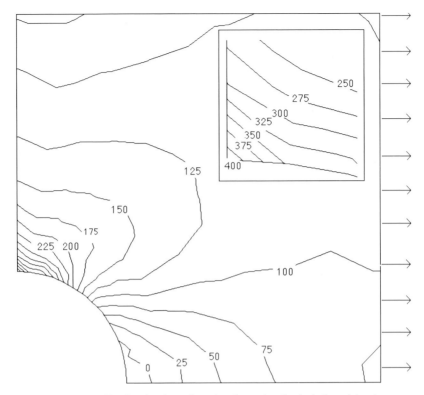

Figure 4.25 Normal stress distribution in x direction for a circular hole in a thin plate under a uniaxial tensile stress of 100 MPa. Inset shows stresses near the hole and indicates a stress concentration factor greater than 3 for this finite-sized plate.

in a fiber-reinforced composite. The fibers are being subjected to a transverse uniaxial stress, which gives rise to a stress concentration in the matrix near the 'poles' of the fiber.

Problems

4.1   A sample of silicon carbide is broken in three-point bending. The speci-
       men dimensions were $60 \times 6 \times 2$ mm and the outer load span was 50 mm.
       The force to break the specimen was 150 N and the maximum deflection
       at failure was 232 μm.
       a) Is the bend test statically determinate? Calculate the maximum
          tensile stress at failure. Where is this stress located? Calculate
          Young's modulus for the SiC.
       b) Calculate the maximum shear stress at failure. Where is the location
          of the maximum shear?
       c) A defect was found on a plane 100 μm from the central loading

Figure 4.26 A phototelastic fringe pattern in a fiber-reinforced composite. The fibers are being subjected to a (vertical) uniaxial stress, which gives rise to a stress concentration in the matrix near the 'poles' of the fiber. (Adapted from Puck, 1967 reproduced courtesy of Carl Hanser Verlag, Munich, Germany.)

cross-section and 100 μm below the tensile surface. Calculate the tensile stress at this point.

d) With a safety and other factors, the design stress for a thick-walled, silicon carbide tube was determined to be 100 MPa. Calculate the minimum thickness of a tube, radius 100 mm, that must withstand an internal pressure of 20 MPa.

4.2   Two hikers walking through a wood, come across a river, 9 m wide. Across the river is a plank of wood, length 10 m, width 200 mm and thickness 100 mm. Knowing the flexural strength of wood to be at least 3 MPa, their weights (50 and 75 kg), they must decide whether they should walk across the plank to cross the river. What is your opinion? Show your calculation. For any hiker that can make it across, calculate the maximum deflection of the plank ($E \sim 1$ GPa for wood).

4.3   a) Show that the relationships between the stresses and the Airy stress function automatically satisfy the stress equilibrium equations, Eq. (4.20).

b) Show that the stress function $\chi = Cr\theta\sin\theta$ satisfies the biharmonic equation, Eq. (4.24). Calculate the stress components σ_{rr}, $\sigma_{r\theta}$ and $\sigma_{\theta\theta}$.

c) For an Airy stress function $\chi = Dx_1x_2^3$, calculate the stress components for plane stress.

4.4   A sharp knife-edge under a linear load of 1 kN/m is placed in contact with a surface. At what distance below the contact point does the stress exceed 1 GPa? At what distance below the contact point does the stress exceed 100 MPa? Sketch the region in which the stresses would exceed 1 GPa.

4.5   Particulate composites with an alumina matrix and SiC particles are of interest for structural applications. For a 30 vol.% SiC particle composite:

a) Calculate the residual stresses in particles after fabrication. Assume the residual stresses arise during cooling from the sintering temperature (1520 °C).

b) If the composite contained a large spherical pore, 100 μm in diameter, calculate the maximum stress at the pore under a uniaxial stress of 100 MPa. What would happen to this stress if the pore surface were not completely smooth?

c) Determine the maximum stress concentration if the applied stress in part b) was changed to an equibiaxial applied stress of 100 MPa.

d) If a plate of the composite contained an elliptically shaped hole, with major and minor axes of 100 μm and 0.2 μm, calculate the maximum tensile stress at the void under the action of a uniaxial tensile stress of 100 MPa.

4.6   The stress field in the vicinity of a crack tip for an infinite plate containing a sharp, through-thickness, internal crack under plane stress conditions and uniaxially-applied tension (σ) is expressed as

$$\sigma_{11} = K_1\cos(\theta/2)[1-\sin(\theta/2)\sin(3\theta/2)]/\sqrt{2\pi r}$$
$$\sigma_{22} = K_1\cos(\theta/2)[1+\sin(\theta/2)\sin(3\theta/2)]/\sqrt{2\pi r}$$
$$\sigma_{12} = K_1\cos(\theta/2)[\sin(\theta/2)\cos(3\theta/2)]/\sqrt{2\pi r}$$

where K_1 is called the stress intensity factor and represents the amplitude of the stress field. K_1 is also related to the applied stress through $K_1 = \sigma\sqrt{\pi a}$. The distance from the crack tip is denoted r and θ is the angle between this vector and the crack plane. Microstructural events occurring near a crack tip are often described by assuming they occur at a critical stress.

a) Draw contours around a crack tip for the critical value of stress$=10\sigma$

for the cases in which the stress is i) the maximum normal stress, ii) the maximum shear stress and iii) the mean stress.

b) By what factor would the radial distances of the contours change if the stress state was plane strain ($\sigma_{33}=v(\sigma_{11}+\sigma_{22})$).

c) Determine the stresses at a point P, 1 μm from the crack tip for the cases where $\theta=0°$ or $90°$, $\sigma=240$ MPa, $a=200$ μm, $v=0.25$. Compare hydrostatic and deviatoric stress components and the principal stresses for these two locations.

4.7 a) Show the change in radius for a thin-walled cylinder under an internal pressure p is given by pr^2/Et, where E is the Young's modulus of the cylinder material, r the cylinder radius and t the wall thickness.

b) A thin-walled steel ($E=210$ GPa) cylinder just fits over a thin-walled copper ($E=100$ GPa) cylinder. The inner and outer radii are 500 and 505 mm for the copper cylinder and 505 and 510 mm for the steel cylinder. Calculate the hoop stresses in the two tubes if the assembly is heated to 100 °C. The thermal expansion coefficients for steel and copper are 6.5 and $9.3\times10^{-6}/°C$, respectively. Ignore the effects caused by the longitudinal expansion difference, i.e., assume plane stress.

c) At 100 °C, the copper cylinder is subjected to an internal pressure of 10 MPa. Compare the hoop stress in the copper tube, with and without the 'shrink-fit' steel tube.

4.8 A sample of alumina is broken in four-point bending. The specimen dimensions were $60\times6\times2$ mm the inner and outer load spans were 30 and 50 mm, respectively. The force to break the specimen was 750 N and the deflection at the inner load point at failure was 425 μm.

a) Use the double-integration method to determine the mathematical equation for the deflection at the inner load points.

b) Calculate maximum tensile stress at failure. Where is this maximum tensile stress located?

c) Calculate Young's modulus for this alumina.

4.9 From the elastic solution for a circular hole in a plate being subjected to uniaxial tension:

a) Plot the stress components (polar coordinates) normalized by the applied stress as a function of distance r normalized by hole radius r_0. Use the distance range from $r/r_0=1$ (hole surface) to $r/r_0=10$ for the case of $\theta=\pi/2$.

b) Suppose the plate was being stressed longitudinally by 500 MPa and transversely by 200 MPa. Determine the maximum stress concentration.

4.10   The surface of a semi-infinite solid is subjected to a concentrated force
       tangential to the surface. An Airy stress function of the same form as a
       concentrated normal force can be used to solve the problem.
       a) Determine the elastic solution in terms of σ_{rr}, $\sigma_{r\theta}$ and $\sigma_{\theta\theta}$.
       b) Does the stress function satisfy the biharmonic equation and the
          equilibrium conditions?

4.11   What type of elastic problems are solved by the Airy stress functions:
       a) $\chi = Ax_1^2 + Bx_1x_2 + Cx_2^2$;
       b) $\chi = Dx_2^3$?

References

F. P. Beer and E. R. Johnston, Jr,
   Mechanics of Materials, McGraw-Hill,
   New York, 1981.

B. Budiansky, J. W. Hutchinson and A. G.
   Evans, Matrix fracture in fiber-reinforced ceramics, *J. Mech. Phys. Solids,*
   34 (1986) 167–89.

A. H. Cottrell, *Mechanical Properties of
   Matter*, Wiley and Sons, New York,
   1964.

R. W. Davidge, *Mechanical Behaviour of
   Ceramics*, Cambridge University Press,
   Cambridge, UK, 1979.

L. J. Gibson and M. F. Ashby, *Cellular
   Solids: Structure and Properties,*
   Pergamon Press, Oxford, UK, 1988.

C-H. Hseuh and A. G. Evans, Residual
   stresses and cracking in metal/ceramic
   systems for microelectronics pack-
   aging, *J. Am. Ceram. Soc.,* **68** (1985)
   120–6.

M-J. Pan, *Microcracking Behavior of
   Particulate Titanium Diboride-Silicon
   Carbide Composites and Its Influence on
   Elastic Properties*, Ph.D. Thesis, The
   Pennsylvania State University, 1994.

A. Puck, Zur Beanspruchung und
   Verformung von GFK-
   Mehrschichtenverbund-Bauelementen,
   Kunststoff., **57** (1967) 965–73.

J. Selsing, Internal stresses in ceramics, *J.
   Am. Ceram. Soc.,* **44** (1961) 419.

S. P. Timoshenko and J. N. Goodier,
   Theory of Elasticity, 3rd Edition,
   McGraw-Hill, New York, 1970.

A. V. Virkar, J. L. Huang and R. A.
   Cutler, Strengthening of oxide ceramics
   by transformation-induced stresses, *J.
   Am. Ceram. Soc.,* **70** (1987) 164–70.

Chapter 5

Viscosity and viscoelasticity

Unlike elastic deformation in which the atoms maintain their nearest neighbors, flow involves changes in nearest neighbors and is a process of shear. This process is also dependent on time, so that one is concerned with the change of strain with time. The ease of flow in a liquid is characterized by its **viscosity**. **Viscous flow** is usually associated with liquids but it can occur in amorphous solids. For such materials, elastic and viscous processes can coexist. This is termed **viscoelasticity** and one can view elastic and viscous deformation as the limiting conditions of such behavior. Flow processes, such as creep, can also occur in crystalline materials. In this situation, the deformation processes involve different mechanisms but they can mimic viscoelastic behavior.

5.1   Newton's Law of viscosity

The shear strain rate $(d\gamma/dt)$ of a viscous material is a function of the applied shear stress τ. The simplest dependence (for small stresses) is a linear function, i.e.,

$$\frac{d\gamma}{dt} = \alpha\gamma\tau \qquad (5.1)$$

or

$$\tau = \eta\frac{d\gamma}{dt} \qquad (5.2)$$

These relationships are known as **Newton's Law of viscous flow**; α is termed the **fluidity** and η the **dynamical shear viscosity**. Newton's Law is analogous to Hooke's Law, except shear strain has been replaced by shear strain *rate* and the shear modulus by shear *viscosity*. As shown later, this analogy is often very important in solving viscoelastic problems. In uniaxial tension, the viscous equivalent to Hooke's Law would be $\sigma = \eta^*(d\varepsilon/dt)$, where η^* is the **uniaxial viscosity**. As $v=0.5$ for many fluids, this equation can be re-written as $\sigma = 3\eta(d\varepsilon/dt)$ using $\eta = \eta^*/[2(1+v)]$, the latter equation being the equivalent of the interrelationship between three engineering elastic constants, ($\mu = E/[2(1+v)]$).

Newtonian viscosity leads to an interesting aspect for materials undergoing a drawing process. The normalized rate of change of the cross-sectional area for a **linear viscous** material is given by $(dA/dt)/A = -(d\varepsilon/dt) = -\sigma/\eta^*$ or $(dA/dt) = -F/\eta^*$, where F is the applied uniaxial force. This derivation shows that, for a constant F, (dA/dt) must be constant, i.e., the body can be reduced in cross-section at a constant rate. Thus, a section with an initially narrow section will not neck down faster than elsewhere.

The ability of a material to flow is usually associated with liquids but whether a stressed body behaves as a solid or a liquid can depend on the period of time over which the stress is applied. With some materials this distinction is obvious. For example, silicone putty can be poured slowly from a container but also be bounced as a rubber ball. In other cases, the flow process can be very slow and the difference is less obvious. The physical distinction between a solid and a liquid is therefore arbitrary but a viscosity of 10^{14} Pa s (100 TPa s) is sometimes used. Figure 5.1 shows schematically the fluidity (=1/viscosity) behavior with changes in temperature of simple molecular materials. The viscosity increases by approximately 30 orders of magnitude as one moves from a gas to a solid. The viscosity of a gas is usually about 10^{-5} Pa s (10 μPa s) and changes to ~10^{-3} Pa s (1 mPa s) on condensing as a liquid. On further cooling, the viscosity rises,

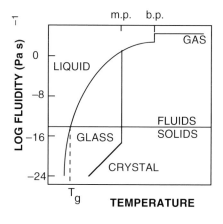

Figure 5.1 Log fluidity (1/viscosity) as a function of temperature of a simple molecular solid.

slowly at first but then more rapidly. If the material crystallizes at the melting point, there is a dramatic increase in 'viscosity' from ~0.1 Pa s to ~10^{17} Pa s (100 TPa ks) and this increases further on cooling. Some materials fail to crystallize and the liquid viscosity continues to increase. At some temperature, however, these materials are no longer considered to be liquids but rather **amorphous solids** or **glasses**. This temperature is known as the **glass transition temperature** T_g.

Silicate glasses are commonly used materials and control of viscosity is critical in their production. Figure 5.2 shows the viscosity of a soda–lime–silica glass as a function of temperature. In the glass melting process, gas bubbles must escape from the molten material. In this process, known as **fining**, the viscosity of the glass must be ~5 Pa s, to ensure the gas-escape process is reasonably rapid. The viscosity must also be controlled in the forming process, its value depending on the particular forming method being used. The viscosity working range of a glass is between 1 kPa s and 10 MPa s. For example, glass blowing is often performed at a viscosity of ~10 MPa s. The **working point** of a glass is defined as the temperature corresponding to a viscosity of 1 kPa s and the **flow point** to a viscosity of 10 kPa s. The viscosity at the **softening point** is not as well defined but it is ~10 MPa s. As a glass cools, stresses due to differential cooling can no longer relax and the body may contain residual stresses. These stresses are removed by a further heating process or by controlled cooling. The relaxation behavior is usually defined by the **annealing point** and the **strain point**. These temperatures correspond to viscosity values of approximately 1 TPa s and 30 TPa s, respectively.

The viscosity of a glass depends on its particular composition. Indeed, the glass composition is chosen, in part, to ensure it has the appropriate viscosity for the various processes it must undergo during its production and use. For example, univalent and divalent ions are often used to decrease the viscosity, as these **network modifiers** tend to break up the silica network. The effect of various ions on the structural modification depends on the field strength and polarizabil-

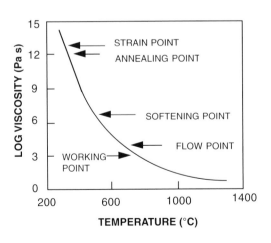

Figure 5.2 Log viscosity as a function of temperature of a soda–lime–silica glass.

ity of the ions. Control of viscosity is important in many other aspects of ceramic processing. For example, in forming processes such as injection molding, slip casting, extrusion, tape casting, etc., it is essential to have a material with the correct flow characteristics. Viscous flow is also important in the sintering of glass powders and in the high-temperature creep behavior of ceramics when liquid phases are present at grain boundaries. Some creep mechanisms in poly-crystalline ceramics are found to exhibit a linear relationship between stress and strain rate. In these cases, the deformation is not strictly viscous flow as it depends on structure. It is, however, sometimes useful to use a viscoelastic formalism to describe the behavior.

5.2   Temperature dependence of viscosity

As indicated by the above discussion, it is important to know the temperature dependence of viscosity. It has become standard industrial practice to use an empirical equation, known as the **Vogel–Fulcher–Tammann (VFT) equation** to describe this temperature dependence. For silicate glasses, the VFT equation often fits the temperature dependence over *ten* orders of magnitude in viscosity. The VFT equation involves three empirical constants (η_0, C and T_0) and can be written

$$\eta = \eta_0 \exp\left(\frac{C}{T - T_0}\right)$$
(5.3)

Originally, the various constants had no particular physical meaning, though one does note that the viscosity becomes infinite when $T = T_0$. The VFT equation runs into some difficulties in fitting data near the glass transition temperature but the success of the empirical approach has often led subsequent theoreticians to interpret their models in terms of the VFT equation.

   A simple theoretical approach to the temperature dependence of viscosity would be to consider it a thermally activated process and utilize the Arrhenius equation, i.e.,

$$\eta = \eta_0 \exp\left(\frac{E_v}{RT}\right)$$
(5.4)

where E_v is the activation energy for viscous flow, η_0 is a constant, R is the gas constant and T is the absolute temperature. There is a clear similarity between Eqs. (5.3) and (5.4) and they become equivalent if one puts $T_0 = 0$ in Eq. (5.3). Unfortunately, the Arrhenius equation only fits liquid viscosity data over a narrow temperature range.

   There is a strong relationship between diffusion and viscosity, in that both involve the movement of atoms in a body. Indeed, viscous flow is often considered in terms of the **free volume** V_F that exists in a liquid or solid and the distances over which the flow must occur. The movement of a molecule in the flow process depends on the availability of an adjoining vacancy. The volume of these diffusing vacancies constitutes the free volume of the liquid and roughly corresponds to the difference in the specific volumes of the liquid V_L and its corresponding crystal V_C. According to the free volume theory

$$\eta = \eta_0 \exp\left(\frac{DV_C}{V_F}\right) \tag{5.5}$$

where D and η_0 are constants. The volumetric thermal expansion coefficient of a liquid β_L is larger than its corresponding crystal value β_C. Thus, if the specific volumes are extrapolated, there is a temperature, T_V, when the free volume becomes zero. The ratio V_F/V_C can be approximated by $(\beta_L - \beta_C)(T - T_V)$ and substitution into Eq. (5.5) gives $T_0 = T_V$, i.e., T_0 can be interpreted as the point at which the free volume disappears. For glasses, some fraction f_g of the free volume will be 'trapped' at the glass transition temperature T_g. For this case, Eq. (5.5) can, therefore, be written as

$$\eta = \eta_0 \exp\left(\frac{D}{f_g + (\beta_L - \beta_C)(T - T_g)}\right) \tag{5.6}$$

This expression is known as the **Williams–Landel–Ferry (WLF) relation**. Comparison of Eqs. (5.3) and (5.6) shows that T_0 is expected to be slightly below T_g.

In the **Adams–Gibbs model**, the approach is to consider the configurational entropy S_c of the liquid. Viscous flow is seen as a cooperative rearrangement of the molecules. As temperature decreases, fewer configurations will be accessible and rearrangement will involve an increasing number of molecules. This leads to a decrease in configurational entropy and an ensuing decrease in mobility. The theory can be expressed as

$$\eta = \eta_0 \exp\left(\frac{E_s}{T S_c}\right) \tag{5.7}$$

where E_s and η_0 are constants. The configurational entropy can be approximated by

$$S_c = \alpha\left(\frac{1}{T_k} - \frac{1}{T}\right) \tag{5.8}$$

where α is a constant and T_k is the temperature at which S_c goes to zero. Comparison of this expression with Eq. (5.3) gives an interpretation of $T_0 = T_k$. As T_k can be determined from thermodynamic data, this is a very useful attribute. The above discussion demonstrates that the VFT equation can be interpreted by various models but experimental viscosity data do not appear to show

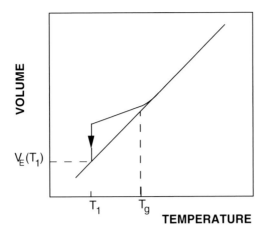

Figure 5.3 Volume change in a liquid near the glass transition temperature.

the divergence in viscosity that is predicted to occur near T_0. The VFT equation is, however, very accurate over a very wide range of temperatures and will continue to be used in many practical applications.

It is important to note that viscosity can apparently become dependent on time, especially near T_g. This is a reflection of the **structural relaxation** that occurs in the material. The physical behavior of glasses can become quite complex in this region, as it is possible to 'freeze-in' structures with differing amount of free volume. Figure 5.3 shows how the liquid volume can depart from the equilibrium curve near the glass transition temperature. The details of this metastable behavior will depend on the cooling rate. If the glass is held at a constant temperature T_1, the volume of the glass will decrease and move towards its equilibrium value, $V_E(T_1)$. The structural relaxation leads to changes in glass properties, such as viscosity, as they move towards their equilibrium values.

The emphasis in this section has been on viscous behavior but both liquids and amorphous solids can show viscoelastic behavior; this will be considered in Section 5.7.

5.3   Simple problems of viscous flow

It is useful to analyze some simple modes of flow in Newtonian fluids. As indicated at the start of this chapter, the analogy between Newton's Law and Hooke's Law can be used to solve some problems of viscous flow. To do this, one assumes the flow is **laminar** and not **turbulent**.[†] Laminar flow usually occurs at small flow rates. For simplicity, it will be assumed that the liquid is incompressible and that no slip occurs at the interface between the fluid and its container.

---

† The transition from laminar to turbulent flow for a Newtonian fluid can be determined from a dimensionless parameter, known as the **Reynolds number**.

Consider two plates separated by a Newtonian fluid under the action of a constant shear stress. One can imagine layers in the liquid being sheared past each other, much like a pack of cards. The relative velocity of the plates will depend on the viscosity of the liquid. Indeed, the strain rate is equivalent to the velocity gradient (dv/dx) and, thus, Newton's Law can be written as

$$\frac{dv}{dx}=\frac{\tau}{\eta} \tag{5.9}$$

Thus, for a constant stress, $v=\tau h/\eta$, where h is the thickness of the fluid. Clearly, this sliding geometry could be used as a technique for measuring viscosity.

A related problem is the flow of a fluid down a cylindrical tube, length L, under the action of a pressure P, as shown in Fig. 5.4. One can again think of layers of fluid sliding down the tube with velocity v in the x direction. In this case, however, the velocity will be dependent on r. Equation (5.9) is, therefore, written as

$$\frac{dv}{dr}=\frac{\tau}{\eta} \tag{5.10}$$

For equilibrium, the applied force, $\pi r^2 P$, must be counterbalanced by the viscous force, $2\pi r L\tau$, acting on the cylindrical surfaces of the flowing liquid. Equation (5.10) can be re-written as

$$dv=\frac{P}{2L\eta}r\,dr \tag{5.11}$$

On integration, one obtains

$$v=\frac{P}{4L\eta}(r_0^2-r^2) \tag{5.12}$$

The velocity profile is found, therefore, to be parabolic, with a maximum velocity at the center of the tube. The volume V_D discharged per unit time from the tube is given by

$$V_D=\int_0^{r_0} 2\pi r v\,dr=\frac{P\pi r_0^4}{8L\eta} \tag{5.13}$$

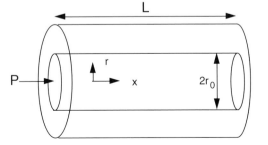

Figure 5.4 Viscous flow down a cylindrical tube.

This is known as the **Poiseuille relation** and can also be used to determine viscosity.

As a third example, consider viscous flow between coaxial cylinders, as shown in Fig. 5.5. The outer cylinder is rotating with a linear velocity ωb, while the inner cylinder is at rest. The 'particles' in the fluid move in concentric circles around the axis of rotation. In polar coordinates, only the tangential component of fluid velocity, u_θ, is non-zero. Using the elastic analogy, from Eq. (4.26) one can write

$$\frac{d\gamma}{dt} = \frac{du_\theta}{dr} - \frac{u_\theta}{r} = \frac{\tau}{\eta} \tag{5.14}$$

The couple exerted on a cylinder, radius r, and length L is $2\pi r^2 L\tau$. This is the same for all values of r and, thus, the couple M exerted on the inner cylinder is

$$\tau = \frac{M}{2\pi r^2 L} \tag{5.15}$$

From Eqs. (5.14) and (5.15)

$$\frac{du_\theta}{dr} - \frac{u_\theta}{r} = \frac{M}{2\pi r^2 L\eta} \tag{5.16}$$

The general solution for this equation is

$$u_\theta = Cr - \frac{M}{4\pi r L\eta} \tag{5.17}$$

Using $u_\theta = 0$ at $r = a$ and $u_\theta = \omega b$ at $r = b$, one obtains

$$M = \frac{4\pi r L\eta\omega a^2 b^2}{b^2 - a^2}; \quad C = \frac{\omega b^2}{b^2 - a^2} \tag{5.18}$$

Thus, the velocity profile is

$$u_\theta = \frac{\omega b^2 (r^2 - a^2)}{r(b^2 - a^2)} \tag{5.19}$$

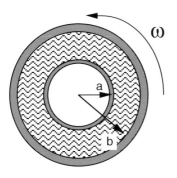

Figure 5.5 Linear viscous flow between rotating coaxial cylinders.

The determination of viscosity from the couple exerted on concentric cylinders is used in many **viscometers**. For silicate glasses, a wide range of techniques are used for measuring viscosity. For low viscosities (1 Pa s to 1 MPa s), a rotating spindle technique, based on the above principle, is often used. For higher viscosities, techniques based on fiber elongation, compression of a cylinder and beam-bending are utilized.

5.4   General equations for slow viscous flow

The simplest approach to analyzing viscous flow is to continue with the previous assumptions. The primary assumption is that the motion is slow, with **negligible inertial forces**. This allows terms involving acceleration, which lead to non-linearities in the equations, to be neglected and thereby preserves the analogy with static elasticity.

In Section 2.4, displacement was considered in terms of the vector u_i at a point x_i. For slow viscous flow, u_i is now the velocity, i.e., the components of u_i are now the time derivatives of displacement. The components of strain rate $\dot{\varepsilon}_{ij}$ (with $\dot{\gamma}_{ij} = 2\dot{\varepsilon}_{ij}$ when $i \neq j$) and rotation rate $\dot{\omega}_{ij}$ are defined in a similar way to Eqs. (2.14) and (2.15). In Section 2.4, it was pointed out that the definition of ε_{ij} was only valid for small deformations. In fluid flow, however, the deformation is usually both finite and large and, thus, this restriction is not needed for $\dot{\varepsilon}_{ij}$. In fluid flow, the deformation is defined at successive times and the time interval can always be chosen such that the changes in the deformation state is infinitesimal. In this sense, the fluid flow has no 'memory' of the previous deformation.

In flow problems, one expects a relationship between the velocity coefficients which expresses the law of matter conservation. For continuous, incompressible fluids, the outflow must equal the inflow, i.e., the 'dilatation' must be zero,

$$\dot{\varepsilon}_{ii} = \frac{\partial u_1}{\partial x_1} + \frac{\partial u_2}{\partial x_2} + \frac{\partial u_3}{\partial x_3} \tag{5.20}$$

This is known as the **Continuity equation**. Equation (2.49) can be used to define the stress equilibrium. Substituting strain rates for strains, η for μ and using $v = 0.5$, the viscous analogy of Hooke's Law can be written by summing the dilatational and deviatoric components (Eq. (2.73))

$$\sigma_{11} = 2\eta\dot{\varepsilon}_{11} - P; \quad \sigma_{12} = 2\eta\dot{\varepsilon}_{12;} \text{ etc.} \tag{5.21}$$

where $P = -\sigma_{ii}/3$. From Eqs. (5.21) and (2.14), Eq. (2.48) can be written as

$$F_1\rho - \frac{\partial P}{\partial x_1} + \eta\left(2\frac{\partial^2 u_1}{\partial x_1^2} + \frac{\partial^2 u_1}{\partial x_2^2} + \frac{\partial^2 u_2}{\partial x_1 \partial x_2} + \frac{\partial^2 u_1}{\partial x_3^2} + \frac{\partial^2 u_3}{\partial x_1 \partial x_3}\right) = 0 \tag{5.22}$$

or

$$F_1\rho-\frac{\partial P}{\partial x_1}+\eta\nabla^2 u_1+\eta\frac{\partial}{\partial x_1}\left(\frac{\partial u_1}{\partial x_1}+\frac{\partial u_2}{\partial x_2}+\frac{\partial u_3}{\partial x_3}\right)=0 \qquad (5.23)$$

where

$$\nabla^2 u_1=\left(\frac{\partial^2 u_1}{\partial x_1^2}+\frac{\partial^2 u_1}{\partial x_2^2}+\frac{\partial^2 u_1}{\partial x_3^2}\right) \qquad (5.24)$$

The final term in Eq. (5.23) vanishes as a result of the continuity equation (Eq. (5.20)). If one generalizes this approach for the other stress equilibrium equations, one obtains

$$\eta\nabla^2 u_i-\eta\frac{\partial P}{\partial x_i}+F_i\rho=0 \qquad (5.25)$$

This is known as the **Navier–Stokes equation** for incompressible flow.

As an example of the use of this equation, consider the flow of a fluid between two parallel planes, separated by a small distance h. Suppose the fluid undergoes planar flow, such that $u_3=0$ and the variation of u_1 and u_2 in the x_3 direction greatly exceeds that in the x_1 and x_2 directions. For small h, one can write

$$\nabla^2 u_1=\frac{\partial^2 u_1}{\partial x_3^2}; \quad \nabla^2 u_2=\frac{\partial^2 u_2}{\partial x_3^2}; \quad \nabla^2 u_3\approx 0 \qquad (5.26)$$

Thus, Eq. (5.25) reduces to

$$\eta\frac{\partial^2 u_1}{\partial x_3^2}=\frac{\partial P}{\partial x_1} \text{ and } \eta\frac{\partial^2 u_2}{\partial x_3^2}=\frac{\partial P}{\partial x_2} \qquad (5.27)$$

As P is independent of x_3, these equations can be integrated to

$$u_1=\frac{x_3(x_3-h)}{2\eta}\frac{\partial P}{\partial x_1} \text{ and } u_2=\frac{x_3(x_3-h)}{2\eta}\frac{\partial P}{\partial x_2} \qquad (5.28)$$

Equation (5.28) gives rise to parabolic velocity profiles, as in Poiseuille flow (Eq. (5.12)). The volume V_1 of fluid flowing through a plane of unit width perpendicular to the x_1 axis is

$$V_1=\frac{\partial P}{\partial x_1}\int_0^h \frac{x_3(x_3-h)}{2\eta}dx_3=-\frac{h^3}{12\eta}\left(\frac{\partial P}{\partial x_1}\right) \qquad (5.29)$$

Consider two circular discs, radius R and spacing h, being pulled apart by a force F, with a fluid between the plates (Fig. 5.6). Assume the fluid is in contact with a reservoir of fluid at a pressure P_0 so that additional fluid can flow to fill the increasing volume. The flow pattern is radial so, for a volume element of height h and sides dr and $rd\theta$, the volume is increasing at a rate$=(dh/dt)rdrd\theta$. From Eq. (5.29), the rate of radial flow through a face of area, $hrd\theta$, is given by

$(h^3 r d\theta/12\eta)(\partial P/\partial r)$. The net flow rate into the element is the change in volume across the radial element, i.e., $(\partial V/\partial r)\, dr$. This can be equated to the volume increase, i.e.,

$$\frac{12\eta}{h^3}\left(\frac{dh}{dt}\right)r=\frac{\partial}{\partial r}\left(r\frac{\partial P}{\partial r}\right)$$
(5.30)

Using $V=0$ (i.e., $\partial P/\partial r=0$) at $r=0$, the equation integrates to

$$\frac{6\eta}{h^3}\left(\frac{dh}{dt}\right)r^2=r\frac{\partial P}{\partial r}$$
(5.31)

Integrating again, with $P=P_0$ at $r=R$, one obtains

$$P=P_0-\frac{3\eta}{h^3}\left(\frac{dh}{dt}\right)(R^2-r^2)$$
(5.32)

The force F must be equal to the pressure difference integrated over the whole disc, so that

$$F=\int_0^R 2\pi r\frac{3\eta}{h^3}\left(\frac{dh}{dt}\right)(R^2-r^2)dr=\frac{3\pi\eta R^4}{h^3}\left(\frac{dh}{dt}\right)$$
(5.33)

The force increases rapidly for small values of h. This principle is exploited in **viscous adhesion**, in which joints can be made using a thin, viscous film. Provided joints made in this manner survive the initial loading, they will fail by viscous flow. For glassy adhesives, high loading rates can give rise to brittle failure. For more fluid adhesives, the joint fails by the nucleation and growth of **spheroidal cavities**. The cavities usually nucleate in non-wetted regions, at gas bubbles or at foreign particles and grow by localized viscous flow.

In the above analysis of slow viscous flow, the materials were assumed to be incompressible. This is not, however, necessary to derive the flow equations. If the compressibility is included, two viscosity coefficients are needed. This is analogous to the need for two constants to describe the elastic behavior of isotropic materials.

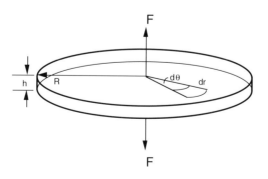

Figure 5.6 Radial viscous flow between circular discs being separated by a tensile force.

5.5   Non-linear viscous flow

In some materials, large shear stresses are required to obtain appreciable flow and the strain rate becomes a non-linear function of stress. As shown in Fig. 5.7, two types of non-linear behavior can be observed: **pseudoplastic** behavior in which flow becomes easier as the stress increases; and **dilatant** behavior in which the opposite occurs. These effects are often related to changes in the structure of the viscous medium. For example, (i) anisometric particles dispersed in a fluid or (ii) large molecules being oriented by the stress. If flow becomes easier then pseudoplasticity follows. In some cases, the re-orientation of the particles can persist and the viscosity becomes time dependent. This effect is termed **thixotropy** and is exploited in many paints. Once vigorously stirred, a thixotropic paint will remain highly fluid but after some time the viscosity will again increase. Dilatant behavior is usually associated with viscous suspensions in which particle interference becomes important at the higher strain rates. For **non-linear** or **non-Newtonian flow**, the strain rate can be written

$$\tau = A\dot{\gamma}^n \tag{5.34}$$

The flow behavior is dilatant for $n>1$, Newtonian for $n=1$ and pseudoplastic for $n<1$. For these materials, the viscosity is still defined as the stress–strain rate ratio. In this way, viscosity becomes dependent on strain rate, i.e.,

$$\eta = A\dot{\gamma}^{n-1} \tag{5.35}$$

For small values of n (e.g., $n<0.1$), the behavior approaches an **ideal pseudoplastic** state, in which the strain rate becomes independent of stress (after reaching a yield stress). The idea of yielding is important for viscous media that contain 'attracting' particles, because a finite stress is needed to initiate flow.

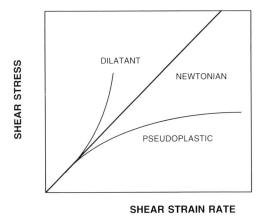

Figure 5.7 Dependence of shear stress on shear strain rate for Newtonian, dilatant and pseudoplastic systems.

Alternatively, particles may be initially in contact but become separated by a fluid at a higher stress. For media in which the yielding behavior is very abrupt, as shown in Fig. 5.8, Eq. (5.34) can be re-written as

$$\tau - \tau_Y = A \dot{\gamma}^n \text{ for } \tau \geq \tau_Y \text{ and } \dot{\gamma} = 0 \text{ for } \tau \leq \tau_Y \qquad (5.36)$$

where τ_Y is the yield stress. For the special case known as **Bingham flow**, the flow becomes linear once the yield stress is reached ($n = 1$ and $A = \eta$). Thixotropic effects are often important in media that exhibit yield behavior. In these cases, the yield strength can become dependent on the time the suspension has been at rest.

5.6　Dispersion of solid particles in a fluid

In ceramics science, one is often concerned with viscous media that consist of solid particles dispersed in a fluid. A **sol** is a solution or suspension of **colloidal** particles (sizes between 1 nm and 1 μm) in a solvent. A **lyophilic sol** contains solutes whose molecules are large enough to be classified as colloidal. A **lyophobic sol** contains insoluble particles but the particles are so small that **Brownian motion** prevents them from settling. Lyophobic sols are usually stabilized by electrical charges or by adsorption of polymeric molecules on the surface of the particles. Concentrated lyophilic sols can undergo **gelation** and this is usually induced by chemical or temperature changes or by loss of solvent. Lyophobic sols can undergo **coagulation** if the repulsive forces between the particles are lost. This leads to particle settling and gelation. Clearly, such changes in the sol structure lead to dramatic changes in viscosity.

For larger particle sizes, one may have to deal with pastes, particle beds and powder dispersions. In these situations, the volume fraction of particles is high

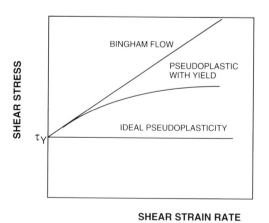

Figure 5.8 Examples of shear stress–shear strain rate behavior for viscous materials with a yield stress.

and viscous flow is usually non-linear. In powder dispersions, chemical forces between particles still play an important role, but in particle beds the important cohesive forces are contact pressures and gravity. The flow behavior for these materials is very sensitive to the volume fraction of particles and this has to be carefully controlled in a technological application. For example, clay becomes 'plastic' at about 30 vol.% water, when there is a continuous film of liquid around the particles. With this consistency the clay can be formed by a range of plastic processes, such as **extrusion** and **molding**. In these situations clays behave more like plastic solids with a distinct yield behavior, such as Bingham flow. At higher water contents, the clay grains flow apart and the resulting **slurry** is used for **slip casting**. This type of behavior is observed for many ceramic powder dispersions. Control of the electrical surface charge or polymer adsorption is used to aid in particle dispersion (**deflocculation**).

The mechanical behavior of wet sand is also sensitive to water content, behaving like a stiff paste when the solids fraction V_p is >0.65 but as a fluid when it is <0.6. In a deformation process, wet sand can therefore undergo a sudden change in behavior if the water content changes. Even dry powders can flow, provided the cohesion between particles can be overcome. It is often useful to agglomerate particles into spherical shapes to help the flow behavior, e.g., by **spray drying**. In other cases, powder beds can be **fluidized** by using a change in pressure to overcome the gravitational contact pressures or by dispersion in a liquid of the same density as the solid. In some particle dispersions, magnetic and electric fields can cause sudden changes in viscosity. These are known as **electro-rheological** and **magneto-rheological** fluids. The viscosity change is associated with the development of particle structures in the liquid.

The viscosity of a fluid is increased by the presence of dispersed particles as they tend to disturb the flow patterns. For dilute suspensions of spherical particles in a Newtonian fluid, the viscosity is described by **Einstein's relation**,

$$\eta = \eta_0 (1 + 2.5 V_p) \qquad (5.37)$$

where η_0 is the viscosity of the pure fluid. Non-spherical particles lead to a greater disturbance in the flow and higher viscous losses occur, often induced by particle rotation. For this case, Eq. (5.37) is often written as

$$\eta = \eta_0 (1 + H V_p) \qquad (5.38)$$

where H is a hydrodynamic shape factor. Although the flow pattern attempts to align rod-shaped particles, the Brownian motion in sols can act to randomize the orientation. For larger particles, however, alignment invariably occurs and pseudoplastic behavior is expected. At high particle concentrations, the above formulae break down as a result of the complex interaction between particles.

For these cases, (empirical) non-linear versions of Eq. (5.38) are sometimes used to describe the variation of viscosity with particle volume fraction.

As indicated earlier, unless particles are small enough to be held in suspension by Brownian motion they will undergo settling as a result of gravity. The force F on a sphere, radius a, moving slowly with a velocity v through a fluid of viscosity η is given by

$$F = 6\pi a v \eta \tag{5.39}$$

a relationship known as **Stokes' Law**. For the settling (or flotation) of particles in a fluid under the action of gravity, $F = 4\pi a^3(\rho_1 - \rho_2)g/3$, where ρ_1 and ρ_2 are the densities of the particle and fluid, respectively, and g is the acceleration due to gravity. The particle velocity is, therefore, given by

$$v = \frac{2(\rho_1 - \rho_2)ga^2}{9\eta} \tag{5.40}$$

Equation (5.40) is often exploited in the measurement of particle size by determining the rate of particle settling. For very fine particles, such a measurement technique can be very time-consuming and, thus, the settling process is often accelerated by centrifuging. Equation (5.40) is also useful in describing the escape of gas bubbles from a glass melt (fining). Indeed, small bubbles have such a low velocity that they can become trapped in the glass. In these cases, chemical processes for dissolving the gas into the glass are often needed.

5.7   **Viscoelastic models**

In many materials, the mechanical response can show both elastic and viscous types of behavior; the combination is known as **viscoelasticity**. In elastic solids, the strain and stress are considered to occur simultaneously, whereas viscosity leads to time-dependent strain effects. Viscoelastic effects are exhibited in many different forms and for a variety of structural reasons. For example, the thermo-elastic effect was shown earlier to give rise to a delayed strain, though **recovery** of the strain was complete on unloading. This 'delayed' elasticity is termed **anelasticity** and can result from various time-dependent mechanisms (**internal friction**). Figure 5.9 shows an example of the behavior that occurs for a material that has a combination of elastic and anelastic behavior. The material is subjected to a constant stress for a time, t. The elastic strain occurs instantaneously but, then, an additional time-dependent strain appears. On unloading, the elastic strain is recovered immediately but the anelastic strain takes some time before it disappears. Viscoelasticity is also important in **creep** but, in this case, the time-dependent strain becomes permanent (Fig. 5.10). In other cases, a strain can be applied to a material and a viscous flow process allows **stress relaxation** (Fig. 5.11).

Viscoelasticity leads to a very wide range of deformation behavior and various models have been developed to account for the differences in behavior. The approach in these models is to combine elastic and viscous components in various configurations. The elastic component is represented by a **spring** and the viscous by a **dashpot**. This latter component consists of a pot that contains a viscous liquid, the flow of which controls the motion of a piston. The deformation behavior of the components is shown in Fig. 5.12. The basic equations for

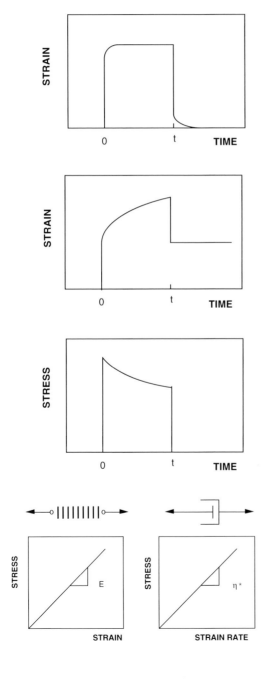

Figure 5.9 Strain response for a material that exhibits a combination of elastic and anelastic behavior when a constant stress is applied for a time *t*.

Figure 5.10 Strain response for a material that exhibits a combination of elastic and viscous creep behavior when a constant stress is applied for a time *t*.

Figure 5.11 Stress response for a material that exhibits a combination of elastic and stress relaxation behavior when a constant strain is applied for a time *t*.

Figure 5.12 Viscoelastic models are constituted from spring and dashpot components. The mechanical response of these components is shown.

the extension of the spring and dashpot, respectively, under the action of a uni-axial tensile stress σ are

$$\sigma = E\varepsilon \text{ and } \sigma = \eta^*(d\varepsilon/dt) \tag{5.41}$$

where ε is the extensional strain, E is Young's modulus and η^* is the tensile viscosity. Alternatively, instead of Eq. (5.41), the spring and dashpot models can be derived for a shear stress τ using

$$\tau = \mu\gamma \text{ and } \tau = \eta(d\gamma/dt) \tag{5.42}$$

where μ is the shear modulus of the spring and η is the shear viscosity of the dashpot.

Consider the mechanical behavior of a model in which a spring and dashpot are placed in series, which is known as the **Maxwell model** (Fig. 5.13). Recognizing that the overall strain rate is the sum of the contributions from the spring and the dashpot, the constitutive equation can be written

$$\dot{\varepsilon} = \frac{\dot{\sigma}}{E} + \frac{\sigma}{\eta^*} \tag{5.43}$$

It is sometimes useful to write this equation in an operator form {O}, using the **linear differential time operator**, $\partial_t = \partial/\partial t$, i.e.,

$$\{\delta_t\}\varepsilon = \left\{\frac{\partial_t}{E} + \frac{1}{\eta^*}\right\}\sigma \tag{5.44}$$

The operator form is useful for setting up the differential equations for more complex spring–dashpot models. If a constant strain ε_0 is applied to the Maxwell model, Eq. (5.44) becomes

$$\left\{\frac{\partial_t}{E} + \frac{1}{\eta^*}\right\}\sigma = 0 \tag{5.45}$$

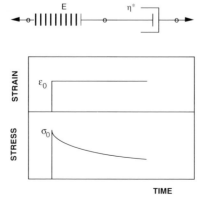

Figure 5.13 The Maxwell model of a spring and dashpot in series. The stress relaxation that occurs for a constant applied strain is shown.

Integration of this equation gives

$$\sigma = \sigma_0 \exp\left(-\frac{t}{T_R}\right) \tag{5.46}$$

where $T_R = \eta^*/E$, is known as the relaxation time and σ_0 is the stress at $t=0$ (i.e., $\sigma_0 = E\varepsilon_0$). As shown in Fig. 5.13, the initial stress is relaxed as the dashpot opens, falling exponentially. The relaxation time is a useful way to classify viscoelastic materials and represents the time for the stress to be reduced by 63.2%. Relaxation times can show a very wide range, from $T_R \sim 10$ ps for simple liquids to $T_R > 10$ ks for solids. If the strain is removed in the Maxwell model before the stress has completely relaxed, the spring unloads and the model will be left with a permanent strain. If a constant stress σ were applied to the Maxwell model, Eq. (5.44) is easily integrated, giving a strain ε that is time dependent, i.e.,

$$\varepsilon = \sigma\left(\frac{1}{E} + \frac{t}{\eta^*}\right) \tag{5.47}$$

The model initially responds elastically but a permanent creep strain develops with time and only the elastic strain is recovered on unloading.

An alternative model is obtained if one places the spring and dashpot in parallel and this is known as the **Voigt** or the **Kelvin model** (Fig. 5.14(a)). For this model, the strains on the two components are the same and the overall stress is the sum of the stresses on the dashpot and spring. The constitutive equation becomes

$$\sigma = \{E + \eta^*\partial_t\}\varepsilon \tag{5.48}$$

For a constant stress σ it is left to the reader to show that the strain is given by

$$\varepsilon = \frac{\sigma}{E}\left[1 - \exp\left(-\frac{t}{T_{R'}}\right)\right] \tag{5.49}$$

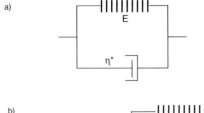

Figure 5.14 Common viscoelastic models: a) Voigt/Kelvin model; b) Zener model/standard linear solid.

where $T_{R'}$ is termed the **retardation time**. For this case, the retardation time is still given by $T_{R'}=\eta^*/E$. The strain approaches an asymptotic limit, σ/E, at long times. On unloading, the spring is forcing the dashpot to close and the strain falls exponentially to zero. The Voigt model exhibits the characteristics of an anelastic strain.

The simplest way to obtain the behavior discussed in connection with Fig. 5.9 is to place a second spring in series with a Voigt model. This is shown in Fig. 5.14(b) and is known as the **Zener model** or **standard linear solid**. The constitutive equation is found by simply adding the strains from the spring and Voigt model (Eqs. (5.41) and (5.48))

$$\varepsilon=\frac{\sigma}{E_1}+\frac{\sigma}{\{E_2+\eta^*\partial_t\}} \tag{5.50}$$

This is a first-order differential equation, which can be solved once the applied stress or strain is specified. For example, under a constant stress ($d\sigma/dt=0$), the solution of the differential equation gives

$$\varepsilon=\frac{\sigma}{E_1}+\frac{\sigma}{E_2}\left[1-\exp\left(-\frac{t}{T_{R'}}\right)\right] \tag{5.51}$$

where $T_{R'}=\eta^*/E_2$. For $t=0$, the modulus of the system is E_1 and this is sometimes termed the **unrelaxed modulus**. As $t\rightarrow\infty$, the anelastic strain, represented by the Voigt model, is now present and the **relaxed modulus** E_R becomes

$$E_R=\frac{E_1 E_2}{E_1+E_2} \tag{5.52}$$

It is sometimes useful to define a term, known as the **modulus defect**,

$$\Delta_0=(E_U-E_R)/E_R \tag{5.53}$$

to show the difference between these moduli. The change of modulus with time, $E(t)$, is shown in Fig. 5.15. In some cases it is useful to consider the **creep compliance** of the system, $\Phi(t)=\varepsilon(t)/\sigma$.

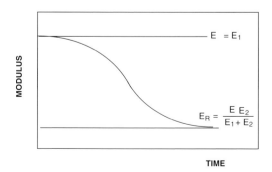

Figure 5.15 Variation of modulus of standard linear solid as a function of time under a constant stress.

If the standard linear solid (SLS) is unloaded from a constant stress, the spring (modulus E_1) closes immediately and the elastic strain is removed. The anelastic strain then decays to zero as the second spring closes the dashpot, i.e., there is complete recovery. Under the action of a *constant strain*, the SLS model will also show stress relaxation but, in this case, the time constant, $T_R = \eta^*/(E_1 + E_2)$. In applying a constant stress to the SLS model, the strain can be considered to lag behind the stress, both on loading and unloading. This lag concept is also very important in considering the effect of a dynamic stress or strain.

Consider a sinusoidal variation of the applied stress on the SLS model, i.e., $\sigma = \sigma_0 \cos \omega t$. The differential equation can be found from Eq. (5.50) and the solution is

$$\varepsilon = A \cos \omega t + B \sin \omega t \qquad (5.54)$$

where

$$A = \sigma_0 \left[\frac{1}{E_1} + \frac{1}{E_2(1 + \omega^2 T_R^2)} \right] \qquad (5.55)$$

and

$$B = \sigma_0 \left[\frac{\omega T_R}{E_2(1 + \omega^2 T_R^2)} \right] \qquad (5.56)$$

where $T_R = \eta^*/E_2$. As B is always positive, the strain can be expressed as

$$\varepsilon = \varepsilon_0 \cos(\omega t - \delta) \qquad (5.57)$$

where

$$\tan \delta = \frac{B}{A} = \left[\frac{E_1 \omega T_R}{E_1 + E_2(1 + \omega^2 T_R^2)} \right] \qquad (5.58)$$

and

$$\varepsilon_0 = \sqrt{A^2 + B^2} \qquad (5.59)$$

In Eq. (5.54), the term $A \cos \omega t$ represents the in-phase component of the strain and $B \sin \omega t$, the out-of-phase component. The 'in-phase' or **storage modulus** E' can be defined as

$$\frac{1}{E'} = \frac{\varepsilon_0}{\sigma_0 \cos \delta} = \frac{1}{E_1} + \frac{1}{E_2(1 + \omega^2 T_R^2)} \qquad (5.60)$$

The 'out-of-phase' or **loss modulus** E'' is defined as

$$\frac{1}{E''}=\frac{\varepsilon_0}{\sigma_0\sin\delta}=\frac{\omega T_R}{E_2(1+\omega^2T_R^2)} \tag{5.61}$$

The 'out-of-phase' modulus is termed the loss modulus because the energy dissipated per unit volume depends on this modulus. An analysis similar to the one above can be performed for strain cycling and different equations are obtained for E' and E''. The strain is, however, still found to lag behind the stress by the same phase angle δ.

Viscoelastic problems can also be considered in terms of the compliance of the system and one can define a **storage compliance**, $\Phi'=\varepsilon_0\cos\delta/\sigma_0$ and a **loss compliance**, $\Phi''=\varepsilon_0\sin\delta/\sigma_0$. The energy dissipated per cycle ΔU can be determined from

$$\Delta U=\int_0^{2\pi/\omega}\sigma_0\sin\omega t\varepsilon_0\sin(\omega t-\delta)\mathrm{d}(\omega t)=\pi\sigma_0^2\Phi'' \tag{5.62}$$

The **specific damping capacity** $\Delta U/U$ of a system is obtained by normalizing to the maximum strain energy of the system ($\sigma_0\varepsilon_0/2$) to give

$$\frac{\Delta U}{U}=2\pi\sin\delta \tag{5.63}$$

In some texts, the average strain energy is used to normalize ΔU, leading to a difference by a factor of two with Eq. (5.63). If δ is small, the relative amount of energy loss in a material is often described by the **loss tangent**, $\tan\delta$ ($\sim\sin\delta$, if δ is small). This is equivalent to the ratios, E''/E' or Φ'/Φ''.

A generalization of these ideas is achieved by expressing the stress and strain in a complex form, i.e.,

$$\sigma=\sigma_0\mathrm{e}^{\mathrm{i}\omega t}\text{ and }\varepsilon=\varepsilon_0\mathrm{e}^{\mathrm{i}(\omega t-\delta)} \tag{5.64}$$

The ratio of the stress to the strain is used to define a **complex modulus** $E^*(\mathrm{i}\omega)$, the real part of which is the storage modulus and the imaginary part the loss modulus, i.e., $E^*(\mathrm{i}\omega)=E'+\mathrm{i}E''$. Alternatively, one can define a **complex compliance** $\Phi^*(\mathrm{i}\omega)$ as the ratio of the strain to the stress. For this case, $\Phi^*(\mathrm{i}\omega)=\Phi'-\mathrm{i}\Phi''$; the real part is the storage compliance and the imaginary part is the negative of the loss compliance (note that $E^*=1/\Phi^*$).

It is useful to consider the behavior of the SLS as a function of dimensionless frequency (ωT_R), as shown in Fig. 5.16. At low frequencies, the dashpot has sufficient time to open and close and elastic behavior is obtained (relaxed modulus). The stress and strain are in phase. At very high frequencies, there is no time for the anelastic strain to develop. The stress and strain are again in phase but the modulus is higher, with only the spring of modulus E_1 opening and closing (unrelaxed modulus). For intermediate frequencies, a lag develops and the stress–strain behavior exhibits a **hysteresis loop** (see Fig. 5.16). The area of this loop is a maximum when the phase angle is a maximum and this represents the maximum energy loss for the system. This condition occurs when $\omega T_R = \sqrt{[E_2/(E_1+E_2)]}$ and, hence, the maximum value of the phase angle δ_{max} is

$$\tan\delta_{max} = \frac{E_1}{2E_2}\sqrt{\frac{E_2}{E_1+E_2}} \tag{5.65}$$

For small values of δ_{max}, this is sometimes expressed as

$$\tan\delta_{max} \sim \frac{E_U-E_R}{2E_R} = \frac{\Delta_0}{2} \tag{5.66}$$

The reciprocal of the phase angle is called the **quality factor** Q and is often used in the description of electric circuits. The inverse of Q, i.e., Q^{-1}, is often called the **loss factor**. The specific damping capacity of the SLS can, therefore, be expressed as $\Delta U/U = 2\pi\delta = 2\pi Q^{-1}$. Although these factors depend on frequency, they should be independent of the stress amplitude. In order to characterize the loss peak in Fig. 5.16, it is common practice to measure the peak width at $1/\sqrt{2}$ of the maximum loss, which is approximately equal to Q^{-1}.

This section has considered only the simplest viscoelastic models and, thus, only touches on the field of viscoelasticity. A strong emphasis was placed on the SLS and this model is very useful in describing anelastic behavior. One approach to describing more complex behavior is to generalize the Voigt and Maxwell models. The **generalized Voigt model** consists of a sequence of Voigt models attached in series. The total strain of this model is simply the sum of the strains of the individual units and the differential equation can be obtained by summing

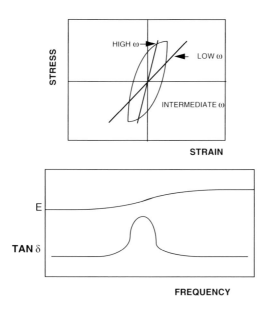

Figure 5.16 Stress–strain behavior for standard linear solid subjected to a sinusoidal stress. The system is elastic with a high modulus at very high frequencies and a lower modulus at low frequencies. At intermediate frequencies, hysteresis develops and the loss passes through a maximum.

the strain operators. Each Voigt unit in the series has a characteristic retardation time. It is sometimes useful to consider an infinite number of Voigt units in series. For this case, the creep compliance $\Phi(t)$, at constant stress, can be written as

$$\Phi(t)=\Phi_U+\int_0^\infty \Phi(T_R)\left[1-\exp\left(-\frac{t}{T_R}\right)\right]dT_R \tag{5.67}$$

where $\Phi(T_R)$ is called the **retardation spectrum** and Φ_U is the unrelaxed compliance. In a similar fashion, one can connect Maxwell models in parallel to form a **generalized Maxwell model**. For the case when there is an infinite number of Maxwell units in parallel, the modulus $E(t)$ can be written

$$E(t)=E_R+\int_0^\infty E(T_{R'})\left[\exp\left(-\frac{t}{T_R}\right)\right]dT_{R'} \tag{5.68}$$

where $E(T_{R'})$ is the **relaxation time spectrum**.

The spring and dashpot models can be formulated in many complex ways. For example, Fig. 5.17 shows Burger's **viscoelastic model** which consists of a Maxwell and Voigt model in series. This model shows all three of the basic viscoelastic response patterns; an instantaneous elastic response from the free spring, a viscous response from the free dashpot and an anelastic response from the Voigt unit. The spring and dashpot models can also be generalized by using a temperature-dependent viscosity or by using Fourier transforms of dynamic data. These approaches are useful in obtaining 'long-time' rheological data from short-term tests. Even with these added complexities, spring and dashpot models clearly only deal with very simple stress and strain states. For a three-dimensional theory there is, however, the strong analogy with linear elasticity. In particular, the stresses and strains can usually be separated into the deviatoric and dilatational parts. These can be used to write the constitutive equations for a viscoelastic material using a **complex bulk modulus** and a **complex shear modulus**. Furthermore, viscoelastic solutions can often be obtained by (Laplacian) transformations of the elastic solutions.

5.8   Anelasticity in ceramics and glasses

Anelastic effects have been studied extensively in glasses. Even in pure glass formers, there are small damping peaks. These are usually associated with the

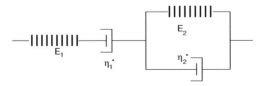

Figure 5.17 Burgers' (four-parameter) viscoelastic model.

motion of oxygen between equivalent positions. Extensive studies have been made of the internal friction in simple alkali–silicate glasses. These glasses often exhibit two distinct loss peaks when measured as a function of temperature at a fixed frequency. The low-temperature peak is usually associated with the motion or diffusion of alkali ions. This peak increases with the alkali content in a fashion similar to the changes in the diffusion coefficient. Moreover, the activation energy is similar to that obtained from diffusivity or electrical conductivity measurements. The higher-temperature peak has been associated with the presence of non-bridging oxygens. The presence of residual glassy phases in poly-crystalline ceramics should also be an important source of anelastic and viscous relaxation.

Time-dependent hysteresis effects can also occur in crystalline materials and these lead to **mechanical damping**. Models, such as the SLS and the generalized Voigt model, have been used extensively to describe anelastic behavior of ceramics. It is, thus, useful to describe the sources of internal friction in these materials that lead to anelasticity. The models discussed in the last section are also capable of describing permanent deformation processes produced by creep or densification in crystalline materials. For polycrystalline ceramics, creep is usually considered from a different perspective and this will be discussed further in Chapter 7.

In Section 2.14, the thermoelastic effect was discussed and it was shown that this leads to a difference in the adiabatic and isothermal elastic constants and is a form of anelasticity. It was also pointed out that the thermoelastic effect can give rise to **thermal currents** in a polycrystalline material if grains are being stressed at different rates or if other stress gradients are rapidly applied. **Thermoelastic damping** is an example of **linear damping**, as the loss is inde-pendent of strain amplitude and the viscous portion obeys Newton's law of viscosity. Another form of linear damping involves the motion of **point defects**. These defects possess a local strain field and can have more than one equivalent orientation in the structure, if the symmetry of the strain is lower than that of the crystal structure. When a stress is applied, these orientations can lose their equiv-alency and the movement of the defect becomes a thermally activated, time-dependent process. Even if substitutional atoms have the same valency as the 'host' atoms, a defect arises that is capable of anelastic motion. For example, sub-stitution of Li^+ for K^+ in the KCl structure leads to a displacement of the Li^+ ion along $<111>$, as a result of its high field strength. Aliovalent (different valence) ion substitution often leads to vacancy-solute pairs with distinct orientations. The defects associated with the substitution of Ca^{2+} into ThO_2, ZrO_2 and CeO_2 have all been studied from an anelastic perspective. In a similar fashion, interstitial-solute pairs can undergo stress-induced motion. For example, when YF_3 is dissolved in CaF_2, the substitution of Y^{3+} onto a Ca^{2+} site leads to a nearby F^- interstitial. The application of a stress can then lead to

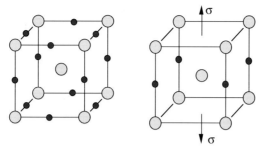

Figure 5.18 In the unstressed state, the interstitials in the octahedral sites of the bcc structure have an equal opportunity of occupying any site. Under a uniaxial stress, however, the interstitials will favor the edge sites parallel to the applied stress.

orientation of these defect pairs. The movement of interstitials in bcc structures is a simple illustration of anelasticity caused by the motion of point defects. Under a uniaxial stress, interstitials can move to the unit cell edges parallel to applied stress direction, as illustrated in Fig. 5.18. Damping is also associated with the motion of line and surface defects. For example, the bowing of a dislocation between pinning points and the motion of grain, twin and domain boundaries have all been considered as sources of internal friction. In many second-order phase transitions, stress-induced ordering gives rise to internal friction.

Problems

5.1   The viscosity of a borosilicate glass was determined to be 150 Pa s, 4.0 MPa s, and 200 TPa Ms at temperatures of 1800, 1000 and 900 K, respectively.

a) Determine the constants in the VFT equation.

b) In a glass-melting tank, the same borosilicate glass is held at 1700 K. The glass contains gas bubbles 1–100 μm in diameter. If the glass is 1 m deep, how long will it take to remove the bubbles by letting them rise to the surface? Assume the glass density is 2000 kg/m^3.

c) The borosilicate glass must be discharged from the tank at a rate of 1 kg s^{-1}. It is to be discharged down a hollow cylinder of length 1 m and inside diameter 100 mm under an applied pressure of 1 MPa. Determine the temperature needed for this discharge rate.

d) Use the VFT equation to find the temperature at which a borosilicate glass fiber would reduce its diameter (initially 10 mm) at a rate of 1 μm/s under a constant uniaxial tensile force of 5 N.

5.2   A glass-bonded alumina consists of the alumina grains (grain size 10 μm) separated by a 0.1 μm layer of glass. Determine the creep strain rate that could be obtained if the grains were allowed to separate by linear viscous flow under a uniaxial tensile stress of 100 MPa. Assume the viscosity of the glass is 1 MPa s.

5.3   The softening point of glass is often determined by measuring the temperature at which a uniform glass fiber, diameter 550–750 μm and length 235 mm, extends at a given rate under its own weight. The upper 100 mm of the fiber is heated at a rate of 5 °C per minute and the critical elongation rate is 1 mm/min. Derive an equation for the viscosity of a glass at its softening point, assuming that the fiber is subjected to a downward force due to its weight and an upward force due to surface tension γ. If the effective length of the fiber is 70 mm, compare the softening point viscosities for a soda–lime silica glass (density=2200 kg/m³, γ=0.3 J/m²) and a high lead silica glass (density=6200 kg/m³, γ=0.2 J/m²).

5.4   The annealing and strain points of a glass are often determined by measuring the temperature at which a uniform glass fiber elongates at a specified rate under a 1 kg weight. Derive an equation for the elongation rate required of a glass at its annealing point, if the viscosities at the annealing and strain points are assumed to be 1 TPa s and 31.6 TPa s, respectively. The following elongation rates were determined experimentally for a glass. Determine the annealing and strain points for the glass assuming the effective fiber length is 500 mm and the fiber diameter is 600 μm.

Temperature (°C)	460	455	450	445	440	435	430	425
Elongation rate (mm/min)	3.8	2.2	1.6	1.0	0.71	0.45	0.30	0.20

5.5   If a beam of a viscous material is loaded in three-point bending show, from first principles, that the deflection rate of the beam, $dy/dt=FL^3/144\eta I$ (ignore the loading from the weight of the beam). The beam length is L, the force is F and I is the second moment of area. It is often said that the bottoms of glass panes thicken over time-spans of a few centuries and that glass rods, a few feet long, if supported horizontally at their ends, will develop a permanent bend in a few months. At ambient temperatures, the viscosity of soda–lime–silica glass has been estimated to be ~100 TPa Ms. Perform some simple calculations to show whether these adages are true.

5.6   A linearly viscous liquid is flowing down an annular tube of inner radius a and outer radius b under a pressure difference P. Derive an expression for the volumetric discharge rate from the tube.

5.7   Two identical circular discs, radius R, are mounted coaxially so that their (parallel) faces are a small distance d apart. If this space is filled with a Newtonian liquid and one of the discs is rotated with an angular velocity ω show that the torque on the discs is $\pi \eta \omega R^4 / 2d$. Assume no slip occurs between the liquid and the discs.

5.8   Some materials demonstrate anelastic behavior and this is often modeled by a spring in series with a parallel spring and dashpot unit (standard linear solid, SLS).
   a) Derive the equations for strain under a constant uniaxial stress for this model.
   b) Plot the strain for the SLS when the stress is 50 MPa. Assume $E=5$ GPa and $\eta^*=50$ GPa s for the parallel unit and $E=10$ GPa for the series spring. What is the effect on the strain of (i) increasing and (ii) decreasing the viscosity by a factor of 10, after a period of 1 s?
   c) Describe the strain behavior of the SLS if the stress were removed. Contrast this behavior with that found in the Maxwell and Voigt models.

5.9   For the Burger's viscoelastic model, derive an expression for the strain under the application of a uniaxial tensile stress.

5.10   A material, which can be described as an SLS, was found to have unrelaxed and relaxed Young's modulus values of 70 and 50 GPa, respectively. Determine the relaxation and retardation times. Plot graphically the compliance of the material as a function of time, under the action of a constant uniaxial tensile stress.

5.11   Derive an expression for the strain of an SLS being subjected to a constant stress rate.

5.12   Assume an SLS is being subjected to sinusoidal stress cycling with $E_1=75$ GPa, $E_2=70$ GPa and $\eta^*=14$ MPa s. Plot the phase difference, storage modulus and loss modulus as a function of frequency. At what frequency does the maximum energy dissipation occur? What is the specific damping capacity of the material at this frequency?

References

R. C. Bradt, *Introduction to the Mechanical Properties of Ceramics*, Class Notes, The Pennsylvania State University, 1980.

A. H. Cottrell, *Mechanical Properties of Matter*, J. Wiley and Sons, New York, 1964.

H. E. Hagy, Rheological behavior of glass, pp. 343–71 in *Introduction to Glass Science*, edited by L. D. Pye *et al.*, Plenum Press, New York, 1972.

W. D. Kingery, H. K. Bowen and D. R. Uhlmann, *Introduction to Ceramics*, J. Wiley and Sons, New York, 1976.

S. I. Krishnamachari, *Applied Stress Analysis of Plastics*, Van Nostrand Rheinhold, New York, 1993.

G. E. Mase, *Theory and Problems of Continuum Mechanics*, Schaum Outline Series, McGraw-Hill, New York, 1970.

F. A. McClintock and A. S. Argon, *Mechanical Behavior of Materials*, Addison-Wesley Publishing Co., Reading, MA, 1966.

J. S. Reed, *Introduction to the Principles of Ceramic Processing*, J. Wiley and Sons, New York, 1988.

G. W. Scherer, Editorial comments on a paper by Gordon S. Fulcher, *J. Am. Ceram. Soc.*, **75** (1992) 1060.

G. W. Scherer, *Relaxation in Glass and Composites*, J. Wiley and Sons, New York, 1986.

I. M. Ward and D. W. Hadley, *An Introduction to the Mechanical Properties of Solid Polymers*, J. Wiley and Sons, New York, 1993.

J. G. Williams, *Stress Analysis of Polymers*, Ellis Horwood Ltd., UK, 1980.

Chapter 6

Plastic deformation

Inelastic deformation can occur in crystalline materials by plastic 'flow'. This behavior can lead to large permanent strains, in some cases, at rapid strain rates. In spite of the large strains, the materials retain crystallinity during the deformation process. Surface observations on single crystals often show the presence of lines and steps, such that it appears one portion of the crystal has slipped over another, as shown schematically in Fig. 6.1(a). The slip occurs on specific crystallographic planes in well-defined directions. Clearly, it is important to understand the mechanisms involved in such deformations and identify structural means to control this process. Permanent deformation can also be accomplished by **twinning** (Fig. 6.1(b)) but the emphasis in this book will be on plastic deformation by **glide (slip)**.

6.1   Theoretical shear strength

Figure 6.2 shows one possible way in which crystal glide could occur, with one plane of atoms being sheared past an adjacent plane. In the perfect crystal, the atoms are assumed to lie directly above each other with a planar spacing d. Clearly, as the atoms are displaced, the stress will rise and pass through a maximum. Once the displacement u reaches a value of $b/2$, i.e., at the mid-shear position, the atoms would be equally as likely to complete the displacement ($u=b$) as to return to their original positions. The stress returns to zero at this equilibrium point. For displacements larger than $b/2$, the atoms will be 'attracted' into their new positions by the interatomic potential and thus the stress becomes negative. The maximum stress τ_{th} is termed the **theoretical shear strength**. The detailed nature of the stress-displacement function must be related to the interatomic potential but, for simplicity, it will be assumed to be approximated by a sine function (Fig. 6.3). For this case, the theoretical shear strength corresponds to a displacement $u=b/4$, i.e.,

$$\tau = \tau_{th}\sin\left(\frac{2\pi u}{b}\right) \tag{6.1}$$

At small displacement values, the material will be elastic and, thus, Hooke's Law can be used to describe the stress-displacement behavior in this region. Using $\sin x \sim x$ for small angles and the shear strain $=u/d$, one obtains

$$\tau = \tau_{th}\left(\frac{2\pi u}{b}\right) = \mu\left(\frac{u}{d}\right) \tag{6.2}$$

Figure 6.1 Permanent deformation of a crystal produced by a) crystal glide and b) twinning. The orientation of the shading represents the crystallographic texture of the crystal.

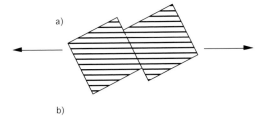

Figure 6.2 Shear deformation of two planes of atoms in a perfect crystal: a) The initial (equilibrium) position of the atoms; and b) the mid-shear position.

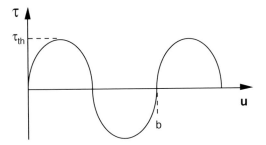

Figure 6.3 The stress variation for shear in a perfect crystal can be assumed to be approximately sinusoidal. The maximum shear stress is termed the theoretical shear strength.

where μ is the shear modulus. Rearranging Eq. (6.2), one finds

$$\tau_{th} = \frac{\mu b}{2\pi d} \tag{6.3}$$

The lowest value of τ_{th} is obtained for planes with low values of b, the slip distance, and high values of d, the spacing between slip planes. Clearly, this does not consider the anisotropy of the shear modulus which, as seen in Chapter 2, can be significant in some cases. For cases in which $b \sim d$, one obtains $\tau_{th} \sim \mu/6$. As pointed out earlier, the use of the sine function is only approximate. More sophisticated calculations lead to theoretical shear strength values in the range $\mu/10$ to $\mu/30$. Experimental yield strengths for materials are often several orders of magnitude less than the theoretical values. Thus, one concludes the deformation does not occur in the way depicted in Fig. 6.2. To explain the discrepancy, the existence of localized structural defects, known as **dislocations**, was postulated in the 1930s. Rather than envisioning planes of atoms sliding past each other in a perfect crystal, plastic deformation is obtained by the motion of these localized defects. In later years, it was found that single-crystal whiskers could be obtained in an essentially dislocation-free state and their shear strength could approach the theoretical values.

6.2   **Dislocations**

Dislocations are **line defects** that separate sheared and 'unsheared' portions of a crystal. There are two fundamental types of dislocations; **edge and screw**, shown schematically in Figs. 6.4 and 6.5, respectively. Edge dislocations are often depicted as 'extra half-planes' of atoms, with the dislocation line being at the termination of the half-plane. For ceramics, which contain two or more types of atoms, edge dislocations are often more complicated than a single half-plane. The application of a shear stress allows an edge dislocation to move. For example, if a shear stress is applied to the edge dislocation as shown in Fig. 6.4, the dislocation could move to the right with the extra half-plane joining the adjacent lower plane and a 'new' half-plane is formed. Thus, edge dislocation lines move parallel to the applied shear stress. In the screw dislocation, shown in Fig. 6.5, the upper plane of atoms is being sheared over the lower plane. If the upper plane of atoms was sheared further (to the right), the screw dislocation would move vertically, i.e., perpendicular to the shear stress. Dislocations often have a mixed character, i.e., they may possess a combination of the 'pure' edge and screw characteristics.

When a **unit dislocation** moves, atoms are displaced from one regular lattice site to an equivalent position. The displacement associated with the dislocation is termed the **Burgers vector**. A technique to identify the Burgers vector is shown in Fig. 6.6. The arrangement of atoms on the left contains a dislocation and a circuit is made around the dislocation. The same circuit is then made in a perfect crystal (atoms on right) and is found to be incomplete. The displacement needed to complete the circuit is the Burgers vector. This is sometimes called the **RHFS rule**, as for a **R**ight-**H**anded circuit, the Burgers vector is the one needed to re-connect the **F**inish of the circuit with the **S**tart. Returning to Figs. 6.4 and 6.5, one should note that an edge dislocation line moves in the same direction as its Burgers vector and a screw dislocation moves perpendicular to its Burgers vector.

If one considers dislocations in more detail, it becomes clear that not just the atoms in the dislocation line are displaced but also the neighboring atoms. For example, consider the edge dislocation shown in Fig. 6.7. The plane of atoms below and above the slip plane are shown. In the initial configuration, several atoms in the upper row (five) are displaced from their normal positions, i.e., there is a **dislocation width**. It is useful to compare this figure with Fig. 6.2 in which the displacement b is obtained by simultaneous motion of all the atoms. The same displacement b is obtained in both cases but for the dislocation it is produced by a localized motion of atoms rather than the simultaneous shear of a 'perfect' plane. Thus, the displacement associated with the dislocation is 'spread' across

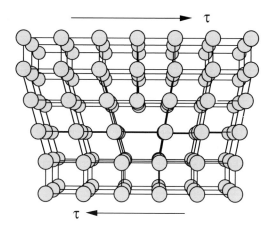

Figure 6.4 Schematic figure showing an edge dislocation in a cubic crystal.

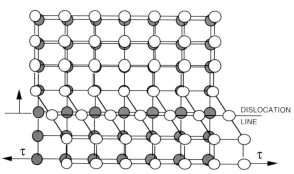

Figure 6.5 Schematic of a screw dislocation. The lower part of the front plane of atoms (light shade) has been sheared with respect to rear plane (dark shade).

several atoms. Moreover, as the shear stress causes the two rows of atoms in Fig. 6.7 to move, each of the individual atoms moves a distance less than b. For these reasons, the stress to move a dislocation along the slip plane, known as the **Peierls–Nabarro stress**, is much less than the theoretical shear strength. A useful analogy is the movement of a carpet; it is much easier to move a 'ripple' in a carpet than slide the whole carpet. The Peierls–Nabarro (or frictional) stress is often expressed in terms of the dislocation width, i.e.,

$$\tau_f = \mu \exp\left(-\frac{2\pi w}{b}\right)$$ (6.4)

The values of τ_f obtained from Eq. (6.4) are usually found to be *less* than the experimental yield strength values in many materials and, thus, one concludes these materials must contain strengthening mechanisms. The frictional stress is clearly very sensitive to the dislocation width and, thus, it is important to identify the material properties that govern this parameter. Dislocation width is governed primarily by the nature of the atomic bonding and crystal structure. In covalent solids, the bonding is strong and directional and, hence, dislocations are very narrow ($w \sim b$). In ionic solids and bcc metals, the dislocations are moderately narrow whereas in fcc metals dislocations are wide ($w > 10b$).

6.3   Stress fields of dislocations

An internal (residual) strain field must exist around a dislocation since atoms associated with the dislocation are displaced from their equilibrium positions. The atoms near the dislocation line have substantial displacements from their normal position but outside this region the stress field will be linearly elastic. The

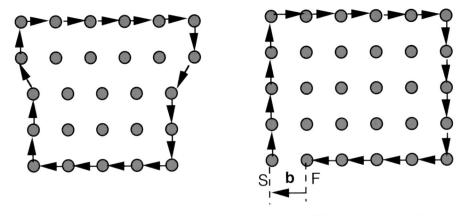

Figure 6.6 The displacement produced by the presence of a dislocation is termed the Burgers vector. This vector can be identified by comparing an interatomic path around the dislocation (*left*) compared to the same path in a perfect crystal (*right*).

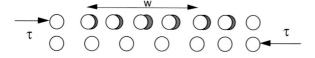

Figure 6.7 Dislocations can have various widths depending on the atomic bonding. The dislocation displacement is 'spread' over several atoms. The motion of the dislocation involves individual atomic displacements less than the Burger's vector. The shaded atoms indicate the position of the atoms after a small displacement.

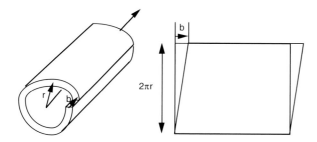

Figure 6.8 A cylindrical element showing the displacement around a screw dislocation. Opening the element into two dimensions shows the stress field to be simple shear.

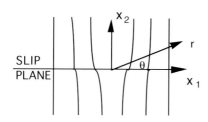

Figure 6.9 Coordinate system for the description of the stress field around an edge dislocation.

non-linear region is termed the **dislocation core**. As dislocations are subjected to external forces and forces from other defects, it is useful to understand the nature of this stress field. The field in the core is difficult to describe but the long-range, linear elastic field can be described in relatively simple terms. For example, a cylindrical element around a screw dislocation is presented in Fig. 6.8, showing the stress field to be simple shear, with the shear strain $= b/2\pi r$. This is obvious if one 'opens' the element into a planar configuration. Using Hooke's Law, the shear stress is

$$\sigma_{\theta z} = \frac{\mu b}{2\pi r} \tag{6.5}$$

The stress field around an edge dislocation is more complex, with both shear and dilatational stresses. For example, from Fig. 6.9, one expects σ_{11} to be compressive in the region above the slip plane due to the insertion of the extra half-plane.

This is compensated by a tensile value of σ_{11} below the slip plane. The stress components around an edge dislocation are given by

$$\sigma_{11} = -\frac{\mu b}{2\pi(1-v)r}\sin\theta(2+\cos\theta) \tag{6.6}$$

$$\sigma_{22} = \frac{\mu b}{2\pi(1-v)r}\sin\theta\cos 2\theta \tag{6.7}$$

$$\sigma_{12} = \frac{\mu b}{2\pi(1-v)r}\cos\theta\cos 2\theta \tag{6.8}$$

$$\sigma_{33} = \frac{\mu b v \sin\theta}{\pi(1-v)r} \tag{6.9}$$

It is useful to consider the energy U associated with the internal stress field of a dislocation. Changes in this energy will control the 'behavior' of the dislocation. For a screw dislocation, it is a straightforward process to derive an expression for the energy of the elastic portion of the field (i.e., the region outside the core). Consider a dislocation line length L; the elastic energy of the line can be obtained by integration from the radius of the core (r_0) to some outer boundary (r_c), i.e.,

$$U = \int_{r_0}^{r_c}\left(\frac{\sigma\varepsilon}{2}\right)dV = \int_{r_0}^{r_c}\left(\frac{\mu b}{2\pi r}\right)\left(\frac{b}{2\pi r}\right)\pi r L\ dr \tag{6.10}$$

Performing the integration and assuming the logarithmic term can be represented by a geometric coefficient α, one obtains an energy per unit length of dislocation

$$\frac{U}{L} = \left(\frac{\mu b^2}{4\pi}\right)\ln\left(\frac{r_c}{r_0}\right) \cong \alpha\mu b^2 \tag{6.11}$$

where α is usually considered to be in the range 0.5–1.0. The analogous expression for the energy of an edge dislocation is

$$\frac{U}{L} = \left(\frac{\mu b^2}{4\pi(1-v)}\right)\ln\left(\frac{r_c}{r_0}\right) \cong \alpha\mu b^2 \tag{6.12}$$

The energy associated with the dislocation core turns out to be a small portion of the total dislocation energy and is often neglected. Thus, $U/L = \alpha\mu b^2$ is commonly used to describe the energy of both dislocation types and it is independent of the applied stress. The energy is high enough, however, to cause the dislocation to be considered a thermodynamically unstable defect, i.e., there will *not* be an equilibrium number present in a material. This does not, however, imply dislocations will spontaneously disappear, as a finite stress is needed for their movement (Eq. (6.4)). The energy per unit length U_L $(= U/L)$ of a dislocation, is called the **line tension** and it acts to straighten a dislocation line, thus lowering the overall energy. An important consequence of the dislocation energy expression is that a dislocation with the smallest Burgers vector has the lowest energy. For this reason, the slip direction in a crystal is often the shortest distance between like atoms. Dislocation energies are also useful in determining whether a particular dislocation 'reaction' will occur. For example, a dislocation with a Burgers vector b_1 would be expected to split into two new dislocations (b_2 and b_3) if $b_1^2 > b_2^2 + b_3^2$.

The energy of a dislocation will change if it interacts with a neighboring dislocation. The change in energy with position must, therefore, give rise to a force F. For example, for two parallel screw dislocations with the same Burgers vector, a repulsive force will occur between the dislocations, which is given by

$$\frac{F}{L} = \left(\frac{1}{L}\right)\frac{dU}{dr} = \frac{\mu b^2}{2\pi r} \tag{6.13}$$

If the Burgers vectors' were of opposite sign (antiparallel), the force becomes attractive. These ideas are useful in rationalizing the behavior of dislocations and their equilibrium geometric configurations. For example, in the absence of stress, edge dislocations of like sign on the same slip plane will repel each other, whereas ones of opposite sign will attract. Similarly dislocations will be attracted to a free surface, as this automatically reduces the dislocation energy by removing part of the stress field. Finally, edge dislocations of like sign on parallel slip planes will be expected to form a linear array (i.e., boundary). With such an alignment, the compressive field for one dislocation will be relaxed by the tensile field of the neighboring dislocation. This array is known as a **tilt boundary**, as the crystal planes on either side of the dislocation array will possess a small misorientation.

6.4   Attributes of dislocations

It is useful to review some of the various attributes of dislocations, especially those that 'complicate' the behavior when compared to the simple pictures described earlier.

6.4.1   Non-planar glide

To this point, it was assumed that dislocations simply glide along a single (slip) plane. For screw dislocations, in which the Burgers vector is parallel to the dislocation line, a unique slip plane is not defined. Thus, if a screw dislocation encounters an obstacle, it may **cross-slip** onto an intersecting slip plane before returning to a parallel primary slip plane. Conversely, an edge dislocation containing a non-planar segment will encounter greater difficulties in movement.

The segment may still be able to move if it lies on a slip plane but it will be subjected to a different shear stress.

6.4.2   Dislocation pinning

In order to overcome obstacles, dislocation lines often assume a curved shape. For example, a dislocation line may be moving between two energy minima and needs to overcome an energy barrier in this process. Instead of moving as a single line, the dislocation can form a **kink** that moves that part to the next energy minimum. Lateral spreading of the kink can then be performed and the whole dislocation line moves to the next minimum. Dislocation lines can also become curved if they encounter obstacles, such as particles or precipitates. For example, consider the situation depicted in Fig. 6.10; the shear stress needed to bow a dislocation segment into a circular arc with radius of curvature R is given by

$$\tau = \frac{U_L}{bR} = \frac{\alpha \mu b}{R} \tag{6.14}$$

As will be seen later, the use of obstacles such as particles is often used to strengthen ductile materials by impeding dislocation motion.

6.4.3   Dislocation structures

The emphasis to this point has been on **glide dislocations** but materials contain many dislocations that cannot move; **sessile dislocations**. For glide dislocations that do move, they are likely to encounter each other in the deformation process. After such intersections, **jogs** will be formed in the dislocation lines. These jogs are likely to be less mobile than the rest of the dislocation and will tend to pin or decrease the velocity of this dislocation. This effect leads to the need for an increasing stress for further deformation and is termed **work hardening**. In other cases, two dislocations can intersect, react and form a new dislocation. The Burgers vector of the new dislocation may no longer lie on the appropriate slip plane and, thus, becomes a **dislocation lock**.

The initial number of dislocations in a material is related to the way in which the material is produced, i.e, dislocations are usually 'grown' into the structure

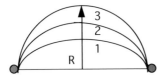

Figure 6.10 Bowing of a dislocation line between two obstacles that are pinning the dislocation. Successive positions of the dislocation line are numbered 1 to 3.

when the material is first processed. Moreover, as just discussed, dislocations can encounter obstacles or produce obstacles by their intersection and interaction. Additionally, under the action of a stress dislocations can also multiply and a well-known example of a multiplication mechanism is the **Frank–Read source**, shown in Fig. 6.11. A dislocation line is pinned between two points. Under the action of a stress, the segment between these points can bow and even extend round the back of the obstacle. At this point, the two segments, marked X and Y, can meet and annihilate each other. The overall process allows the formation of a new dislocation loop and the 'return' of the pinned segment. The dislocation loop can then expand, causing slip, and another loop can be generated. In summary, plastic deformation leads to an increase in dislocation density and complex dislocation structures. These structures usually make further plastic deformation difficult unless relaxation processes can be induced, e.g., by increasing temperature.

6.4.4   Dislocation climb

The movement of edge dislocations by glide was discussed earlier and this is referred to as **conservative motion**. Edge dislocations can also move by **non-conservative motion**. For example, the atoms along portions of the dislocation line can exchange with nearby vacancies, leading to **dislocation climb**. An example is shown in Fig. 6.12. Alternatively, an atom could move into an edge dislocation and cause **reverse climb**. These processes lead to jogs in the dislocation line, which can act to pin the dislocation. Such jogs are expected to occur naturally by thermal activation but, under the action of an applied stress, climb can also be used to circumvent obstacles in the path of the dislocation. Climb is expected to occur more easily at moderate to high temperatures since it involves vacancy diffusion. For this reason it is often associated with creep processes, as will be discussed later.

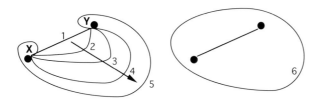

Figure 6.11 A pinned segment of a dislocation undergoing bowing under the action of a shear stress. A dislocation loop forms as the bowing segments (X and Y) merge and the pinned segment returns. This mechanism allows dislocation multiplication and is known as a Frank–Read source.

6.4.5   Dislocation velocity

The velocity v of dislocations is limited at high stress by the velocity of sound but at lower stresses it is often expressed as

$$v = \left(\frac{\tau}{\tau_m}\right)^p \qquad (6.15)$$

where p and τ_m are material constants. The extent of macroscopic shear strain γ in a crystal must depend on the number of dislocations and the distance they move, i.e., $\gamma = \rho b u$, where u is the average displacement and ρ is the dislocation density. If the dislocation density is constant, differentiation of the shear strain with respect to time gives an expression for the shear strain rate

$$\dot{\gamma} = b\rho v \qquad (6.16)$$

This value can be very large, e.g., if $b = 0.4$ nm, $\rho = 10^{10}$ /m^2 and $v = 1$ km/s, then $\dot{\gamma} = 4000$ /s. As pointed out earlier, dislocations can multiply and under some circumstances the increase in ρ can cause **strain softening**. In ionic single crystals, such a process can lead to a yield 'drop' (see Section 6.7), i.e., the stress may fall suddenly at a fixed strain rate (cf. Eq. (6.16)). Conversely, as dislocations move and interact to form tangled structures within a material, the average dislocation velocity will drop and, thus, an increase in stress is needed to produce a given strain rate, giving rise to **work (strain) hardening** (again, see Section 6.7).

6.5   The geometry of slip

To this point, the motion of dislocations on slip planes in particular slip directions has been discussed. The factors that control the choice of these **slip systems** have not, however, been clearly identified, particularly with respect to crystal structure. It was established earlier (Eq. (6.3)) that the easiest slip process should be one that involves the smallest (unit) displacement on planes that are most

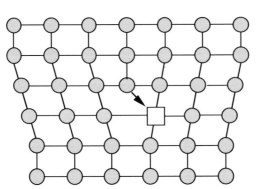

Figure 6.12 Dislocation climb produced by atom on the dislocation line moving to a nearby vacancy.

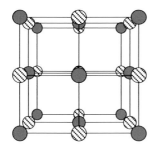

Figure 6.13 Ion positions in the rock-salt structure.

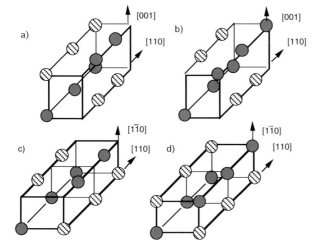

Figure 6.14 Comparison of slip on {100} and {110} in the rock-salt structure. The left-hand figures a) and c) show the initial atom positions for slip on {100} and {110}. The right-hand figures b) and d) show the mid-shear positions.

widely spaced, i.e., small values of b/d. Let us consider the application of this idea to two crystal structures of interest to ceramics science; the rock-salt and corundum structures.

6.5.1 Rock-salt structures

The rock-salt structure is shown in Fig. 6.13. In crystals having this structure, the smallest spacing between ions of the same type is along <110> and the most widely spaced planes with these closely packed directions are the {100} planes. Experimental observations confirm the slip direction as <110> but the slip planes are usually found to be the {110} planes. The systems for which slip is easiest are termed the **primary slip systems** and, thus, for rock-salt structures they are usually {110}<1$\bar{1}$0>. Slip may occur with greater difficulty on other systems and these are termed **secondary slip systems**. Slip does not occur on {100} planes because of the electrostatic interaction that occurs between the ions in this process. This is depicted in Fig. 6.14, in which the initial (a, c) and mid-shear (b, d) positions of the ions are shown. For slip on {100} planes (a to b), the distance between like ions is increased and between opposite ions, it is decreased. For slip

Table 6.1 *Comparison of primary glide planes in crystals having the rock-salt structure*

(After Davidge, 1979.)

Crystal	Primary glide plane	Polarizability (10^{-30} /m³)			Lattice constant (nm)
		Anion	Cation	Total	
LiF	{110}	0.03	1.0	1.03	0.401
MgO	{110}	0.09	3.1	3.19	0.420
NaCl	{110}	0.18	3.7	3.88	0.563
PbS	{100}	3.1	10.2	13.3	0.597
PbTe	{100}	3.1	14.0	17.1	0.634

on {110} (b to d), the distance between like ions is also increased but opposite ions are brought closer together, decreasing the overall electrostatic interaction energy. Not all structures with the rock-salt structure have {110}$<1\bar{1}0>$ as the primary slip systems, as shown in Table 6.1. For PbS and PbTe, the polarizability ('deformability') of these large ions is such that the electrostatic interaction is greatly reduced and, thus, {100} become the slip planes. The atomic arrangement on the {110} planes of the rock-salt structure is shown in Fig. 6.15 with the Burgers vector identified. This notation uses the magnitude of *each* vector component (*a*/2) outside the parentheses, so the crystallographic slip direction remains identified, e.g., (*a*/2)[110].

Dislocations in ionic ceramics are often charged and this can influence their behavior. For example, the charge can affect the dislocation bowing process and it is thought to enhance core diffusion. Consider the edge dislocation in Fig. 6.16. To maintain a perfect crystal structure on either side of the dislocation, it is necessary to introduce two extra half-planes. The line of ions at the bottom of these half-planes consists of alternating positive and negative ions. Thus, if a jog is formed in the dislocation line by the loss of an ion, it will be charged. The intrinsic formation of a cation vacancy is usually easier than an anion vacancy. Thus, if a cation moves into a jog to leave a vacancy, the dislocation acquires a positive charge. This is then compensated by a 'cloud' of negatively charged cation vacancies. Conversely, a dislocation line can act as a sink for extrinsic cation vacancies, e.g., as a result of the addition of divalent solutes. In this case, the dislocation line acquires a negative charge and an 'atmosphere' of positively charged impurities. This can significantly reduce the mobility of the dislocation. As temperature increases, it is sometimes possible for the charge on the dislocation to change from negative to positive. The temperature at which this switch occurs is called the **isoelectric temperature**. Dislocations are also thought to acquire a transient charge by picking up vacancies as they glide through a crystal.

6.5.2 Alumina

Alumina in its α-phase form has a hexagonal structure and, thus, the basal plane is expected to be the primary slip plane, which is confirmed experimentally. For the oxygen ions, <1100> are the close-packed directions but these are *not* the slip directions. This is a result of the stacking sequence in the aluminum layers, in which only two-thirds of the sites are occupied. In order to preserve the same stacking sequence of the aluminum ions, unit slip must occur along the <1120>, as shown in Fig. 6.17. It is important to note, however, that temperatures >1300 °C are needed before plastic deformation becomes appreciable in α-alumina single crystals. This is a result of the high Peierls stress needed for dislocation motion in this material. Figure 6.18 shows a micrograph of the dislocation structure in α-alumina after deformation at high temperature, obtained using trans-

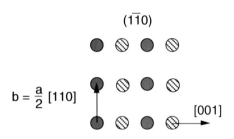

Figure 6.15 The atomic arrangement on the {110} slip planes in the rock-salt structure with the Burgers vector identified.

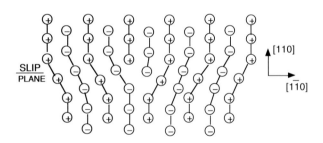

Figure 6.16 Schematic of a (110)[1$\bar{1}$0] edge dislocation in the rock-salt structure. (Adapted from Cook and Pharr, 1994.)

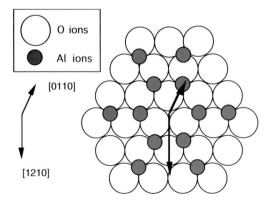

Figure 6.17 Atomic arrangement on primary slip plane of α-alumina (corundum).

mission electron microscopy. For some single crystals, dislocation behavior can also be studied using etch pits. The slip systems for various ceramics are given in Table 6.2.

6.6   Partial dislocations

To this point, only unit dislocations have been discussed but, in some cases, unit slip can be caused by the combined motion of **partial dislocations**. This has been

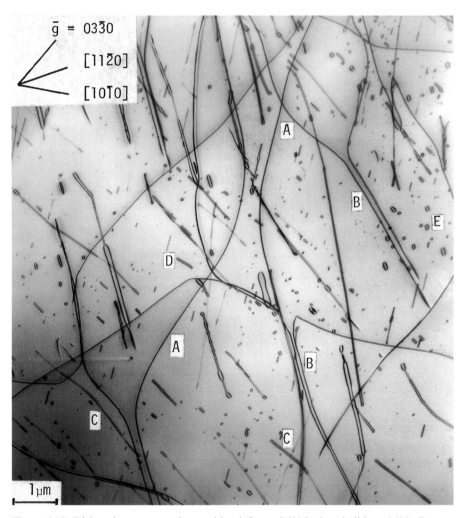

Figure 6.18 Dislocation structure in sapphire deformed 4% by basal glide at 1400 °C, consisting of A) glide dislocations, B) edge dipoles, C) faulted dipoles and D, E) dislocation loops. Basal foil, 650 kV. (Micrograph from B. J. Pletka and T. E. Mitchell, Case Western Reserve University, reproduced courtesy of The American Ceramic Society, Westerville, OH.)

Table 6.2 *Primary and secondary slip systems of ceramics*

Material	Crystal structure	Slip systems		Activation temperature (°C)	
		Primary	Secondary	Primary	Secondary
Al_2O_3	hexagonal	$\{0001\}\langle11\bar{2}0\rangle$	several	1200	
BeO	hexagonal	$\{0001\}\langle11\bar{2}0\rangle$	several	1000	1700
MgO	cubic (NaCl)	$\{110\}\langle1\bar{1}0\rangle$	$\{001\}\langle110\rangle$	0	
$MgO.Al_2O_3$	cubic (spinel)	$\{111\}\langle1\bar{1}0\rangle$	$\{110\}\langle1\bar{1}0\rangle$	1650	
β-SiC	cubic (ZnS)	$\{111\}\langle1\bar{1}0\rangle$		>2000	
$β-Si_3N_4$	hexagonal	$\{10\bar{1}0\}\langle0001\rangle$		>1800	
TiC, (ZrC, HfC, etc.)	cubic (NaCl)	$\{111\}\langle1\bar{1}0\rangle$	$\{110\}\langle1\bar{1}0\rangle$	900	
UO_2, (ThO_2)	cubic (CaF_2)	$\{001\}\langle110\rangle$	$\{110\}\langle1\bar{1}0\rangle$	700	1200
ZrB_2, (TiB_2)	hexagonal	$\{0001\}\langle11\bar{2}0\rangle$		2100	
C (diamond)	cubic	$\{111\}\langle1\bar{1}0\rangle$			
C (graphite)	hexagonal	$\{0001\}\langle11\bar{2}0\rangle$			
$β-SiO_2$	hexagonal	$\{0001\}\langle11\bar{2}0\rangle$			
CaF_2, (BaF_2, etc.)	cubic	$\{001\}\langle110\rangle$			
CsBr	cubic (CsCl)	$\{110\}\langle001\rangle$			
TiO_2	tetragonal	$\{110\}\langle1\bar{1}0\rangle$	$\{110\}\langle001\rangle$		
WC	hexagonal	$\{10\bar{1}0\}\langle0001\rangle$	$\{10\bar{1}0\}\langle11\bar{2}0\rangle$		

observed experimentally in graphite. The structure of graphite is shown schematically in Fig. 6.19. The shaded atoms represent the carbon atoms in a single basal
plane. In an adjacent basal plane, however, the carbon atoms are displaced from
these positions and some of these are shown as open circles. Unit dislocations in
this structure are shown as the vectors AB, AC or AD. In order to obtain the
unit slip AC, there is the possibility that slip could occur by the motion of two
partial dislocations, i.e, AO plus OC. This can be written as a dislocation reaction, i.e.,

$$\frac{a}{3}[2\bar{1}\bar{1}0]=\frac{a}{3}[1\bar{1}00]+\frac{a}{3}[10\bar{1}0] \tag{6.17}$$

Notice that each component of the direction vector is balanced in the equation.
If one determines the energy of these dislocations (proportional to b^2), one finds
that the reaction favored is

$$a^2 \rightarrow \frac{a^2}{3}+\frac{a^2}{3} \tag{6.18}$$

Such a process does, however, lead to a **stacking fault** for the carbon atoms. When
the first partial moves through the crystal, the carbon at position A now lies
above the carbon atom at position O. The decrease in energy produced by the
dissociation into two partials is therefore offset by the stacking fault energy γ.
The partial dislocations move cooperatively through the crystal, separated by a
distance d such that

$$d=\frac{\mu\bar{b}_1\cdot\bar{b}_2}{2\pi\gamma} \tag{6.19}$$

Thus, measurement of d allows an estimation of the stacking fault energy.

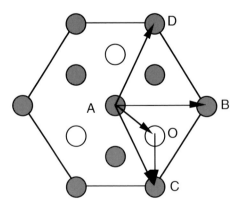

Figure 6.19 The arrangement of atoms in the basal
plane of graphite (shaded
atoms). The open circles
represent atomic positions
in an adjacent basal plane.
Unit dislocations, AB, AC
and AD are shown as well
as the dissociation of AC
into two partial disloca-
tions AO and OC.

6.7   **Plasticity in single crystals and polycrystalline materials**

Consider a single crystal being subjected to uniaxial tension or compression, as shown in Fig. 6.20. Clearly, the ease with which plastic deformation is activated will depend not only on the ease of dislocation glide for a particular slip system but also the shear stress acting on each system. This is similar to the problem discussed in Section 2.10 (Eq. (2.44)) though one should note the plane normal, the stress direction and the slip direction are not necessarily coplanar, $(\phi+\lambda)\neq90°$. In other words, slip may not occur in the direction of the maximum shear stress. The **resolved shear stress** acting on the slip plane in the slip direction is

$$\tau=\sigma\cos\phi\cos\lambda \tag{6.20}$$

The uniaxial yield stress σ_Y will determine when a **critical resolved shear stress** τ_c is obtained for slip on a particular plane and direction (**slip system**), i.e.,

$$\sigma_Y=\frac{\tau_c}{\cos\phi\cos\lambda} \tag{6.21}$$

The stress–strain behavior of single crystals is shown in Fig. 6.21. Following yielding, work hardening is initially low and slip occurs on the (single) slip system for which the resolved shear stress is the highest. Stage I is termed **easy glide** and is usually terminated when slip is initiated on multiple systems. The work hardening observed in Stage II is associated with dislocations interacting on intersecting slip planes until, in Stage III, the hardening is exhausted and processes such as cross-slip alleviate the hardening process. The type of behavior

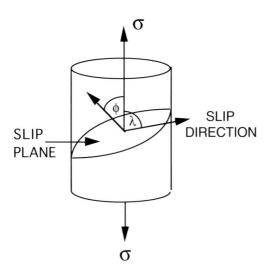

Figure 6.20 Geometry of the slip plane and slip direction in a single crystal, defined with respect to a uniaxial applied stress.

shown in Fig. 6.21 has been observed and studied in many ceramic single crystals, notably the alkali halides and MgO at room temperature. For favorably oriented single crystals, large strains (~20%) are possible. Ductility can, however, be
limited by the tendency for **crack nucleation** at slip band intersections. Thus,
although plastic deformation can occur in ceramic single crystals, this process is
in strong competition with fracture.

The onset of yielding is not always as 'smooth' as that depicted in Fig. 6.21
and it can be accompanied by a stress drop (strain softening). In some cases, the
drop occurs over a range of strain whereas, in the other cases, it is instantaneous.
The former behavior occurs when yielding is associated with a rapid increase in
the dislocation density. This allows the dislocation velocity to drop (Eq. (6.16))
and, hence, yielding can continue at a lower stress. A sharper **yield drop** is usually
associated with dislocations being pinned by impurity 'atmospheres' and, once
the dislocations escape, yielding continues at a lower stress.

The effect of temperature and strain rate on the critical resolved shear stress
is shown in Fig. 6.22. In the low-temperature region (Region I), τ_c decreases with
increasing temperature and decreasing strain rate. In this domain, thermal
fluctuations are sufficient to allow dislocations to overcome the short-range
stress fields of obstacles in the slip plane. In Region II, however, τ_c becomes
athermal and independent of strain rate. Dislocations now interact with the
long-range internal stress fields of other dislocations, precipitates, etc. Finally, in
Region III, plastic deformation can combine with diffusion such that τ_c
decreases with increasing temperature and decreasing strain rate. This behavior
is important in creep deformation and will be discussed further in Chapter 7.
Figure 6.23 shows the temperature dependence of τ_c for several ceramic single
crystals. The effect of the bonding on the yield stress is clear with the more ionic

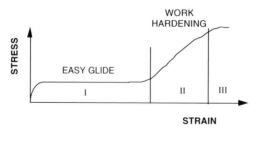

Figure 6.21 Stress–strain
behavior for a single
crystal that is favorably
oriented for plastic flow.

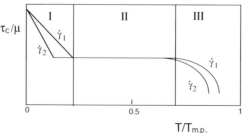

Figure 6.22 Schematic
illustration of the dependence of normalized critical resolved shear stress
on temperature for two
strain rates, $\dot{\gamma}_1 > \dot{\gamma}_2$.

solids (NaCl and MgO) having the lower yield stresses. In addition, the effect of impurity atoms is shown; an effect that can be exploited in **solid solution strengthening**. For this latter behavior, the increase in yield stress is a result of the dislocations interacting with the stress and electric fields of the impurity atoms (see Section 6.8.1).

The stress–strain behavior of ceramic polycrystals is substantially different from single crystals. The same dislocation processes proceed within the individual grains but these must be constrained by the deformation of the adjacent grains. This constraint increases the difficulty of plastic deformation in polycrystals compared to the respective single crystals. As seen in Chapter 2, a general strain must involve six components, but only five will be independent at constant volume (ε_{ii}=constant). This implies that a material must have at least five **independent slip systems** before it can undergo an arbitrary strain. A slip system is independent if the same strain cannot be obtained from a combination of slip on other systems. The lack of a sufficient number of independent slip systems is the reason why ceramics that are ductile when stressed in certain orientations as single crystals are often brittle as polycrystals. This scarcity of slip systems also leads to the formation of stress concentrations and subsequent crack formation. Various mechanisms have been postulated for crack nucleation by the 'pile-up' of dislocations, as shown in Fig. 6.24. In these examples, the dislocation pile-up at a boundary or slip-band intersection leads to a stress concentration that is sufficient to nucleate a crack.

For NaCl, the activation of the secondary slip systems at temperatures >200 °C is required before ductility in polycrystals is obtained. A similar **brittle to ductile transition** occurs in KCl at 250 °C. For MgO, this transition occurs ~ 1700 °C. Some cubic materials, such as TiC, β-SiC and MgO.Al$_2$O$_3$ have sufficient independent primary systems but, unfortunately, the dislocations tend to be immobile in these materials. Thus, overall it is found that most ceramic polycrystals lack sufficient slip systems or have such a high Peierls stress that they are brittle except under extreme conditions of stress and temperature.

The constraint in polycrystals also removes the presence of the easy glide region observed in single crystals and, thus, work hardening immediately

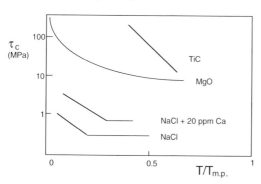

Figure 6.23 Schematic illustration of variation in critical resolved shear stress with temperature for various ceramic single crystals.

follows yielding. Figure 6.25 shows a stress–strain curve for a polycrystalline material that undergoes extensive plastic deformation. The stress apparently reaches a maximum value (**ultimate tensile strength**) and then decreases (curve 1). The decrease is associated with a geometric instability, known as **necking**, in which the plastic deformation becomes localized to one section of the test specimen. If one considers the true stress–true strain behavior for this section of the specimen, the stress maximum is eliminated (curve 2) and **true fracture stress** and **true failure strain** can be defined. For the engineering stress–strain curve, it can be difficult to define the exact stress at which yielding occurs. To overcome this problem, an **offset yield stress** is often used. As shown in Fig. 6.25, a straight line is drawn with the same slope as the elastic portion but offset by a given strain, e.g., 0.2%. The intersection of the line with the stress–strain curve is then used to define the yield stress. Even in ductile ceramic polycrystals, one often finds that failure coincides or shortly follows the onset of yielding. Thus, necking and a maximum in the stress–strain curve is rarely observed in ceramic polycrystals.

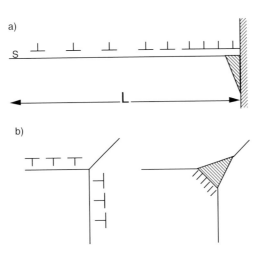

Figure 6.24 Crack nucleation as a result of dislocation pile-up: a) Zener model; b) Cottrell model.

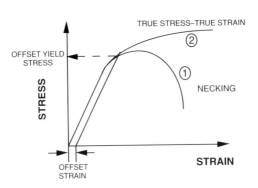

Figure 6.25 Stress–strain behavior (curve 1) for a polycrystalline material that demonstrates significant plastic deformation and necking. In addition, the true stress–true strain (curve 2) is shown schematically.

6.8   Obstacles to dislocation motion

For materials in which dislocation motion is easy, one is often concerned with means to introduce obstacles to such motion, thereby utilizing **strengthening mechanisms**. Conversely, for materials in which dislocations are difficult to move, one is more concerned with removing such obstacles. Figure 6.26 shows a schematic of a dislocation interacting with an array of obstacles. This interaction causes bowing of the dislocation under the action of an increasing stress. This process continues until a critical angle ϕ_c is reached and the obstacles can then be by-passed. For 'strong' obstacles (large ϕ_c), the maximum shear stress to accomplish this process is given by

$$\Delta\tau_Y = \frac{\mu b}{L}\cos\frac{\phi_c}{2}$$

(6.22)

It is now worth summarizing some of the mechanisms that lead to the formation of such obstacles.

6.8.1   Point defects

Point defects, such as vacancies, interstitial and substitutional atoms, even at low concentrations (~100 ppm), will give rise to localized stress fields that reduce dislocation mobility, as shown in Fig. 6.27. Indeed, the use of solute atoms is often used to strengthen ductile materials and this is termed **solid solution strengthening**. The nature of the interaction between a solute atom and the dislocation depends on the volume misfit (spherical distortion) and modulus of the solute. For edge dislocations, solute atoms with a positive volume misfit would be attracted to the region below the extra half-plane, as this region is in tension (see Section 6.3). The interaction between point defects and dislocations is controlled by the stress field though for ionic crystals the presence of charged dislocations may also play a minor role. It is found that solutes with a different valence

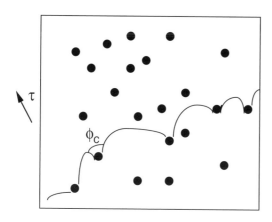

Figure 6.26 Obstacles to dislocation motion leads to bowing of dislocations as they glide through the material.

(**aliovalent**) than the host atoms are more effective in increasing the yield stress than ions with the same valence (**isovalent**). For aliovalent ions, the stress field of the impurity and its associated vacancy produces an asymmetric elastic distortion which interacts strongly with dislocations. The yield stress contribution $\Delta\tau_Y$ by solid solution strengthening has been analyzed (e.g., see Cook and Pharr, 1994) and can be described by

$$\left(\frac{\Delta\tau_Y}{\tau_0}\right)^{1/2} = 1 - \left(\frac{T}{T_0}\right)^{1/2} \tag{6.23}$$

where T is the absolute temperature, $\tau_0 = \mu\Delta\varepsilon c^{1/2}/3.3$, $T_0 = \mu\Delta\varepsilon b^3/(3.86\alpha k)$, c is the defect concentration, $\Delta\varepsilon$ is a measure of the strain misfit, b is the Burgers vector, α is a numerical constant and k is Boltzmann's constant.

6.8.2 Work hardening

As pointed out earlier, dislocations can interact strongly with each other and, thus, techniques to increase the dislocation density will also act to strengthen the material. As dislocations intersect, jogs are often formed and these act to pin dislocations. In addition, reactions can occur between dislocations to form sessile dislocations, which then act as barriers to the other glide dislocations.

After the various dislocation interactions, the stress needed for further plastic deformation will depend on the mean free dislocation length L. The dislocation density should be proportional to $1/L^2$ and thus Eq. (6.22) can be used to estimate the shear stress needed to overcome the obstacles, i.e.,

$$\Delta\tau_Y = \alpha\mu b\sqrt{\rho} \tag{6.24}$$

Dislocation structures at high strains can become rather complex, often forming cellular structures. In these cases, the process is closer to a boundary interaction, which will be described next.

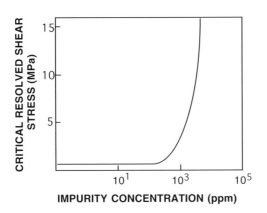

Figure 6.27 Effect of various divalent impurities on the critical resolved shear stress in various alkali halides. (After Cook and Pharr, 1994.)

6.8.3 Boundary interactions

Internal boundaries in a material act as obstacles and can be used in **boundary strengthening**. Boundaries are expected to be more potent obstacles than point or line defects, as they will impede a dislocation over a larger length. Grain boundaries are particularly effective obstacles, as the change in crystal orientation at the boundary will not allow passage of dislocations. Yielding needs to be activated in the adjacent grain before the plastic deformation can continue. If the adjacent grain is not favorably oriented, dislocations will pile up against the boundary, in a similar fashion to those depicted in Fig. 6.24(a). Assuming crack nucleation does not occur, slip can be initiated in the adjacent grain under the action of the stress concentration produced by the pile-up. Analysis of this process leads to the **Hall–Petch relation,** in which the contribution to the yield stress is given by

$$\Delta \tau_Y = \frac{k_y}{\sqrt{D}} \tag{6.25}$$

where D is grain size and k_y is the locking parameter, which describes the strength of the boundary interaction. The above relationship recognizes that the stress concentration at the boundary increases with increasing slip distance (proportional to grain size), which aids in initiating yielding in the adjacent grain. Boundary strengthening can also arise due to interaction with cell boundaries, sub-boundaries and stacking faults but these are usually less effective obstacles.

6.8.4 Particles

The introduction of small particles into a ductile material can substantially increase the yield strength, even if the volume fraction is low (< 10 vol.%). The particles can be introduced by *precipitation* (**precipitation hardening**) or by *physical addition* (**dispersion strengthening**). For example, Fig. 6.28 shows the effect of precipitation of $MgO.Fe_2O_3$ on the stress–strain behavior of MgO. The extent of the strengthening is determined by several factors, including volume fraction,

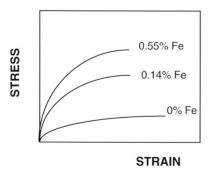

Figure 6.28 Stress–strain curves for MgO containing precipitates of $MgO.Fe_2O_3$. (Adapted from McColm, 1983, reproduced courtesy of Chapman and Hall, Andover, UK.)

particle size and the shape and nature of the particle–matrix interface. For this latter factor, an important distinction is whether the dislocations shear the particles or bow past them. For deformable ('shearable') coherent particles, factors similar to solid solution strengthening play a role, e.g., particle misfit and modulus. In addition, however, one is also concerned with the energy needed to create new particle–matrix interface during shear. This is termed **chemical strengthening** and, in some material systems, it can be a major contributor to the strength increase. If the particle–matrix interface is disordered or a coherent particle exceeds a critical size, the particles can be by-passed by dislocation bowing. In this process, the dislocation bows round the obstacle until the bowing segments join. The dislocation can then proceed, leaving a dislocation (**Orowan**) **loop** around the particle. As expected from Eq. (6.22), for a fixed particle volume fraction, the stress required for by-passing decreases as the interparticle spacing or particle size increases. For precipitation-hardened systems, the yield stress is often found to pass through a maximum as particle size increases (at fixed volume fraction).

In recent times, it has been shown that particles at much higher concentrations can also lead to strengthening and this has been exploited in **metal-matrix composites**. In these materials, particles are introduced with a high elastic modulus, e.g., SiC; the matrix is often a lightweight metal, e.g., Al or Ti, of lower modulus. The strengthening is often considered to be a result of the modulus difference, with the high-modulus particles carrying most of the stress. Thus, strength is expected to increase with increasing particle volume fraction. This is termed **modulus strengthening**. It has, however, also been suggested that the thermal expansion mismatch can substantially increase the dislocation density, thereby giving rise to dislocation strengthening. Reinforcement in metal-matrix composites can also be accomplished using fibers or single-crystal whiskers.

6.9   Plasticity mechanics

For linear elastic materials, Hooke's Law is a constitutive relationship between stress and strain. There have been substantial efforts in identifying similar relationships for plastic solids. In uniaxial tests, the portion of the true stress–true strain curve beyond yielding is often described by

$$\sigma_T = A\varepsilon_T^n \dot{\varepsilon}^m \tag{6.26}$$

where A is a material constant, n is the strain-hardening coefficient and m is the strain rate sensitivity. For $n=1$, the material is elastic and for $n=0$, ideally plastic. In this way, uniaxial testing is often used to define the plastic deformation behavior of a material but, clearly, the behavior under multiaxial stresses must also be understood. The competition between plastic deformation and fracture is highly sensitive to the stress state. For instance, one would expect that the deviatoric stresses would be most important in driving shear deformation processes.

One simple criterion for yielding under multiaxial stresses is known as the **Tresca yield criterion**. This approach recognizes that the maximum shear stress is one-half the difference between the maximum and minimum principal stresses. In terms of the uniaxial yield stress $\sigma_Y (=2\tau_c)$ this can be written as

$$\sigma_{max} - \sigma_{min} = \sigma_Y \tag{6.27}$$

An alternative approach is the **von Mises yield criterion**, in which the role of all three principal stresses is introduced, i.e.,

$$\frac{1}{\sqrt{2}}\left[(\sigma_1-\sigma_2)^2+(\sigma_1-\sigma_3)^2+(\sigma_2-\sigma_3)^2\right]^{1/2}=\sigma_Y \tag{6.28}$$

If the left-hand sides of Eqs. (6.27) and (6.28) are less than the uniaxial yield stress, yielding does not occur. Thus, the equations can be seen as defining a yield surface, within which yielding does not occur. The two criteria are shown schematically in Fig. 6.29 for the case of biaxial stresses. The Tresca yield locus is given by horizontal and vertical lines in the first and third quadrants (where $\sigma_3=0$ is the minimum or maximum stress) and by two 45° lines in the second and fourth quadrants (where the minimum and maximum stresses are σ_1 or σ_2). The fact that yielding is controlled by only one of the stresses in the first and third quadrants is unexpected and does not agree with experimental data. For biaxial loading, the von Mises yield locus is an ellipse and the values of both σ_1 and σ_2 play a role in defining the yield criterion. With $\sigma_3=0$, Eq. (6.28) becomes

$$(\sigma_1^2+\sigma_2^2-\sigma_1\sigma_2)^{1/2}=\sigma_Y \tag{6.29}$$

The two criteria agree for uniaxial and equibiaxial loading. The von Mises criterion is essentially an empirical approach, though it does equate the deviatoric

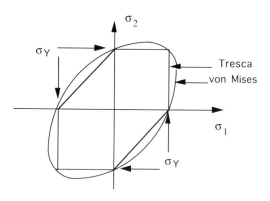

Figure 6.29 Comparison of von Mises and Tresca yield criteria for biaxial stress states.

elastic energy for multiaxial and uniaxial stressing. In mechanical design, the Tresca criterion is often used as it is more conservative.

The idea of 'plastic' yielding is also applied to granular materials, such as soils and powders. In this case, the shear stress required for deformation depends on the packing density, and the particle shape and surface characteristics. The shear resistance is commonly described by the **Mohr–Coulomb yield criterion**, i.e.,

$$\tau_{mc} = \tau_0 + \mu_f P = \tau_0 + P\tan\phi \tag{6.30}$$

where τ_0 is the shear strength due to cohesive forces, P is the normal stress (assumed compressive), μ_f is the friction coefficient and ϕ is the friction angle. For strong cohesion with no friction, Eq. (6.30) reduces to the Tresca criterion. For weak cohesion ($\tau_0 = 0$), e.g., sand, $\tau = P\tan\phi = \mu_f P$, where ϕ is equated to the **angle of repose**. In this case, the friction angle is the steepest angle a bank of the material can sustain without slipping. For loosely packed sand, $\phi \sim 30°$ but for interlocked particles or higher packing densities, ϕ may be in the range 40–45°. In densely packed powders, an increase in volume is often needed before slip can occur (dilatancy).

6.10   **Hardness**

The resistance of a material to the formation of a permanent surface impression by an indenter is termed **hardness**. The deformation process must be inelastic and, hence, it is inherently related to the resistance of a material to such a deformation (indentation). Hardness impressions can be formed even in brittle materials, though at higher loads this is usually accompanied by localized crack-ing. For more ductile materials, however, one would expect hardness to be related to the yield stress of a material. In order to create the surface impression, various geometries are used for the indenter (Fig. 6.30). In most tests, hardness is defined as the applied load divided by the actual or projected area of the impression and, thus, the units are the same as stress (Pa).

One common hardness test is the **Brinell test**, in which a hardened steel ball is used as the indenter. The **Brinell Hardness Number** (BHN) is the applied force F divided by the surface area of the indentation, i.e.,

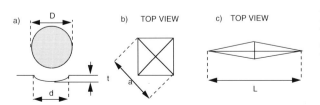

Figure 6.30 Geometries of various hardness impressions: a) Brinell; b) Vickers; and c) Knoop.

$$BHN = \frac{F}{\pi D t} \tag{6.31}$$

where D is the diameter of the ball and t is the indentation depth (see Fig. 6.30). The **Meyer Hardness Number** (MHN) can be defined using the same test with the area term being the projected surface area, i.e.,

$$MHN = \frac{4F}{\pi d^2} \tag{6.32}$$

For ductile materials, which do not work harden, one can show that MHN $\sim 3\sigma_Y$, where σ_Y is the uniaxial yield stress. The hardness value is higher than the yield stress because all the material below an indenter must yield before indentation occurs.

Other common hardness tests involve the use of diamond pyramids. In the **Vickers hardness test**, a square pyramid is used and in the **Knoop hardness test**, the pyramid is elongated. The area term in the former test is the actual indentation area and in the latter, the projected area. From the impression geometries, shown in Fig. 6.30, the **Vickers Hardness Number** (VHN) and **Knoop Hardness Number** (KHN) can be shown to be VHN$=1.854F/a^2$ and KHN$=14.2F/L^2$, respectively. A common hardness test in the USA is the **Rockwell hardness test**, which uses various indenter types and loads. The result of these tests is a dimensionless number and leads to the use of various hardness scales (e.g., Rockwell B, Rockwell C).

Problems

6.1   For plastic deformation in CsCl, the primary slip system is $\{110\}<001>$.

     a) Use a drawing of the unit cell to indicate *one* primary slip system and all the associated slip direction(s) within this plane (no need to show ions).

     b) Draw the two-dimensional arrangement of ions on the slip plane identified in part a) and indicate the Burgers vector(s) for primary slip within this plane (show orientation of plane with two orthogonal arrows).

     c) Sketch the atomic arrangement around an edge dislocation in CsCl.

     d) For some materials with the CsCl structure, the slip directions change to $<111>$, especially if the bonding has a metallic character. If the critical resolved shear stress for both types of slip is 10 MPa, compare the magnitude of the normal stress on the (100) plane needed to initiate yielding on these two types of slip systems. On which specific plane or planes will yielding first occur?

e) Estimate the theoretical shear strength of the primary slip plane (lattice parameter CsCl=0.411 nm, $\mu\{110\}<001>=8$ GPa).

6.2   For plastic deformation in TiC, the primary slip system is $\{111\}<110>$.

a) Use a drawing of the unit cell to indicate one primary slip system and associated slip direction(s) within the slip plane.

b) Draw the arrangement of ions on the slip plane identified in a) and indicate the various Burgers vector(s) for primary slip within this plane. Label the Burgers vector using notation $m[hkl]$, where m is the magnitude of each component.

c) Estimate the frictional stress for primary slip in TiC ($v=0.25$).

d) If the critical resolved shear stress for primary slip in a TiC single crystal is 100 MPa, determine the uniaxial stress required to initiate yielding if stress is applied along [001]. On which specific plane(s) and in which directions will yielding first occur?

e) Estimate the theoretical shear strength of the primary slip plane (lattice parameter TiC=0.432 nm, $\mu=175$ GPa).

6.3   Determine an expression for the mean stress ($\sigma_{ii}/3$) around an edge dislocation in units of $\mu b/r$ and plot these values as a function of the angle θ. At which angle will the mean stress be a maximum? (r is the radial distance from the dislocation line.)

6.4   In ionic crystals, the following dislocation reaction can occur.

$$\frac{a}{2}[011]+\frac{a}{2}[10\bar{1}]=\frac{a}{2}[110]$$

Show that the reaction is energetically favorable. For NaCl, calculate the energy of the final dislocation in units (L/b) and compare it to the cohesive binding energy for this material. The cohesive binding energy is the depth of the interatomic potential 'well'.

6.5   Derive an expression for the total energy required to fail, at constant strain rate, a ductile material exhibiting a stress–strain behavior described by Eq. (6.26).

6.6   From the expression for the elastic energy of an isotropic solid in terms of the principal stresses and strains, show that the von Mises yield criterion is obtained when the dilatational part is removed, leaving the elastic shear strain energy.

6.7   Derive the equation VHN=$1.854F/a^2$, given that the angle between opposing faces at the apex of the Vickers indenter is 136°. Indicate any approximations in derivation. How would the equation change if the projected indentation area was used?

6.8 Suggest an energetically favorable dislocation reaction for the splitting of a unit dislocation into partial dislocations for basal slip in sapphire (α-alumina) single crystals.

6.9 For a uniaxial compression test, determine the plane on which the maximum shear stress will occur for a material that obeys the Mohr-Coulomb yield criterion. Suggest an approach for measuring the material parameters (ϕ, τ_0) in the Mohr–Coulomb yield criterion. (Hint: construct the Mohr's circle (see Fig. 2.23) for various values of the normal stress.)

6.10 The hardness test can be considered to be similar to uniaxial compression ($\sigma_1 = \sigma$) but with the constraint that $\sigma_2 = \sigma_3 \neq 0$. If the applied stress required to yield the material is three times the uniaxial yield stress, determine σ_2 and σ_3 using the von Mises and Tresca yield criteria. Illustrate your answer using the Mohr's circle construction (see Fig. 2.23).

6.11 Show that Eq. (6.24) follows directly from Eq. (6.22) by using the appropriate relationship between dislocation density and spacing.

6.12 It has been suggested that edge dislocations in α-alumina should dissociate into two or four partial dislocations for basal slip. Sketch the ionic structure of α-alumina and discuss the displacement paths that could be associated with these two dislocation dissociation mechanisms.

References

M. F. Ashby and D. R. H. Jones, *Engineering Materials 1*, Pergamon Press, Oxford, 1980.

R. C. Bradt and R. E. Tressler, *Deformation of Ceramic Materials*, Plenum Press, New York, 1975.

R. F. Cook and G. M. Pharr, Mechanical properties of ceramics, pp. 339–407 in *Materials Science and Technology*, Vol. 11, edited by M. Swain, VCH Publishers, Germany, 1994.

A. H. Cottrell, *Mechanical Properties of Matter*, J. Wiley and Sons, New York, 1964.

T. H. Courtney, *Mechanical Behavior of Materials*, McGraw-Hill, 1990.

R. W. Davidge, *Mechanical Behaviour of Ceramics*, Cambridge University Press, Cambridge, UK, 1979.

R. W. Hertzberg, *Deformation and Fracture of Engineering Materials*, J. Wiley and Sons, New York, 1989.

D. Hull, *Introduction to Dislocations*, Pergamon Press, Oxford, 1965.

W. D. Kingery, H. K. Bowen and D. R. Uhlmann, *Introduction to Ceramics*, J. Wiley and Sons, New York, 1976.

F. A. McClintock and A. S. Argon, *Mechanical Behavior of Materials*, Addison-Wesley Publishing Co., Reading, MA, 1966.

I. J. McColm, *Ceramic Science for Technologists*, Blackie and Sons Ltd, Glasgow, UK, 1983.

T. E. Mitchell, Application of transmission electron microscopy to the study of deformation in ceramic oxides, *J. Am. Ceram. Soc.*, **62** (1979) 254–66.

Chapter 7

Creep deformation

As indicated in the last chapter, plasticity can become time dependent at high temperatures and this is termed **creep**. This behavior is common in ceramic materials and becomes an important factor in the use of these materials in high-temperature structural applications. For the design process, the amount of creep strain must usually be kept below a particular value, otherwise design tolerances will be lost or creep failure may occur. The type of strain behavior associated with creep at constant stress is shown in Fig. 7.1. It is generally divided into three regions: **primary**; **secondary**; **and tertiary creep**. This behavior is sometimes discussed qualitatively in terms of hardening and softening (recovery) processes. In the primary stage, the strain rate is decreasing and the softening process is usually associated with changes in structure (e.g., grain size, dislocation density) or with the redistribution of stresses. In secondary or **steady-state creep**, the strain rate is constant and a balance appears to be occurring between hardening

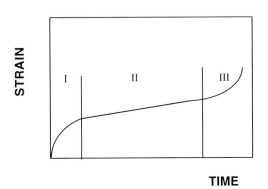

Figure 7.1 Three stages of creep deformation: I primary; II secondary; III tertiary.

and softening processes. For this case, creep is considered to be a result of the deformation of an invariant microstructure. In tertiary creep, the strain rate is increasing and this stage is associated with the initiation of the failure process, i.e., the formation of cracks or voids. In ceramics, the tertiary creep stage is often very short or can be entirely missing. The creep failure strain in ceramics is usually very small, typically <2%. The effect of increasing stress on the creep behavior is shown in Fig. 7.2. It gives rise to increases in strain rate, a shortening of the steady-state creep period and lower failure strains. Similar behavior occurs with increasing temperature. Recently, creep in some ceramics has been shown to consist entirely of a primary creep stage. For example, Fig. 7.3 shows the creep curves of SiC fibers exhibiting this behavior, tested in uniaxial tension.

In order to control creep behavior in ceramics, the mechanisms involved in this process need to be identified and analyzed. For this text, the emphasis will be on the mechanisms involved in steady-state creep.

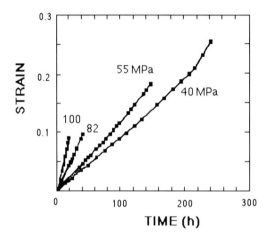

Figure 7.2 Effect of increasing stress on the uniaxial tensile creep deformation of a hot-pressed alumina. (Adapted from Robertson et al., 1991, reproduced courtesy of The American Ceramic Society, Westerville, OH.)

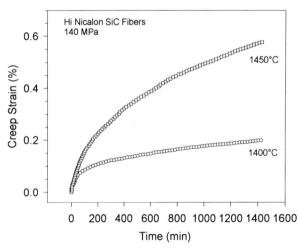

Figure 7.3 Creep curves for SiC fibers (Hi Nicalon™) at 1400 and 1450 °C in argon at a stress of 140 MPa. The inelastic deformation shows only the presence of primary creep. (From Rugg and Tressler, 1997, reproduced courtesy of The American Ceramic Society, Westerville, OH.)

7.1   Creep in single crystals

7.1.1   Dislocation creep

As mentioned in Chapter 6, dislocation motion is impeded by a variety of obstacles. If the range of the stress field associated with these obstacles is short, thermal activation can allow the dislocation to by-pass the obstacle. For the creep of single crystals at high temperature, this occurs primarily by **dislocation climb** (see Section 6.4.4 and Fig. 6.12). When combined with dislocation glide, this process gives rise to **dislocation creep**. Dislocation climb requires a short-range diffusion process, which allows the dislocation to move to an adjacent slip plane. The creep strain rate (referred to hereafter as the creep rate) can be controlled by the climb or glide process. It is usually considered to be the former but in some cases, solute 'atmospheres' are thought to exert a significant drag on the glide process, making this the rate-limiting step. The climb-controlled process has been analyzed and leads to a strain-rate equation of the form

$$\dot{\varepsilon} = \frac{\alpha D_{\mathrm{L}} \mu b}{kT} \left(\frac{\sigma}{\mu} \right)^{n} \tag{7.1}$$

where α is a constant, D_{L} is the diffusion coefficient (lattice diffusivity), k is Boltzmann's constant, b is the Burgers vector, T is absolute temperature, σ is the applied stress, μ is the shear modulus and n is the stress exponent. The value of n was in the range $4-5$. For the glide-controlled process, an equation of the same form is obtained but with $n = 3$. Other dislocation creep processes, such as dislocation climb (without glide) or the dissolution of dislocation loops, have been analyzed. These mechanisms also give rise to stress exponents in the range $3-5$.

7.1.2   Stress-induced diffusion

The creep of materials can also occur solely by diffusion, i.e., without the motion of dislocations. Consider a crystal under the action of a combination of tensile and compressive stresses, as shown in Fig. 7.4. The action of these stresses will be to respectively increase and decrease the equilibrium number of vacancies in the vicinity of the boundaries. (The boundaries are acting as sources or sinks for the vacancies.) Thus, if the temperature is high enough to allow significant vacancy diffusion, vacancies will move from boundaries under tension to those under compression. There will, of course, be a counter flow of atoms. As shown in Fig. 7.4, this mass flow gives rise to a permanent strain in the crystal. For lattice diffusion, this mechanism is known as **Nabarro–Herring creep**. The analysis showed that the creep rate $\dot{\varepsilon}$ is given by

$$\dot{\varepsilon} = \frac{\alpha D_{\mathrm{L}} \sigma \Omega}{d^2 kT} \tag{7.2}$$

where α is a constant, Ω is the atomic volume and d is the crystal size. The inverse-square dependence on crystal size makes this mechanism more important for polycrystals than single crystals. For the former, this implies that the creep rate will increase with decreasing grain size (shorter diffusion distance). An interesting feature of Nabarro–Herring creep is the linear dependence between stress and strain rate. Thus, diffusional creep gives rise to equations with the same form as Newton's Law for viscous flow, implying deformation can occur without necking (see Section 5.1).

7.2   Creep in polycrystals

Steady-state creep in polycrystals can involve both dislocation mechanisms (**dislocation creep**) and purely diffusional mechanisms (**diffusional creep**). In this latter category, Nabarro–Herring creep was discussed in the last section and the analysis considered diffusion through the grains. Another possibility in polycrystals is diffusion along grain boundaries and this is termed **Coble creep**. The alternative diffusion paths are shown schematically in Fig. 7.5. Nabarro–Herring creep is controlled by the lattice diffusivity D_L, and Coble creep by the grain boundary diffusivity D_G. In this latter process, it is necessary to define the grain boundary width d_G and it is found that the creep rate depends on d^{-3}. This higher

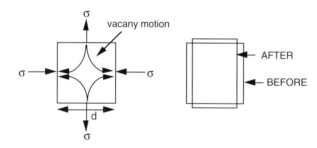

Figure 7.4 Stress-induced lattice diffusion in a single crystal: a) Crystal under the action of tensile and compressive stresses (the arrows in the crystal show directions of vacancy motion); b) Strains produced by the stress-assisted diffusion process.

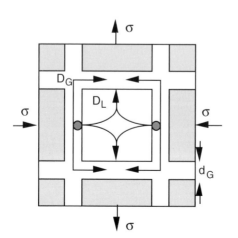

Figure 7.5 In polycrystals, diffusional creep may occur by diffusion through the grains (diffusivity D_L) or along the grain boundaries (diffusivity D_G). The two possible atom diffusion paths are shown by the arrows. (Adapted from Cook and Pharr, 1994, reproduced courtesy of VCH Publishers, Weinheim, Germany.)

Figure 7.6 Transmission electron micrograph showing the presence of a grain boundary glassy phase in silicon nitride. (Courtesy of X. G. Ning and D. S. Wilkinson, McMaster University.)

sensitivity to grain size allows Coble creep to dominate over Nabarro-Herring creep at lower temperatures. For both creep mechanisms, it is assumed that the grain boundaries are perfect sources or sinks for vacancies. It has, however, been suggested by Ashby (see Cook and Pharr, 1994) that these processes may be controlled by a reaction at the interface, leading to **interface (boundary) reaction-controlled diffusional creep**.

A third group of creep mechanisms are those that involve **grain boundary sliding**. Some ceramics possess a grain boundary glassy phase that is used to aid densification in the fabrication process (see Fig. 7.6). The softening of this phase at high temperatures allows creep to occur by grain boundary sliding. It is not, however, the sliding process that controls the creep rate but, rather, the glass viscosity. For large amounts of liquid, the grains could be thought to be 'floating freely' in the matrix. For this case, the material can be considered to have an effective viscosity η_E given by

$$\eta_E = \frac{\eta_0}{\left(1 - \dfrac{V_G}{V_C}\right)^2} \tag{7.3}$$

where η_0 is the liquid viscosity, V_G is the volume fraction of grains and V_C is the percolation threshold (i.e., the critical volume fraction at which a continuous network of grains is formed). Above this threshold, the creep behavior is more complex (see Fig. 7.7). One mechanism that has been analyzed involves a dissolution–precipitation reaction. In this mechanism, grains dissolve into the liquid at points of high stress, and this solute then diffuses through the liquid and precipitates at low stress regions. The creep rate can be stated as

$$\dot{\varepsilon} = \frac{\alpha w \sigma \Omega^{2/3}}{\eta_0 d^3} \qquad\qquad (7.4)$$

where w is the thickness of the glassy layer and Ω is the atomic volume of the diffusing species. An alternative mechanism is to consider the redistribution of the glass by viscous flow. Liquid is 'squeezed' out of boundaries under compression, flowing to those under tension. This problem is similar to that considered in Fig. 5.6 and analyzed as Eq. (5.33). For this mechanism, it has been shown that

$$\dot{\varepsilon} = \frac{\alpha w^3 \sigma}{\eta_0 d^3} \qquad\qquad (7.5)$$

Figure 7.8 shows a high-resolution transmission electron micrograph of the grain boundary liquid phase in silicon nitride. Using such observations, Wilkinson *et al.* (1991) have verified that redistribution of the liquid does occur. Grain boundary sliding can also be coupled with diffusional creep, as shown in Fig. 7.9. As grains elongate in the tensile direction by diffusional flow, adjacent

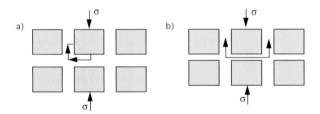

Figure 7.7 Schematic illustration of a) dissolution-precipitation and b) liquid redistribution mechanisms in a material containing a grain boundary glassy phase.

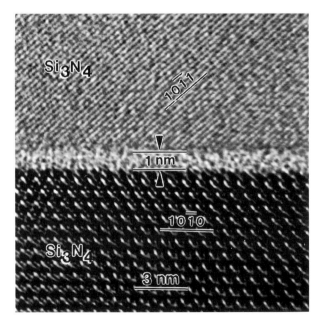

Figure 7.8 High-resolution transmission electron micrograph showing the presence of a grain boundary glassy phase in silicon nitride. These observations have shown the glassy phase changes its thickness during creep. (Courtesy of Q. Jin and D. S. Wilkinson, McMaster University.)

grains must slide relative to each other but diffusional creep can be the rate-controlling step.

In summary, a large number of creep mechanisms have been analyzed and these can be expressed by a single equation, i.e.,

$$\dot{\varepsilon}=\frac{AD\mu b}{kT}\left(\frac{b}{d}\right)^{m}\left(\frac{\sigma}{\mu}\right)^{n} \tag{7.6}$$

where A is a dimensionless constant, D is the diffusion coefficient (diffusivity) associated with the creep process, μ is the shear modulus, b is the Burgers vector, d is the grain size, m is the grain size exponent and n is the stress exponent. The various creep mechanisms give rise to different m and n values and some of these are summarized in Table 7.1. From the analysis of creep data, values of m and n can be obtained and thus, in principle, one can identify the predominant mechanism.

Equation (7.6) is symmetric with respect to the sign of the applied stress, implying the creep behavior should be the same in uniaxial tension or compression. This has been confirmed for several single-phase materials but often breaks down for two-phase materials. For the latter case, cavitation usually occurs more readily in tension. This is thought to be a result of cavity formation at interfaces, so the cavity volume adds to the creep strain. Cavitation is suppressed by compressive stresses and, hence, leads to the stress asymmetry. Figure 7.10 shows evidence of cavitation damage in alumina. It is also possible that contact between grains could be involved in the stress asymmetry. This is because such contacts would be more likely to form under the action of a compressive stress. It has been suggested that the liquid redistribution mechanism might depend on the sign of the applied stress.

In addition to temperature and stress, creep is controlled by the microstructure. Diffusional creep is prevalent for low stresses and at small grain sizes. For these cases, the inverse grain size dependence (Eqs. (7.2) or (7.4)) implies the creep rate can be decreased by increasing grain size. For dislocation creep, the creep rate becomes independent of grain size ($m=0$), e.g., Eq. (7.1). Porosity and second-phase particles can enhance creep by giving rise to stress concentrations. These high stresses, if tensile, also tend to promote cavitation and crack growth.

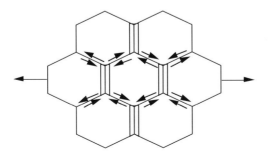

Figure 7.9 An array of grains showing the coupling between grain boundary sliding and diffusional elongation. (Adapted from Cook and Pharr, 1994, reproduced courtesy of VCH Publishers, Weinheim, Germany.)

Table 7.1 *Creep equation exponents and diffusion paths for various creep mechanisms*

Creep mechanism	m	n	Diffusion path
Dislocation creep mechanisms			
Dislocation glide climb, climb controlled	0	4–5	Lattice
Dislocation glide climb, glide controlled	0	3	Lattice
Dissolution of dislocation loops	0	4	Lattice
Dislocation climb without glide	0	3	Lattice
Dislocation climb by pipe diffusion	0	5	Dislocation core
Diffusional creep mechanisms			
Vacancy flow through grains	2	1	Lattice
Vacancy flow along boundaries	3	1	Grain boundary
Interface reaction control	1	2	Lattice/grain boundary
Grain boundary sliding mechanisms			
Sliding with liquid	3	1	Liquid
Sliding without liquid (diffusion control)	2–3	1	Lattice/grain boundary

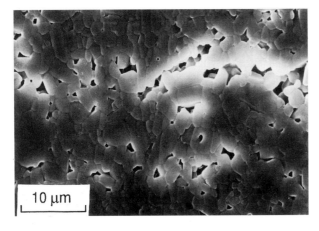

10 µm

Figure 7.10 Scanning electron micrograph showing cavitation damage in alumina after high-temperature deformation. (From D. S. Wilkinson *et al.*, 1991, reproduced courtesy of The American Ceramic Society, Westerville, OH.)

For ceramics that possess a grain boundary glassy phase, the composition of the glass is very important in determining the overall creep rate. In this respect, as a sintering aid in silicon nitride ceramics, Y_2O_3 has been found superior to MgO in improving creep resistance. In some cases, attempts are made to crystallize the glass, thereby increasing the shear resistance. As indicated in Eqs. (7.4) and (7.5) by the dependence on w and d, the liquid volume fraction and grain size are also important in determining the creep rate. Thus, decreasing the amount of liquid or increasing the grain size should decrease the creep rate. It has also been suggested that changing the aspect ratio of the grains could be used to control creep.

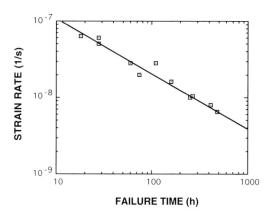

Figure 7.11 Creep rupture behavior of a siliconized silicon carbide. The data were obtained over a range of stresses and temperatures and show a reasonable fit to the Monkman–Grant relationship. (Adapted from Wiederhorn, 1992, reproduced courtesy of CNR/IRTEC, Faenza, Italy.)

Adding ceramic whiskers in volume fractions above the percolation threshold has been found to improve creep resistance, often increasing the creep resistance by two orders of magnitude. One would expect a similar effect with fibers but, in some cases, the fibers have such a small grain size (for high strength) that they can show very poor creep resistance. Other important factors that can affect the creep rate of a material are composition, stoichiometry, defect density and environment, often through their dependence on diffusivity.

The emphasis to this point has been on steady-state creep but clearly information is also needed on creep rupture. For ceramics, an empirical approach suggested by Monkman and Grant is often used (see Wiederhorn, 1992). In this approach the failure time t_f is given as a power function of the steady-state creep rate $\dot{\varepsilon}_s$, i.e.,

$$t_f = C\dot{\varepsilon}_s^{-p} \tag{7.7}$$

where p and C are constants for a given material. If the failure strain is independent of the creep rate $p=1$ and this is found in some metals. For ceramics, however, p is often greater than unity and the failure strain decreases with increasing steady-state creep rate. Figure 7.11 shows a log–log plot of strain rate and failure time for a siliconized silicon carbide. The strain rate exponent, $p=1.45$, was obtained from the slope of the data $(=-p)$. Equation (7.7) suggests that the creep rate is the prime determinant of lifetime and, thus, improvements in creep resistance will also increase lifetime.

7.3 Deformation mechanism maps

For a given material, different creep mechanisms may dominate in different temperature and stress regions. This information can be conveniently given in the form of a **deformation mechanism map**, as shown schematically in Fig. 7.12. This

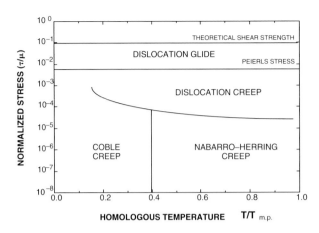

Figure 7.12 Schematic of a creep deformation map for a polycrystalline material. (After Frost and Ashby, 1982.)

particular map is for a constant grain size. In some cases, the map may be given with grain size as a variable at constant temperature. The boundaries in these maps are obtained by equating the creep rate for the various mechanisms. One can also superimpose strain-rate contours on these deformation maps. In order to generate these maps, a large amount of experimental information is required to confirm the various boundaries and strain-rate contours.

At low stresses and temperature, materials are expected to be elastic and some authors include an elastic field on the creep deformation maps. The difficulty with this approach is that such a boundary moves with time. As shown in Fig. 7.12, dislocation glide occurs at the highest stresses. This region is bounded at high stress by the theoretical shear strength (see Section 6.1) and at low stress by the Peierls stress (see Section 6.2). For the creep mechanisms, low stresses favor diffusional creep and high stresses promote dislocation creep. The creep rates associated with the various mechanisms are considered to be **independent processes**, so that the overall creep rate is that associated with the dominant mechanism. From these diagrams, one notes that Coble creep dominates over Nabarro–Herring creep at the lower temperatures (and finer grain sizes). It is important to note that diffusion in ceramics involves more than one ion and, thus, the overall creep rate will depend on the diffusion behavior of the slowest moving ion (**sequential process**). Moreover, it is necessary to consider the various diffusion paths for the ions (lattice or boundary). In MgO, for example, diffusion of oxygen ions through the lattice often controls the steady-state creep rate. As discussed earlier, dislocation climb-glide is a sequential process and may be controlled by either glide or climb, depending on the conditions.

7.4   Measurement of creep mechanisms

The mechanisms in steady-state creep are usually studied experimentally using Eq. (7.6). There are four main variables: strain rate; stress; temperature; and

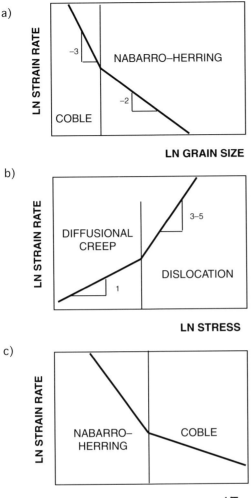

Figure 7.13 Various experimental approaches to determine mechanisms involved in steady-state creep.

grain size. By keeping two parameters constant, one can determine the relationship between the other two, as illustrated in Fig. 7.13. For example, if the strain rate is measured as a function of grain size at constant temperature and stress, the grain size exponent m can be obtained. Taking the (natural) logarithm of Eq. (7.6), the slope of the lines will be equal to $-m$. In Fig. 7.13(a), a transition between Coble and Nabarro–Herring creep is shown, thus the slope changes from -3 to -2. The other two examples in Fig. 7.13 show the other combinations of experimental conditions. For constant temperature and grain size, the slope is the stress exponent n and here a transition from diffusional to dislocation creep is shown (Fig. 7.13(b)). In the final example, grain size and stress are constant. Equation (7.6) includes an exponential dependence on temperature. For example, diffusivity is usually considered to follow the Arrhenius relationship. Thus, Eq. (7.6) can be written as

$$\dot{\varepsilon}=\dot{\varepsilon}_0 \exp\left(-\frac{Q}{kT}\right) \qquad\qquad (7.8)$$

where Q is the activation energy associated with a particular mechanism. On a ln–ln plot, the slope equals $-Q/k$. For the case in Fig. 7.13(c), a transition between Nabarro–Herring and Coble creep is shown.

Experimental creep data for ceramics have been obtained using mainly flexural or uniaxial compression loading modes. Both approaches can present some important difficulties in the interpretation of the data. For example, in uniaxial compression it is very difficult to perform a test without the presence of friction between the sample and the loading rams. This effect causes specimens to 'barrel' and leads to the presence of a non-uniform stress field. As mentioned in Section 4.3, the bend test is statically indeterminate. Thus, the actual stress distribution depends on the (unknown) deformation behavior of the material. Some experimental approaches have been suggested for dealing with this problem. Unfortunately, the situation can become even more intractable if asymmetric creep occurs. This effect will lead to a shift in the neutral axis during deformation. It is now recommended that creep data be obtained in uniaxial tension and more workers are taking this approach.

Problems

7.1   Locate a deformation mechanism map for MgO (grain size 10 μm).
   a) If the maximum allowable creep rate is 10^{-9}/s, determine the maximum normalized shear stress that can be applied to the material at 1000 °C and 1600 °C.
   b) For the same strain rate, name the creep mechanisms occurring at 1000 °C and 1600 °C. If the mechanism is diffusional creep, indicate whether it is Coble or Nabarro–Herring and the type of ion controlling the creep rate.
   c) What is the theoretical shear strength of MgO?
   d) What is the yield shear stress at room temperature at a strain rate of 10^{-2}/s?
   e) What is the maximum creep rate that can be obtained by Coble creep for any temperature and stress?
   f) Estimate the activation energy for O^{2-} diffusion. (Hint: for a single stress, determine the temperatures at which creep rates are 10^{-8} and 10^{-9} /s, $k=8.31$ J/mol K.)
   g) Estimate the creep stress exponent(s) at 1600 °C.
   h) Repeat part a) but assume the grain size of the MgO is now 5 μm.

7.2   The following steady-state creep data were obtained for polycrystalline
        UO_2.

Temperature (°C)	Stress(MPa)	Strain rate (10^{-3} /h)
1666	14	2.0
1666	28	3.1
1666	35	4.0
1666	52	25
1666	78	100
1535	14	0.30
1535	20	0.35
1535	35	0.5
1535	52	4.0
1535	70	10.5
1430	14	0.09
1430	20	0.12
1430	28	0.18
1430	35	0.21
1430	52	1.0
1430	68	3.5
1430	75	4.5

a) Plot the creep rate as a function of applied stress. Which creep mech-
    anisms could be occurring under the experimental conditions?
    Microscopy did not reveal any grain boundary glassy phase.
b) Determine the activation energies for steady-state creep.
c) Predict the creep rate for stresses of 20 and 52 MPa at 1600 °C, if the
    grain size was doubled?

7.3   The following data were gathered for the steady-state creep rate of poly-
        crystalline Al_2O_3. All grain sizes 10 μm except* which are 30 μm.

Temperature (°C)	Stress (MPa)	Strain rate (10^{-4}/h)
1400	10	0.30
1400	10	0.011*
1425	10	0.80
1450	10	1.6
1475	10	3.0
1500	1	0.5
1500	2	1.1
1500	5	2.6
1500	10	5.1

Table (*cont.*)

Temperature (°C)	Stress (MPa)	Strain rate (10^{-4}/h)
1500	10	0.19*
1500	15	8.0
1525	10	11.0
1550	10	24.5
1575	10	54.6
1600	10	121.5
1600	10	13.5*

For steady-state creep, determine:
a) the stress exponent;
b) the grain size exponent;
c) the activation energy; and
d) the probable mechanism(s).
e) Estimate the steady-state creep rate for the same material at 1400 °C with grain sizes of 5 and 20 µm. State any assumptions.

7.4   The following data were gathered by R. T. Tremper *et al.* (1974), for the steady-state creep rate of polycrystalline MgO doped with 2.65 at.% Fe (grain size 14 µm)

Temperature (°C)	Stress (MPa)	Strain rate (10^{-7}/s)
1400	7.2	3.9
1400	14.7	7.8
1400	16.7	8.3
1400	19.1	10.0
1375	14.7	3.9
1350	14.7	2.8
1325	14.7	1.6
1300	14.7	1.1

a) Does the creep behavior agree with a linear viscous law?
b) Determine the activation energy for steady-state creep.
c) Assume creep is controlled by lattice diffusion of Mg^{2+} ions. For an applied stress of 14.7 MPa, plot the steady-state creep rate at 1400 °C as a function of grain size ($1 - 100$ µm).
d) Calculate the diffusion coefficient for Mg^{2+} ions (lattice) at 1400 °C (assume the atomic volume $= 1.86 \times 10^{-29}$ m^3).

7.5.   The following data were gathered for the steady-state creep rate of poly-crystalline Al_2O_3 doped with 1 at.% Fe. All grain sizes 10 µm except * which is 30 µm.

Temperature (°C)	Stress (MPa)	Strain rate (10^{-4}/h)
1400	10	0.3
1425	10	0.8
1450	10	1.6
1475	10	3.0
1500	1	0.5
1500	2	1.1
1500	5	2.6
1500	10	5.1
1500	10	0.6*
1500	15	8.0
1525	10	9.9

For steady-state creep in this material, determine:
a) the probable mechanism; and
b) the activation energy.
c) Estimate the steady-state creep rate at 1550 °C with grain sizes of 5, 10 and 20 μm. State any assumptions.

7.6   At 1750 °C, the steady-state creep rate of Al_2O_3 is approximately four orders of magnitude faster in the polycrystal compared to the single crystal. Explain the source of this difference. Would you expect the activation energy to be the same for both forms of the material?

7.7   Derive an expression for the critical grain size below which Coble creep dominates and above which Nabarro–Herring creep dominates (at constant temperature).

7.8   Creep rupture data for siliconized silicon carbide were fitted to the Monkman–Grant equation to give $p=1.45$. For example, for a strain rate of 2×10^{-8}/s the failure time is 100 hours. Determine the steady-state strain rate that would give a failure time of 10 000 hours. Discuss the problems that might be associated with such an extrapolation. Describe briefly the creep failure observed in this material (need to check references for this chapter).

7.9   Using the creep deformation map for alumina (grain size, 100 μm) given in Fig. 7.14, determine:
a) the shear stress (in MPa) that will give rise to a creep rate of 10^{-10}/s at 1400 °C;
b) the creep mechanism under these conditions;
c) the stresses at which power law creep becomes dominant in the temperature range 1400–2000 °C; and
d) the Peierls–Nabarro stress.

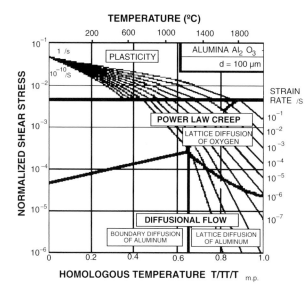

Figure 7.14 Creep deformation map for polycrystalline alumina (grain size, 100 μm). Shaded region represents dynamic recrystallization. (Adapted from Frost and Ashby, 1982, reproduced courtesy of H. J. Frost, Dartmouth College, NH.)

References

M. F. Ashby and D. R. H. Jones, *Engineering Materials 1*, Pergamon Press, Oxford, 1980.

W. R. Cannon and T. G. Langdon, Creep of ceramics: Part 1, *J. Mater. Sci.*, **18** (1983) 1–80, Part 2, *J. Mater. Sci.*, **23** (1988) 1–20.

R. F. Cook and G. M. Pharr, Mechanical properties of ceramics, pp. 339–407 in *Materials Science and Technology*, Vol. 11, edited by M. Swain, VCH Publishers, Germany, 1994.

A. H. Cottrell, *Mechanical Properties of Matter*, J. Wiley and Sons, New York, 1964.

T. H. Courtney, *Mechanical Behavior of Materials*, McGraw-Hill, 1990.

R. W. Davidge, *Mechanical Behaviour of Ceramics*, Cambridge University Press, Cambridge, UK, 1979.

A. G. Evans and T. G. Langdon, Structural ceramics, *Prog. Mater. Sci.*, **21** (1976) 171–441.

H. J. Frost and M. F. Ashby, *Deformation Mechanism Maps, the Plasticity and Creep of Metals and Ceramics*, Pergamon Press, Oxford, UK, 1982.

R. W. Hertzberg, *Deformation and Fracture of Engineering Materials*, J. Wiley and Sons, New York, 1989.

W. D. Kingery, H . K. Bowen and D. R. Uhlmann, *Introduction to Ceramics*, J. Wiley and Sons, New York, 1976.

T. G. Langdon, Grain boundary deformation processes, pp. 101–26 in *Deformation of Ceramic Materials*, edited by R. C. Bradt and R. E. Tressler, Plenum Press, New York, 1975.

A. G. Robertson, D. S. Wilkinson and C. H. Caceres, Creep and creep fracture in hot-pressed alumina, *J. Am. Ceram. Soc.*, **74** (1991) 915–21.

K. L. Rugg and R. E. Tressler, Comparison of the creep behavior of silicon carbide fibers, pp. 27–36 in

Advances in Ceramic-Matrix Composites III, Ceramic Transactions, Vol. 74, edited by N. P. Bansal and J. P. Singh, The American Ceramic Society, Westerville, OH. 1996.

R. T. Tremper, R. A. Giddings, J. D. Hodge and R. S. Gordon, Creep of polycrystalline MgO–FeO–FeO solid solutions, *J. Am. Ceram. Soc.,* **57** (1974) 421–7.

S. M. Wiederhorn, Creep of ceramics, pp. 123–36 in *Introduction to Mechanical Behaviour of Ceramics,* edited by G. de Portu, CNR/IRTEC, Faenza, Italy, 1992.

D. S. Wilkinson, C. H. Caceres and A. G. Robertson, Damage and fracture mechanisms during high-temperature creep in hot-pressed alumina, *J. Am. Ceram. Soc.,* **74** (1991) 922–33.

Chapter 8

Brittle fracture

Fracture has been of importance to humankind since the shaping of primitive tools was first introduced and yet the scientific understanding of this subject has only been developed during this century. Ceramics are prone to brittle fracture, which usually occurs in a rapid and catastrophic manner. Clearly, such behavior is unacceptable for many technological applications of these materials. In response to this challenge, there has been a substantial research effort in the last 25 years aimed at improving the reliability and safety of these materials.

8.1   Theoretical cleavage strength

It is first useful to establish the maximum strength expected from a material, based on the strength of the atomic bonding. Consider two planes of atoms being pulled apart by a tensile stress σ as shown in Fig. 8.1. From consideration of the interatomic potential, the stress is expected to increase initially as a function of the interplanar spacing d. At some point, however, the stress will pass

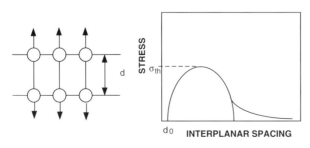

Figure 8.1 Cleavage of atomic planes and the related stress-displacement function.

through a maximum as the interatomic forces are overcome. This maximum stress is termed the **theoretical cleavage stress** σ_{th} and many different approaches have been used to calculate the magnitude of this stress. For our purposes, the stress-displacement function will be assumed to be sinusoidal, so that one can write

$$\sigma = \sigma_{th}\sin\left(\frac{2\pi u}{\lambda}\right) \tag{8.1}$$

where u is the displacement and λ is the wavelength of the sine function. For fracture to occur, one must be able to expend enough energy to create **two** new surfaces. This implies that the energy associated with the overall process must at least equal 2γ, where γ is the surface energy. Thus, one can equate the area under the stress-displacement curve with this energy, i.e.,

$$\int_0^{\lambda/2} \sigma_{th}\sin\left(\frac{2\pi u}{\lambda}\right)du = \frac{\lambda\sigma_{th}}{\pi} = 2\gamma \tag{8.2}$$

In order to determine λ, one recognizes that the material will be elastic at low stresses. For this situation $\sin x \sim x$ and, using Hooke's Law, one can write

$$\sigma = \sigma_{th}\left(\frac{2\pi u}{\lambda}\right) = \frac{Eu}{d_0} \tag{8.3}$$

because the longitudinal strain is given by u/d_0, where d_0 is the equilibrium interplanar spacing (under zero stress). Combining Eqs. (8.2) and (8.3) to eliminate λ, one obtains

$$\sigma_{th} = \sqrt{\frac{E\gamma}{d_0}} \tag{8.4}$$

The theoretical cleavage strength is found to be a material property. Using typical values of γ, Eq. (8.4) can be approximated as $E/10$. More sophisticated theoretical approaches have been used to estimate σ_{th} but the results are very similar. For polycrystalline ceramics, values of E are typically in the range 100–500 GPa. Comparison of the theoretical cleavage stress with experimental data indicates actual strengths are often a factor of 100 or more less. The above discussion is similar to that in Chapter 6 with the calculation of the theoretical shear strength (Section 6.1). Thus, one might similarly expect the low strengths to be related to the presence of a defect in the crystal structure. In pioneering work, Griffith (1920) postulated that materials already contain (pre-existing) cracks and it is the stress concentration associated with these cracks that gives strengths less than σ_{th}. Thus, fracture is seen not as the separation of two perfect crystal planes but as the extension of a pre-existing crack. For small diameter whiskers and fibers, the probability of the presence of such cracks must decrease. Figure 8.2 shows data on the strength of glass fibers from the work of Griffith (see Section 8.3). As the fiber diameter decreases, the strengths approach the theoretical cleavage stress (for silicate glass $E/10 \sim 7$ GPa). Similar observations on other fibers have been made and **fiber composites** have been developed, in part to exploit the high strengths associated with fine-diameter fibers.

8.2   Stress concentrations at cracks

The simplest approach to developing a fracture criterion would be to equate the highest tensile stress associated with a crack to the theoretical cleavage stress. For the geometry shown in Fig. 8.3, the crack could be considered to be an elliptical hole and the elastic solution discussed in Section 4.9 can be utilized, i.e.,

$$\sigma_{\text{tip}} = \sigma\left(1 + 2\sqrt{\frac{c}{\rho}}\right) \approx 2\sigma\sqrt{\frac{c}{\rho}} \tag{8.5}$$

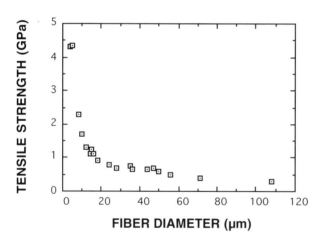

Figure 8.2 Tensile strength of glass fibers as a function of fiber diameter. (After Griffith, 1920.)

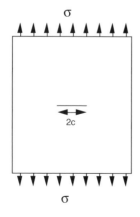

Figure 8.3 An internal crack in a large plate under the action of a uniaxial stress.

where ρ is the radius of curvature at the ends of the major axis of the crack. In most cases, one would expect $\rho << c$ and Eq. (8.5) can be simplified as shown above. If the stress at the crack tip is equated to σ_{th} (Eq. (8.4)), one obtains

$$\sigma_f = \sqrt{\frac{E\gamma\rho}{4cd_0}} \tag{8.6}$$

as a failure criterion. For atomically sharp cracks this approach runs into difficulty, as the stresses at the crack tip would be so high that the stress field is not expected to be linear elastic. Moreover, at these dimensions, treating the material as a continuum must also run into problems and, thus, a different approach is needed.

8.3   The Griffith concept

The work by A. A. Griffith (1920) is generally considered the first breakthrough in developing a sound, scientific basis for fracture. Griffith reasoned that the **free energy** of a cracked body under stress should decrease during crack extension. Consider an isolated system in which a cracked body is being loaded by a set of surface tractions, as shown in Fig. 8.4. The total energy U of the system may be written as

$$U = U_0 + U_E - W_L + U_S \tag{8.7}$$

where U_0 is the elastic energy of the (uncracked) loaded plate, U_E is the change in elastic energy caused by the introduction of the crack, W_L is the work performed by the external forces and U_S is the energy associated with the formation of new surfaces. Utilizing an elastic solution put forward by Inglis (1913) for the loading geometry shown in Fig. 8.3, Griffith was able to show that, for an atomically sharp crack

$$U = \left(\frac{-\pi c^2 \sigma^2}{E}\right) + 4c\gamma \tag{8.8}$$

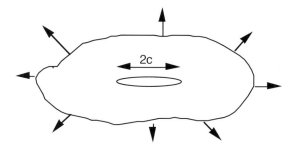

Figure 8.4 A cracked body in an isolated system under the action of applied tractions.

2c

For equilibrium conditions, the term U_S is given by the product of the crack length, the number of surfaces formed (two) and the thermodynamic surface energy per unit area γ, i.e., $4c\gamma$. If the total energy of the system decreases during crack extension, the failure condition is simply given by $dU/dc=0$ and $d^2U/dc^2<0$ (i.e., U is a maximum). For the geometry in Fig. 8.3, the total energy of the system is shown in Fig. 8.5. From Eqs. (8.7) and (8.8), this energy is seen to consist of two parts (Fig. 8.5(a)): the mechanical energy term $(U_M=U_0-W_L+U_E)$ that depends on c^2; and the surface energy term (U_S) that is a linear function of c. When these two parts are combined, the total energy possesses a maximum. Denoting the crack size associated with this maximum as c^*, cracks with sizes $\geq c^*$ should propagate because any increment in crack length decreases the total energy of the system. In many real situations, the stress on a body is increased monotonically. This leads to a series of total energy curves, in which the maximum moves to smaller crack sizes as the stress increases (see Fig. 8.5(b)). For low stresses, the crack size c' is to the left of the maximum, so fracture does not occur. As the stress increases, it reaches its fracture value and c' is then termed the **critical crack size**.

An alternative way of expressing the Griffith approach is to equate the energy differential of the mechanical terms with that associated with the creation of new surface, i.e.,

$$-\frac{d(-W_L+U_E)}{dc}=\frac{dU_S}{dc} \tag{8.9}$$

This alternative will be developed further in Section 8.5. If the failure criterion $dU/dc=0$ is applied to Eq. (8.8), the critical stress for crack extension (the fracture stress σ_f) is obtained, i.e.,

$$\sigma_f=\sqrt{\frac{2E\gamma}{\pi c}} \tag{8.10}$$

a)

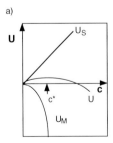

b)

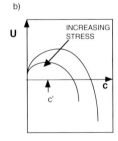

Figure 8.5 The total energy change for body shown in Figure 8.3. In a) the energy is shown as the sum of two parts: the elastic energy decrease that 'drives' the crack (U_M); and the energy needed to create new surface (U_S). In b) the maximum in the total energy moves to smaller crack sizes with increasing applied stress.

This approach indicates that the fracture stress depends on the material para-meters, E and γ and on the crack size c. Equation (8.10) is commonly termed the **Griffith equation**. In order to confirm this equation, Griffith introduced cracks of known size into glass bodies and measured their effect on the strength of the glass. The data obtained by Griffith are shown in Fig. 8.6, plotted to show that fracture stress is inversely proportional to $\sqrt{c}$, as predicted by Eq. (8.10).

It is useful at this point to compare the equation developed by Griffith with the one developed earlier based solely on the maximum stress at a crack tip. For ease of comparison, Eq. (8.6) can be re-arranged into the form

$$\sigma_f = \sqrt{\left(\frac{2E\gamma}{\pi c}\right)}\sqrt{\left(\frac{\pi\rho}{8d_0}\right)} \tag{8.11}$$

For cases where $\sqrt{(\pi\rho/8d_0)}<1$, the fracture stress from this equation is less than that obtained from the Griffith equation. Fracture is not, however, predicted to occur under these conditions because the energy approach is a more global approach. Equation (8.11) will, however, be useful for cases when $\sqrt{(\pi\rho/8d_0)}>1$.

It is interesting to consider a different loading geometry (Fig. 8.7) and some results obtained by Obreimoff (1930) when this geometry was used to cleave mica. Using the Griffith energy balance approach for constant displacement, one can write

$$W_L=0; \quad U_E=\left(\frac{E\delta^3h^2}{8c^3}\right) \tag{8.12}$$

For rigid wedge loading, the work is zero because the loading force suffers no displacement. The elastic energy is estimated using beam-bending theory by

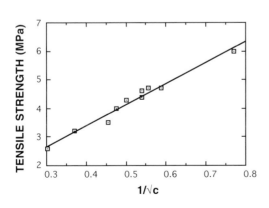

Figure 8.6 The tensile strength of glass is shown to increase linearly with with the inverse of $\sqrt{c}$. (Data from Griffith, 1920.)

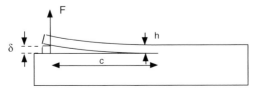

Figure 8.7 Fracture of a mica flake induced by wedging action. (After Obreimoff, 1930.)

assuming the flake being removed is equivalent to a cantilever beam with a fixed support (see Section 4.3). For this case, the total energy is found to possess a minimum and cracks will propagate to the length c_0 associated with this minimum. Setting $dU/dc=0$ to determine the minimum, one finds

$$c_0 = \left(\frac{3E\delta^3 h^2}{16\gamma}\right)^{1/4} \tag{8.13}$$

For this geometry, after the **crack arrest**, failure occurs in a stable fashion as the displacement is further increased. For each displacement increment, the crack will incrementally increase in length and this is termed stable growth. Obreimoff made some interesting observations on the fracture of mica. For example, the value of γ obtained from Eq. (8.13) was found to be dependent on the test environment. Moreover, crack growth was not always instantaneous, which implies kinetic effects may be involved in fracture. Obreimoff also observed that crack growth in mica can be accompanied by a visible electrostatic discharge (**triboluminescence**) and that, although the cracks appeared to heal as the wedge was removed, re-propagation of the crack was easier. These observations indicate that fracture is not really an equilibrium process, as treated by Griffith, but that other **energy dissipation processes** must be present.

8.4   Nucleation and formation of cracks

It has been assumed, to this point, that materials 'inherently' contain cracks. In some cases, cracks may indeed be present, but it is also clear that cracks can form under the action of stress. For ceramics, it is now generally considered that there are a large number of possible microstructural sources for cracks. Thus, one must consider the conditions that lead to a particular type of flaw. For brittle materials, cracks are generally assumed to form by the cleavage of atomic bonds in highly stressed regions. These stresses may be due to stress concentrations or residual stress and they will be particularly effective in producing cracks if weak interfaces are available, i.e., ones that are easy to cleave. The presence of high stresses is generally associated with the heterogeneous nature of the material at the microstructural level or inelastic deformation at localized contacts. For example, voids may be present in a material as a result of the processing. During service, the stress concentration at a sharp corner of these voids may be sufficient to produce a crack. Thus, cracks can be a result of the processed microstructure and may be present before use or they may form during subsequent service. In this latter category, it is known that high stresses can occur in contact events (e.g., impact, erosion, wear, etc.), leading to crack formation in the vicinity of the contact site. Sudden changes in temperature can also lead to stresses, known as **thermal stresses** or **thermal shock** (see Chapter 9). In addition to possessing a variety of **flaw populations** that compete to be fracture origins, the crack sizes within any one population will form a distribution. The fracture stress of a brittle material is, therefore, best considered as a distribution rather than a fixed number. Figure 8.8 shows schematically a possible scenario for a group of flaw populations that could exist in a material. In this example, the most severe flaws are surface cracks, perhaps present from machining. If these flaws can be removed or reduced in severity, the next most severe population is voids and these would become the main failure origins. Finally, if the pores can be reduced in severity, failure from impurity inclusions can dominate. Within this scenario, there is also the possibility of mixed failure origins. The complexity of failure sources in brittle materials has led to empirical statistical approaches in describing strength distributions.

In highly brittle materials, it is useful to identify sources of high stress in the microstructure. For example, stress concentrations will be present if a region exists with a different elastic constant (Section 4.9). Thus, voids and second-phase inclusions are likely sites for crack nucleation. Regions of the micro-structure with a size misfit can lead to microscopic residual stresses. For example, inclusions that undergo a phase transformation or that have a thermal expansion mismatch may also lead to crack nucleation. Anisotropic thermal expansion in polycrystalline materials can also lead to crack nucleation (see Section 2.9). For these microscopic residual stresses, there is a **critical grain size** or **inclusion size** below which microcracking can be suppressed and this will be discussed later. In some ceramics, limited plastic deformation is possible and these materials are considered to be **semi-brittle**. The dislocation motion is usually impeded by grain boundaries or other slip bands and the stress concentration associated with the dislocation pile-up is sufficient to nucleate a crack (see Fig. 6.24). Ductile failure of polycrystalline ceramics, in which cracks do not cause failure, is rare. Once cracks are nucleated in a ceramic, they may undergo further growth in a **formation stage**, especially if there is a localized stress field associated with the nucleation process or if growth is impeded by microstructural obstacles.

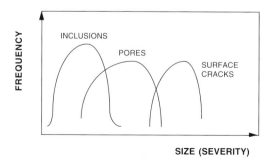

Figure 8.8 Brittle materials contain various types of strength-controlling flaws.

8.5   Linear elastic fracture mechanics

After Griffith's work, it became clear that the theory really applied only to 'ideally' brittle materials. For example, in metals the energy to form fracture surfaces must include a large plastic energy term. Even in ceramics, attempts to measure 'γ' from fracture tests gave values that were generally much higher than the thermodynamic surface energy. The implication is that other energy dissipative mechanisms are occurring when one forms a fracture surface in a ceramic. These may include acoustic emission, heat generation, inelastic deformation or microstructural interactions. It is these last two groups that have been of particular interest to ceramics scientists because, if controlled, materials could be designed in which the resistance to crack propagation would be higher. In any case, it was clear that the Griffith approach was in need of modification.

8.5.1   Energy approach

In the late forties, Irwin suggested the energy balance of Eq. (8.7) could be considered in a slightly different way. In Eq. (8.9), the mechanical energy terms, involving W_L and U_E, were coupled because they involve terms that act to promote crack extension, whereas the surface term U_S was treated separately because it represents the resistance of the material to fracture. One can, therefore, define a parameter

$$G = -\frac{d(-W_L + U_E)}{d(2c)} \tag{8.14}$$

for a sample of unit thickness. This parameter G represents the elastic energy per unit crack area that is available for infinitesimal crack extension. It has been given a variety of names in the literature but in this text it will be termed the **crack extension force** or the **strain energy release rate**. The units of G are those of an energy per unit area but they can also be considered a force per unit length of crack front. Consider an isotropic solid subjected to an arbitrary set of forces, as shown in Fig. 8.9. The energy terms can be seen as a series of steps. From (a) to (b), a crack is introduced into the solid but a set of 'imaginary' tractions is placed on the crack such that the overall stress field is unchanged. The only energy involved is U_S. If the crack surface tractions are now slowly relaxed, an equilibrium crack is formed and some mechanical energy, $U_M = -W_L + U_E$, is released. If the process is reversed, (c) to (b), one sees that this mechanical energy must be uniquely related to the prior stress state in the uncracked body. This process implies that the energy involved in the crack formation is solely related to, and *predetermined* by, the prior stress field.

It is useful at this point to consider the energetics involved when a crack increases its length by a small amount. Consider the loading geometry shown in

Fig. 8.10. The crack can extend by an amount dc under two extreme types of loading: constant load; or constant displacement. In both cases, as the crack extends, the compliance λ of the system increases. It can be shown that the mechanical energy released is the same for both types of loading ($=-F^2\mathrm{d}\lambda/2$). Thus, the value of G can be determined from the change in compliance of a cracked body as the crack extends by a small amount. One can show that for constant load

$$G=\frac{F^2}{2b}\left(\frac{\mathrm{d}\lambda}{\mathrm{d}c}\right) \qquad (8.15)$$

and, for fixed grips (constant displacement)

$$G=\frac{u^2}{2b\lambda^2}\left(\frac{\mathrm{d}\lambda}{\mathrm{d}c}\right) \qquad (8.16)$$

These equations can be shown to be equivalent by setting $u=\lambda F$. Equations (8.15) and (8.16) also allow G to be determined experimentally from compliance measurements. This is accomplished by measuring the compliance of bodies with various crack sizes which allows dλ/dc to be evaluated. It is important to

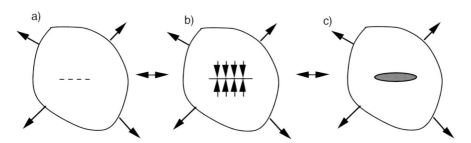

Figure 8.9 The energy changes induced in a body by the introduction of a crack: a) prospective crack site; b) crack introduced but tractions applied to keep original stress field; and c) tractions released.

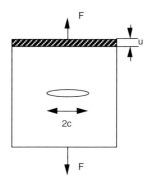

Figure 8.10 The compliance of a cracked body can be defined from the displacement u and force F. The compliance increases as a crack extends.

note that the definition of G does *not* relate to the failure process. It only indicates how the mechanical energy changes *if* a crack were to propagate.

8.5.2   Stress approach

At the same time as the strain energy release rate concept was being developed, it became clear that an alternative approach could be used to describe fracture. This alternative was based on the idea that a variety of elastic problems involving cracks could be solved. All stress fields in the vicinity of a crack can be derived from three modes of loading, which are illustrated in Fig. 8.11. In the linear elastic solutions for all three modes, the stresses σ_{ij} and displacements u_i in the vicinity of a crack tip take the form

$$\sigma_{ij} = \frac{Kf_{ij}(\theta)}{\sqrt{2\pi r}} \tag{8.17}$$

$$u_i = \frac{Kf_i(\theta)}{2E}\sqrt{\frac{r}{2\pi}} \tag{8.18}$$

where r and θ are cylindrical polar coordinates of a point with respect to the crack tip (Fig. 8.12). The parameter K is called the **stress intensity factor** and $f_{ij}(\theta)$ and $f_i(\theta)$ are angular functions. The stress intensity factor is therefore an indicator of the magnitude of the stresses near a crack tip or the *amplitude* of the elastic

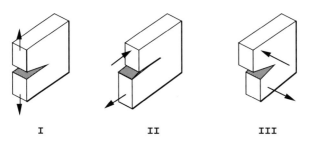

I   II   III

Figure 8.11 The three basic crack loading modes used in linear elastic fracture mechanics: I, uniaxial tensile (opening) mode; II in-plane shear mode; III, out-of-plane shear (tearing) mode.

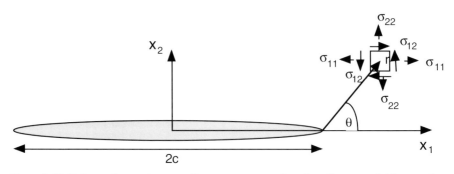

Figure 8.12 Polar and cartesian coordinate systems used to describe stress field around a crack.

field. Equations (8.17) and (8.18) show only the leading term from the general elastic solution but it is the largest term for the region near sharp crack tips. The parameter K is also related to the boundary conditions of the elastic problem. It can be shown for a given mode that

$$K_i = \sigma Y \sqrt{c} \tag{8.19}$$

where Y is a dimensionless parameter that depends on the crack and loading geometries and i represents the loading mode (I, II or III). Solutions for Y have been tabulated for many different types of situations and are available in handbooks. For example, for the geometry considered by Griffith (Fig. 8.3) it can be shown that $Y = \sqrt{\pi}$.

8.5.3 Relation between energy and stress approaches

One might expect that K or the magnitude of the crack tip stresses would be related to the energy-based parameter G. Irwin (1958) was able to show that for plane stress

$$G = \frac{K_I^2}{E} + \frac{K_{II}^2}{E} + \frac{K_{III}^2(1+v)}{E} \tag{8.20}$$

and, for plane strain

$$G = \frac{K_I^2(1-v^2)}{E} + \frac{K_{II}^2(1-v^2)}{E} + \frac{K_{III}^2(1+v)}{E} \tag{8.21}$$

The subscript on K identifies the mode of loading and v is Poisson's ratio. Ceramics generally fail and are tested in mode I loading, and for these conditions the relationship between G and K is given by $G = (K_I^2/E)$ and $G = K_I^2(1-v^2)/E$ for plane stress and plane strain, respectively. To derive Eqs. (8.20) and (8.21), the crack propagation process occurring in Fig. 8.13 is considered. As the crack extends from A to A', the mechanical energy associated with G is released. If, however, the crack could be closed by tractions to return to A, the near-tip stress and displacement fields associated with K would need to be re-constituted. It is

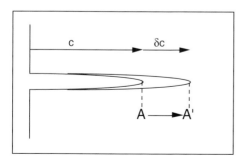

Figure 8.13 Relationship between G and K is found by considering changes in stress–displacement behavior at a crack tip and relating these changes to energy changes.

the integration of the near-tip stresses and displacements that allows the relationship between G and K to be derived.

8.5.4 Failure criterion

Failure is considered to occur when G reaches a critical value (G_C) which is equal to dU_S/dc at the moment of propagation. The term dU_S/dc now contains all the dissipative terms that are characteristic of the formation of the fracture surfaces by crack-tip motion in a particular material. It represents *all* the crack resistance processes available in a material. It is called the **crack resistance (force)** R and, thus, failure occurs when

$$G_C = \frac{dU_S}{dc} = R \tag{8.22}$$

i.e., when the crack *extension* force is equal to the crack *resistance* force. In the ceramics literature, it was popular for some time to keep the formalism similar to Griffith and, instead of R, a term 2Γ was used to replace 2γ. The parameter Γ is termed the **fracture surface energy** and is defined by $2\Gamma = R = dU_S/dc$. It should be noted that there is a factor of two difference between the dU_S/dc term used by Griffith ($4c\Gamma$) and that considered here. This is because G is associated with the (virtual) motion of a single crack tip. For the uniaxial tension geometry considered by Griffith, the failure condition can be written $G_C = \pi \sigma_f^2 c/E = R$.

If fracture occurs at a critical value of G, then this implies fracture occurs at a critical value of K for a given mode of loading. For fracture under pure mode I loading, the failure criterion is given by $K_{IC} = T$, where T is the **fracture toughness** of the material. Thus, the failure criterion can be described in a variety of ways, e.g.,

$$K_{IC} = \sigma_f Y \sqrt{c} = T \tag{8.23}$$

In terms of the fracture stress, failure can be written as

$$\sigma_f = \frac{K_{IC}}{Y\sqrt{c}} = \frac{1}{Y}\sqrt{\frac{E'R}{c}} = \frac{1}{Y}\sqrt{\frac{2E'\Gamma}{c}} \tag{8.24}$$

where $E' = E$ for plane stress, $E' = E/(1-v^2)$ for plane strain and $Y = \sqrt{\pi}$ for the uniaxial loading geometry. To determine the fracture toughness of a material experimentally, the general approach is to introduce a crack of known size and measure the strength of the cracked material. Thus, as long as Y is known for the test geometry, the value of K_{IC} or T can be obtained from Eq. (8.23).

Using the type of formalism outlined above, if the value of T is known for a material, the size of flaw it can tolerate at a given stress can be calculated. Alternatively, if both the strength and fracture toughness are known, it is possi-

ble to calculate the critical flaw size. Another important aspect of the relationship between fracture toughness and strength is that in situations where new
flaws are created, such as in contact damage, the higher the toughness of a
material, the more difficult it is to create flaws.

In the Irwin approach, as with the Griffith approach, strength is found to
depend on a combination of a material property (intrinsic) and a flaw size
(extrinsic). In the linear elastic fracture mechanics approach, however, the
material property is T or R and it has a component that depends on the microstructure of the material. Thus, if the mechanisms that increase T for a material
can be identified, an approach is available to increase the reliability of brittle
materials. It is this philosophy that has been a major driving force in the recent
production of ceramics with higher strengths and toughnesses than had previously been considered possible.

8.5.5 Crack stability

For the uniaxial tension crack geometry (Fig. 8.3), the interplay between G and
R is shown in Fig. 8.14. In this figure, the failure condition is when the G and R
curves intersect. The value of G at this point is G_C, the critical crack extension
force. It was assumed initially that R was independent of crack length. This idea
was prevalent in the early use of fracture mechanics for brittle ceramic materials. Consider Fig. 8.14, in which a body contains a crack of length c_1. If a stress
σ_1 is applied, $G<R$ for this crack length, so crack propagation will not occur. If,
however, a stress σ_2 is applied, $G=R$ for this crack length, initiating fracture.
Moreover, as the crack propagates (i.e., c increases), the crack extension force G
is increasing more rapidly than the crack resistance force R. The crack propagation will therefore be unstable. Another way of stating the failure condition in
Fig. 8.14 is to say that c_1 is the critical crack length for the applied stress σ_2.
Alternatively, one can say that c_1 is the maximum size of crack the body can
tolerate at this stress. For this situation, increasing R increases the critical crack
size.

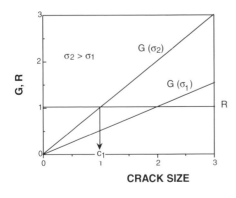

Figure 8.14 Relationship
between G and R for
geometry shown in Fig.
8.3. Failure occurs in an
unstable fashion.

It is worth considering some further, more subtle, details concerning crack stability. The value of G for the geometry described in Fig. 8.3 is linearly proportional to crack length (Fig. 8.14). For other geometries this may not be the case. Consider the crack geometry shown in Fig. 8.7 and used by Obreimoff (1930). From Eq. (8.12), one finds that, for fixed grips, G decreases with increasing crack length. Figure 8.15 shows the relationship between G and R. For this case, the crack propagates to an equilibrium length but immediately arrests as $G<R$ for any further crack increment. For further propagation it is necessary to increase the displacement. As indicated earlier, this causes stable crack growth. Thus, for unstable crack growth, the requirement is that $G=R$ and $dG/dc>dR/dc$. Conversely, if $G=R$ and $dG/dc \leq dR/dc$, the growth is stable. Depending on the shape of the G curve, crack growth behavior in some geometries can become complex with mixed stability regions. An important feature of unstable crack growth in brittle materials is the tendency for the crack to branch. Indeed, this tendency is often used to identify the region in which a dynamic crack initiated its growth, i.e., it is the region prior to first branching.

8.6   Stress intensity factor solutions

Various approaches have been developed to obtain analytical and numerical expressions for the stress intensity factor associated with a wide variety of crack and loading geometries. These solutions are useful not only in developing fracture toughness testing techniques, but also in understanding the interaction of cracks with structure at all scale levels.

8.6.1   Stress intensity factor solutions and test geometries

It is useful to consider some stress intensity factor solutions for basic situations, especially those that relate to typical testing geometries used for fracture tough-

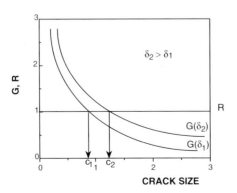

Figure 8.15 For some loading geometries, such as that shown in Fig. 8.7 (fixed grips), failure can be stable since G decreases with crack length.

ness evaluation. In these test geometries, a crack of known size is introduced into a specimen that is subsequently stressed to failure. With respect to Eq. (8.23), one is often considering the expression needed to determine the geometric parameter Y.

(A) Through-thickness internal crack

The geometry given in Fig. 8.3, a crack loaded in uniaxial tension (mode I), has been discussed several times in this chapter. It is also shown in Fig. 8.16(a), emphasizing the presence of infinite boundaries (small crack size). For this geometry

$$Y = \sqrt{\pi} \tag{8.25}$$

(B) Through-thickness surface crack

Cracks are often found at the outside surface of a material. For mode I loading (Fig. 8.16(b)), the solution is expected to be similar to that given in Eq. (8.25). Indeed, it is found that

$$Y = 1.12\sqrt{\pi} \tag{8.26}$$

The presence of the surface gives rise to a **free surface correction factor**. In this case, it is a correction of 12%. The effect of the free surface is to remove some of the constraint on crack opening, thereby increasing the stresses at the crack tip. One should note here that internal cracks are given the dimension $2c$, whereas the surface cracks have a dimension c. The surface crack has strong geometric (and mathematical) similarities to 'one half' of an internal crack.

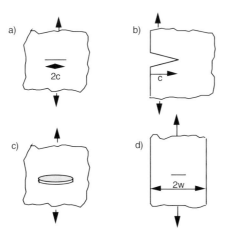

Figure 8.16 Various mode I loading geometries associated with uniaxial tension:
a) 'through-thickness' internal crack in an infinite body;
b) 'through-thickness' surface crack in a semi-infinite body;
c) internal crack with a circular crack front ('penny' crack); and
d) 'through-thickness' internal crack in a finite-width body.

(C) Internal circular ('penny') crack

The crack shown in Fig. 8.16(c) is similar to that in Fig. 8.16(a), except that the crack front is not linear but circular, radius c. For this case

$$Y = \frac{2}{\sqrt{\pi}} \tag{8.27}$$

The effect of crack geometry will be discussed further in Section 8.6.3.

(D) Finite width specimen

To this point, it has been assumed that the external boundaries are infinite or semi-infinite. For the geometry shown in Fig. 8.16(d), this is not the case and the crack front can now 'sense' the boundaries of the body. The crack tip stress field must be 'carried' by a smaller amount of material and, thus, the stress intensity factor is enhanced. The external surfaces can be considered to 'attract' the crack and the geometric factor can be written as

$$Y = \left[\frac{2w}{c} \tan\left(\frac{\pi c}{2w} \right) \right]^{1/2} \tag{8.28}$$

For $c \ll w$, $\tan(\pi c/2w) \sim (\pi c/2w)$, which gives $Y = \sqrt{\pi}$, the same as Eq. (8.25). As the crack size approaches the boundary ($c \to w$), Y increases towards infinity.

(E) Double cantilever beam geometry

A more practical fracture mechanics testing geometry, known as the double cantilever beam (DCB), is shown in Fig. 8.17. It is useful to consider this geometry in some detail, as a simple expression for K can be derived using an elastic beam analysis. It is assumed that the specimen can be treated as two cantilever beams with rigid supports. The deflection of a single beam was derived in Section 4.3. For two beams, one can write

$$2\delta = \frac{2Pc^3}{3EI} \tag{8.29}$$

In terms of the compliance of the specimen, $\lambda = 2\delta/P$, one can write, for beams of rectangular cross-section ($I = bh^3/12$),

$$\lambda = \frac{8c^3}{Ebh^3} \tag{8.30}$$

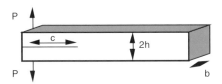

Figure 8.17 Double cantilever beam crack geometry.

If a constant load is applied to the DCB geometry, an expression for G can be obtained using Eq. (8.15), i.e.,

$$G=\frac{P^2}{2b}\left(\frac{d\lambda}{dc}\right)=\frac{12P^2c^2}{Eb^2h^3} \tag{8.31}$$

If one now utilizes Eq. (8.20) for mode I loading only, one can write

$$K_1=\frac{\sqrt{12}Pc}{bh^{3/2}} \tag{8.32}$$

In reality, the cantilever beams are not rigidly supported, but rather they are attached to the rest of the unbroken specimen, i.e., the support is actually elastic. For this situation, one can derive a more exact solution,

$$K_1=\frac{\sqrt{12}Pc}{bh^{3/2}}\left(1+0.64\frac{h}{c}\right) \tag{8.33}$$

The beam height h is usually less than c, and, thus, the expression in Eq. (8.32) is seen to be a good approximation. The DCB specimen can also be loaded using a fixed displacement and for this case G can be determined from Eqs. (8.16) and (8.30)

$$G=\frac{2\delta^2}{\lambda^2b}\left(\frac{d\lambda}{dc}\right)=\frac{3E\delta^2h^3}{4c^4} \tag{8.34}$$

If one compares Eqs. (8.31) and (8.34), one notes that G increases with crack length for constant load but decreases for constant displacement. Thus, crack extension in the DCB geometry can be stable or unstable, depending on the mode of loading. A third variant of the DCB test is to load the cantilever arms using a constant moment arrangement, as depicted in Fig. 8.18. For this case, it has been shown that

$$G=\frac{12M^2}{Eb^2h^3} \tag{8.35}$$

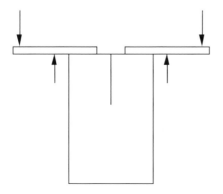

Figure 8.18 Constant moment double cantilever beam crack geometry.

and G is found to be *independent* of crack length. This 'constant K or G' loading mode is sometimes very useful because crack length measurements are not needed. For example, at high temperatures or in aggressive environments it may be difficult to measure crack length.

(F) Single edge-notched beam

Bending geometries are often used in the mechanical testing of ceramics, as they allow simple specimen preparation. For example, Fig. 8.19 shows a three-point bend test with a single edge notch. For this geometry, Y is often given graphically or a polynomial expression is fitted to the numerical data. For the case when $L/h=8$,

$$Y=1.93-3.07\left(\frac{c}{h}\right)+14.53\left(\frac{c}{h}\right)^2-25.11\left(\frac{c}{h}\right)^3+25.8\left(\frac{c}{h}\right)^4 \tag{8.36}$$

For a four-point bending geometry

$$Y=1.99-2.47\left(\frac{c}{h}\right)+12.97\left(\frac{c}{h}\right)^2-23.17\left(\frac{c}{h}\right)^3+24.8\left(\frac{c}{h}\right)^4 \tag{8.37}$$

(G) Chevron-notched beam

An important aspect of fracture toughness testing is that the artificially introduced cracks must be sharp enough to simulate real cracks. This is often accomplished by cyclic fatigue or by pre-loading a machined notch in a testing geometry that promotes stable growth or crack arrest. Another approach is to design the notch so that stable crack growth can occur. For example, Fig. 8.20 shows a chevron-notch geometry. This configuration can replace the straight notch configuration in bend tests. A crack is easily initiated at the tip of the chevron but is arrested by the increasing cross-section. Equations are then available to relate the maximum load to the stress intensity factor at the onset of unstable growth.

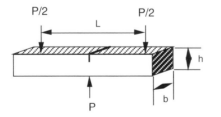

Figure 8.19 Single edge-notched beam geometry. Loading is usually either three- (shown here) or four-point bending.

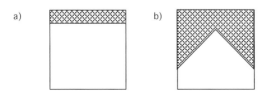

Figure 8.20 Various notch geometries can be used: a) 'through-thickness' crack; and b) chevron notch. The shaded areas are the machined slots.

(H) Double torsion

Another geometry that has been found useful in some situations is the double torsion specimen, which is shown in Fig. 8.21. The crack front is rather complicated in this geometry as the upper surface is in compression. This geometry is, however, a 'constant K' specimen, with K being independent of crack length, i.e.,

$$K_I = Pa\left(\frac{3(1+v)}{bth^3}\right)^{1/2} \tag{8.38}$$

where a is the moment arm and t is the plate thickness in the reduced section. As pointed out earlier, these 'constant K' *loading* geometries are useful when measurement of crack length is difficult.

8.6.2   Fracture toughness measurement

As mentioned in Section 8.6.1, fracture toughness can be determined by introducing an artificial crack into a specimen, which is then loaded to failure. It is important to ensure that the tip of the artificial crack is sharp enough to simulate natural cracks that form in the material. Using the value of Y for the chosen test geometry combined with experimental measurements of strength and crack length, Eq. (8.23) allows K_{IC} to be determined. The theory has been developed for linear elastic materials. For some materials, the high stresses at a crack tip can lead to inelastic processes but if this **process zone** is small compared to the specimen size, linear elastic fracture mechanics can still be utilized. For brittle materials, one expects that $K_{IC}(=T)$ should be independent of crack length. It will be shown later that some microstructures can lead to T values that change with crack length. This can promote stable crack growth even in loading geometries that promote unstable growth. In such materials, it usually becomes necessary to monitor the crack size during the fracture test. Another important aspect of testing is the loading rate. It will be shown later that cracks can grow at values of $K < K_{IC}$, a phenomenon known as **sub-critical crack growth**. This is a kinetic process and, thus, fracture toughness testing is often performed at high loading rates so that the sub-critical crack growth can be neglected. The above remarks

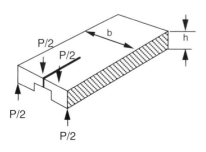

Figure 8.21 Double torsion fracture mechanics test geometry.

assume that there is a single contribution to K from the applied stress. In some cases, however, there may be contributions to K from other sources, such as residual stress (see Section 8.8).

8.6.3   Elliptical cracks

To this point, crack fronts have been assumed to be straight or circular. Another important crack shape is the ellipse and analytical expressions for K are available for such cracks. In this case, Eq. (8.19) can be written as

$$K_1=\sigma\left(\frac{Y}{Z}\right)\sqrt{a} \tag{8.39}$$

where Y is a factor related to the loading geometry, Z is a crack shape factor and $2a$ is the minor axis length. For an internal crack with an elliptical crack front (Fig. 8.22(a))

$$Z=\frac{\Phi(a/c)\sqrt{c}}{[a^2\cos^2\theta+c^2\sin^2\theta]^{1/4}} \tag{8.40}$$

where $\Phi(a/c)$ is the elliptic integral of the second kind, a parameter available in mathematical tables. A useful approximation for $\Phi(a/c)$ is

$$\Phi(a/c)=\left(\frac{3\pi}{8}\right)+\left(\frac{\pi a^2}{8c^2}\right) \tag{8.41}$$

Evaluation of Eq. (8.40) shows that K_1 varies with position along the crack front, being a maximum at the end of the minor axis. For the case of $a=c$, Eq. (8.40) reduces to that for a circular crack (Eq. (8.27)). For surface cracks, the presence of the free surface will influence K_1. For semi-elliptical surface cracks (Fig. 8.22(b)) and quarter-elliptical cracks (Fig. 8.22(c)), numerical solutions are

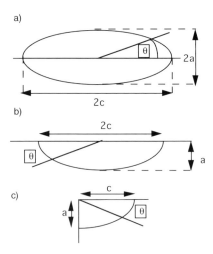

a)

b)

c)

Figure 8.22 Various elliptical crack front configurations: a) internal ellipse; b) surface semi-ellipse; and c) corner quarter-ellipse.

available (e.g., Raju and Newman, 1979). In these cases, the presence of the free surface enhances K in the vicinity of the surface. For shallow semi-elliptical cracks, the correction factor is ~1.12, similar to that given in Eq. (8.26). For quarter-elliptical cracks, the correction factor is ~1.2.

8.7 Methods of determining stress intensity factors

A knowledge of the stress intensity factor associated with a cracked body is critical not only in the safe design of structures but also in understanding the interaction of cracks with the structure of a material. There are many methods to determine stress intensity factors. Figure 8.23 lists these various approaches in terms of the time needed to find a required solution. Many K solutions are now available in handbooks and these can be 'evaluated' in a very short time. Clearly the number of possible geometries of cracked configurations is immense and, thus, it may sometimes be necessary to determine K from first principles. As shown in Fig. 8.23, the time involved for some of these techniques can be long and, moreover, the mathematical tools needed are often complex. There are, however, a set of (short time) techniques that require less effort and, although the solutions

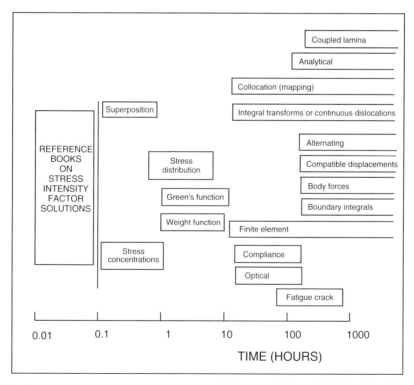

Figure 8.23 Summary of techniques to determine stress intensity factors. (Adapted from Rooke *et al.*, 1981, reproduced courtesy of Elsevier Science Ltd, Kidlington, UK.)

are sometimes only approximate, they give useful insight into the 'behavior' of cracks. It is the aim of this section to review some of the easier techniques.

8.7.1   Superposition

One of the simplest techniques to determine K in a complex configuration is to use **superposition** to 'build' up the solution from a set of simpler and known solutions. Clearly, the precision with which the superposed geometries replicate the final, more complex, structure will impact the accuracy of the final solution. Consider the situation shown in Fig. 8.24, in which cracks emanating from a circular hole is subjected to a biaxial stress. This solution can be broken down into two uniaxial stress solutions, K_a and K_b. Thus, the total stress intensity factor is found by superposition, $K=K_a+K_b$. A somewhat more complex configuration is shown in Fig. 8.25. The problem again involves a cracked circular hole but, in this case, it is being loaded along a semi-circular portion of the hole. The problem is asymmetric but, as shown, it can be found from the superposition of two symmetric solutions, i.e., $K_a=(K_b+K_d)/2$.

One superposition technique that is used extensively is shown in Fig. 8.26. The K solution on the left is found by superposing the two configurations on the right. In the final configuration, the crack surfaces must be free of stress. For the uncracked configuration $K_A=0$, but there is a stress concentration that occurs over the superposed location of the crack. This stress is then removed by the crack surface tractions applied to the cracked body without an external load and, thus, the superposed $K=K_B$.

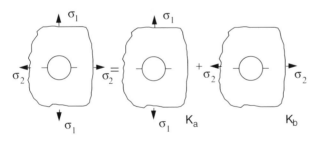

Figure 8.24 Simple example of superposition for biaxial stress from two uniaxial stress solutions. (Adapted from Rooke *et al.*, 1981, reproduced courtesy of Elsevier Science Ltd, Kidlington, UK.)

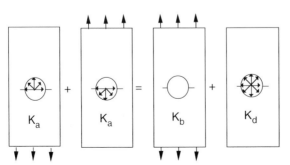

Figure 8.25 Example of superposition for pin-loaded holes in uniaxial tension. (Adapted from Rooke *et al.*, 1981, reproduced courtesy of Elsevier Science Ltd, Kidlington, UK.)

Superposition of K solutions is subjected to the same restrictions as those used for stresses and displacements. For example, the stress intensity factors must be associated with a single loading mode, often mode I, and the body geometry should be the same. An additional restriction is that the crack surfaces must be separated along their entire length in the final configuration. This can be a problem if one of the basic solutions involves compressive stresses that push the crack surfaces together.

8.7.2   Stress concentrations

Elastic solutions are often available for notches with small flank angles and small root radii. In the limit of zero root radius, the notch will be equivalent to a crack, i.e.,

$$K_1 = \text{limit } r \to 0 \ \sigma_{22} \sqrt{2\pi r} \tag{8.42}$$

where σ_{22} is the stress ahead of the notch, perpendicular to the crack surfaces. In terms of the maximum stress σ_{22}^* one can write

$$K_1 = \text{limit } \rho \to 0 \ \frac{\sigma_{22}^*}{2} \sqrt{\pi\rho} \tag{8.43}$$

This approach often utilizes analytical or numerical solutions for σ_{22}^* but one can also use experimental data.

As an example, consider the solution for the maximum stress at the end of an elliptical hole in a large plate under uniaxial tension. Substituting Eq. (8.5) into Eq. (8.43), one finds $K_1 = \sigma\sqrt{(\pi c)}$, which is in agreement with Eq. (8.25). This approach can also be used in an inverse mode to determine the maximum stress at a notch. For example, consider the same elliptical hole but assume it is loaded by a point force as shown in Fig. 8.27. The stress intensity factor for this situation is given as $K_1 = P/\sqrt{(\pi c)}$. Using $a^2 = \rho c$, one finds, from Eq. (8.43), that

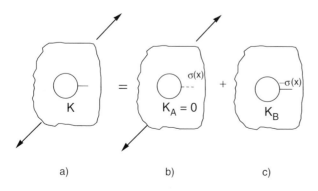

a)   b)   c)

Figure 8.26 Superposition for the equivalence of K for general loading and crack surface tractions: a) unknown crack configuration; b) uncracked configuration; and c) initial configuration with external loading removed. (Adapted from Rooke *et al.*, 1981, reproduced courtesy of Elsevier Science Ltd, Kidlington, UK.)

$\sigma_{22}^*=2P/(\pi a)$. As a final example, consider a single edge notch in a plate under uniaxial tension, as shown in Fig. 8.28. The ordinate of this graph will be equivalent to $K_1/[\sigma\sqrt{(\pi c)}]$ in the limit of $\rho=0$ (K_t is the stress concentration factor, σ_{22}^*/σ). As shown schematically in Fig. 8.28, extrapolation of the data gives an ordinate value of ~1.13, in good agreement with Eq. (8.26).

8.7.3 Stress distributions

As shown in Fig. 8.26 and discussed earlier, if the stress distribution is known in an uncracked body, one can determine K by superposition. If the crack length is small, then, to a first approximation, the stress could be considered constant. Three possible choices for this stress, denoted as σ^*, would be a) the maximum stress, b) the stress at the crack tip location and c) the mean stress. Consider the problem shown in Fig. 8.29. If the surface tractions can be considered constant, the stress intensity factor for a small crack is given as

$$K_1=1.12\sigma^*\sqrt{\pi L} \tag{8.44}$$

The stresses at a circular hole were discussed in Section 4.9 and Eq. (4.53) can be re-arranged to obtain

$$\sigma_{22}=\sigma\left(1+\frac{R^2}{2(R+x)^2}+\frac{3R^4}{2(R+x)^4}\right) \tag{8.45}$$

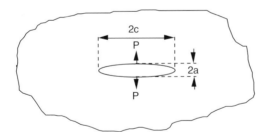

Figure 8.27 Elliptical hole subjected to a central point force.

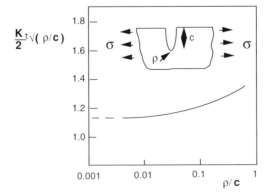

Figure 8.28 Stress concentration factor as a function of notch radius for a single edge-notched beam. (Adapted from Rooke *et al.*, 1981, reproduced courtesy of Elsevier Science Ltd, Kidlington, UK.)

The maximum stress is 3σ at $x=0$. Using this value for σ^* gives $K=3.36\sigma\sqrt{(\pi L)}$ from Eq. (8.44). If the 'crack tip stress' is used for σ^*, one can simply substitute $x=L$ into Eq. (8.45). For the mean stress approach, using

$$\sigma_m = \frac{1}{L}\int_0^L \sigma(x)dx \tag{8.46}$$

one finds

$$\sigma_m = \sigma\left(1 - \frac{R^2}{2L(R+L)} - \frac{R^4}{2L(R+L)^3} + \frac{R}{L}\right) \tag{8.47}$$

The three approaches are compared in Fig. 8.30 and some more exact K calculations for this configuration are included. Of the approximate approaches, the crack tip stress approach gives the most accurate solution. For long cracks, the limiting solution would be $K_1 = \sigma\sqrt{[\pi(R+L)]}$, as there is no need for the surface correction factor. Various procedures have been suggested for interpolating between the long and short crack solutions. For example, it has been proposed that the surface correction factor M can be given by

$$M = 1.0 + 0.12\left(1 - \frac{L}{0.3R}\right) \text{ for } 0 \le L/R \le 0.3, \text{ and}$$
$$M = 1.0 \qquad\qquad \text{ for } L/R \ge 0.3 \tag{8.48}$$

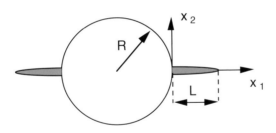

Figure 8.29 Cracks emanating from a circular hole in a large body, which is subjected to uniaxial tension in the x_2 direction.

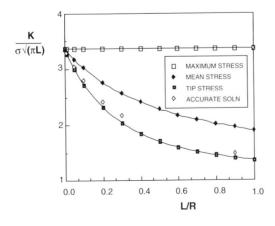

Figure 8.30 Stress distribution approach solutions for cracks emanating from a circular hole.

This allows M to vary between 1.12 and 1 for increasing crack size (compare Eqs. (8.25) and (8.26)).

8.7.4   Green's function approach

In the previous section, the idea of surface tractions in obtaining a superposed K solution was used but the tractions were assumed constant over the whole crack. Clearly, if it was known how to 'weight' the tractions, it would not be necessary to assume a constant stress. One approach to identify this weighting is to use functions derived from the point-force solutions (see Fig. 8.31). These are called **Green's functions**, $g(x)$. For example, the stress intensity factor can be written in terms of the stress σ_{22} along the crack site in the uncracked body,

$$K_1 = \sqrt{\frac{1}{\pi a}} \int_{-a}^{a} \sigma_{22} g(x) dx \tag{8.49}$$

Similar equations can be written for K_{II} and K_{III}. The asymmetric point-force solution for an internal, through-thickness crack (Fig. 8.32) is given as

$$K_{I,\,II,\,III,} = \sqrt{\frac{1}{\pi a}} (P, Q, R) \left(\frac{a \pm b}{a \mp b}\right)^{1/2} \tag{8.50}$$

and, for the symmetric case (Fig. 8.33), as

$$K_{I,\,II,\,III,} = \sqrt{\frac{1}{\pi a}} (P, Q, R) \left(\frac{2a}{\sqrt{a^2 - b^2}}\right) \tag{8.51}$$

In these expressions P, Q and R are the point forces for the various loading modes, $2a$ is the crack size and b is the position of the point force. Allowing b to

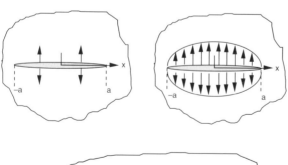

Figure 8.31 If the point-force solution is known (*left*), it can be integrated for more complex distributions (e.g., *right*).

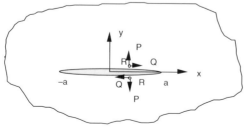

Figure 8.32 Single point-force configuration for an internal crack (asymmetric case).

become the variable ($b=x$), the Green's functions for these two cases can be written

$$g_1(x)=\left(\frac{a+x}{a-x}\right)^{1/2} \text{ and } g_2(x)=\left(\frac{2a}{\sqrt{a^2-x^2}}\right) \tag{8.52}$$

For the case when σ_{22} is constant, it is easy to show that substitution of $g_1(x)$ and $g_2(x)$ into Eq. (8.49) gives $K_1=\sigma\sqrt{(\pi a)}$. Figures 8.34 to 8.36 show the point-force geometry for circular and edge cracks. For a single point force acting on a circular crack

$$K_1=\left(\frac{P}{\pi\sqrt{\pi a}}\right)\frac{\sqrt{a^2-r^2}}{a^2+r^2-2ar\cos\theta} \tag{8.53}$$

For a circular line of forces on a 'penny' crack

$$K_1=\left(\frac{P}{\sqrt{\pi a}}\right)\frac{2r}{\sqrt{a^2-r^2}} \tag{8.54}$$

For an edge crack subjected to a point force

$$K_1=\left(\frac{P}{\sqrt{\pi a}}\right)\frac{2Ma}{\sqrt{a^2-b^2}} \tag{8.55}$$

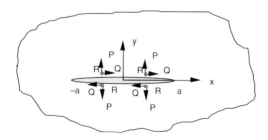

Figure 8.33 Single point-force configuration for an internal crack (symmetric case).

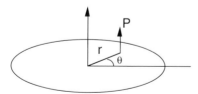

Figure 8.34 Circular crack under action of a single point force. The position of the point force is designated by the coordinates r and θ.

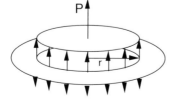

Figure 8.35 Circular crack under the action of a circular line of point forces acting at a distance r from the center of the crack.

where

$$M=1.294-0.6857\left(\frac{b}{a}\right)^2+1.1597\left(\frac{b}{a}\right)^4-1.7627\left(\frac{b}{a}\right)^6+1.5036\left(\frac{b}{a}\right)^8-0.5094\left(\frac{b}{a}\right)^{10}$$

The configuration shown in Fig. 8.35 was used by the present author to determine the stress intensity factor for an annular crack located at the equator of a spherical void (Green, 1980). This geometry was meant to simulate the presence of pores in a linear elastic continuum. The solution is shown schematically in Fig. 8.37. As expected, the solution is similar to the edge crack solution for short cracks ($a/R<0.1$) and the internal circular crack solution (radius ($R+a$)) for long cracks ($a/R>0.5$).

This type of approach has also been used to understand the formation of microcracks at residually stressed inclusions. Figure 8.38 illustrates three possi-

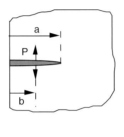

Figure 8.36 Through-thickness edge crack subjected to a point force at a position $x=b$.

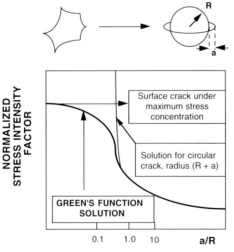

Figure 8.37 Complex void shapes have been modeled by a circular crack emanating from a spherical void. This allows the stress intensity factor to be calculated. For short cracks, the solution approaches that of a surface crack, while for long cracks that for a circular crack of radius ($R+a$).

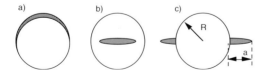

Figure 8.38 Possible microcracking configurations for an inclusion under residual stress: a) interfacial failure; b) inclusion failure; and c) radial cracking.

ble microcracking geometries. The residual stress field for a spherical inclusion was discussed in Section 4.8. For the configurations in Fig. 8.38(a) and (b), the inclusion is under hydrostatic tension and microcracks can form by failure of the interface (see Fig. 3.21) or the inclusion. In the configuration in Fig. 8.38(c), the inclusion is under hydrostatic compression and the maximum tensile stresses are tangential in the matrix. This promotes radial cracking in the matrix; an example is shown in Fig. 8.39. For all three configurations, the stress intensity factors exhibit a maximum. This reflects the localized nature of the stress field, which rises initially with crack length but decreases at long crack length where the stresses are relaxed or diminish to zero. This is shown schematically in Fig. 8.40,

Figure 8.39 Radial cracking at an alumina inclusion in a brittle polymer matrix, observed using a dye penetrant. The inclusion had a lower expansion coefficient than the matrix; width of field 5 mm. (Courtesy of F. F. Lange.)

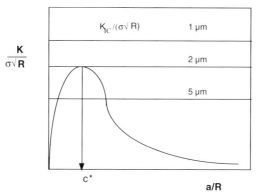

Figure 8.40 The normalized stress intensity factor for cracks at residually stressed inclusions show a maximum value such that cracks cannot form if the inclusion is below a certain (critical) size.

in which the stress intensity factor is normalized by $\sigma\sqrt{R}$, where σ is the stress in the inclusion and R is its radius. The condition for fracture can be found by setting the normalized stress intensity factor equal to $K_{1C}/(\sigma\sqrt{R})$. It is seen immediately that cracks cannot form if the inclusions are below a certain size (2 μm in the example). This behavior is an explanation of the **critical size effect** that is associated with microcracking events. For inclusions above the critical size, the cracking behavior depends on the size of the pre-existing defect that acts as the nucleus for the microcrack. For defects with sizes less than c^*, microcracks will form in an unstable fashion once $K=K_{1C}$. The cracks will arrest in the region where $K<K_{1C}$, as expected for a localized field. The region in which the stress intensity factor decreases with crack length is analogous to that observed for indentations cracks (discussed in Section 8.8). The ability to define the failure conditions at residually stressed inclusions is very useful in understanding and interpreting fracture-initiating events. Evans (1982) has analyzed failure from a wide range of inclusions and other defects in silicon nitride, as shown in Fig. 8.41. An important aspect of these models is that they relate defect size to fracture stress. In the **non-destructive evaluation** of components, defect sizes are often obtained using techniques such as ultrasonic scattering. Once the size is identified, the strength of individual components can be estimated and those possessing strengths below the design stress can be discarded. Figure 8.42 shows the fracture surface of a silicon nitride specimen that failed from a surface crack. The study of the features on a fracture surface is called **fractography** (discussed in Section 8.13).

8.7.5   Weight function approach

Two-dimensional problems concerning a crack of length L in an infinite body subjected to an arbitrary symmetrical loading stress $\sigma(x)$ can be solved using a weight function approach. For this approach, the stress intensity factor K'_1 and the crack face displacements $v'(L, x)$ must be known for another symmetric case. The stress intensity factor K''_1 for the unknown configuration is

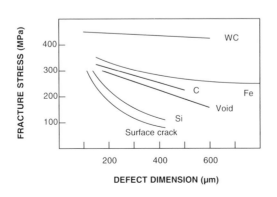

Figure 8.41 Effect of various inclusions on the strength of silicon nitride. (Adapted from Evans, 1982, reproduced courtesy of The American Ceramic Society, Westerville, OH.)

$$K_I'' = \int_0^L \sigma(x)h(L,x)dx \tag{8.56}$$

where $h(L, x)$ is the weight function and is defined as

$$h(L,x) = \frac{8\mu\left(\frac{\partial v'}{\partial L}\right)}{(1+\kappa)K_I'} \tag{8.57}$$

where $\kappa = (3-4v)$ for plane strain, $\kappa = (3-v)/(1+v)$ for plane stress and μ is the shear modulus. The weight function approach is exact provided the correct weight functions are used. As a simple example, consider a crack subjected to an internal pressure $p(x)$. For a crack subjected to a uniform pressure p_0,

$$v(L,x) = \frac{p_0(1+\kappa)}{4\mu}\sqrt{x(L-x)} \tag{8.58}$$

and

$$K_I' = \frac{p_0\sqrt{\pi L}}{\sqrt{2}} \tag{8.59}$$

If one now utilizes Eqs. (8.56) and (8.57), one finds

$$K_I'' = \left(\frac{2}{\pi L}\right)^{1/2} \int_0^L p(x)\left(\frac{x}{L-x}\right)^{1/2} dx \tag{8.60}$$

Figure 8.42 Fracture surface of silicon nitride; failure was initiated from a contact-damage surface crack.

If one uses $L=2a$ and changes the coordinates to the crack center, the expression is equivalent to Eqs. (8.49) and (8.52), the Green's function solution.

8.7.6   Compounding

The geometry of cracked components often contains several boundaries, such as holes, edges, interfaces, etc., and each of these will influence the stress intensity factor. **Compounding** is a type of superposition approach, in which the complex configuration is broken down into N simpler configurations that have known solutions. Each of the simpler geometries usually contains one of the boundaries that is interacting with the crack. The contributions from each of the ancillary configurations is then 'compounded' as follows

$$K_T=\bar{K}+\sum_{n=1}^{N}(K_N-\bar{K})+K_E \tag{8.61}$$

where K_T is the resultant stress intensity factor, K_N is the stress intensity factor for the configuration with only the nth boundary present, $\bar{K}$ is the stress intensity factor in the absence of all boundaries and K_E is the stress intensity factor due to boundary–boundary interactions. Equation (8.61) is often expressed in a normalized form by dividing by $\bar{K}$,

$$Q_T=1+\sum_{n=1}^{N}(Q_N-1)+Q_E \tag{8.62}$$

As an example, consider Fig. 8.43, in which a complex geometry is broken into three simpler configurations for which the stress intensity factor solutions are known. For the first of these configurations, it is found that $Q'=Q''=0.88$. The prime and double prime notation will be used to denote the right and left crack tips. For the second geometry, $Q'=1.201$ and $Q''=1.561$ and for the final geometry $Q'=1.091$ and $Q''=1.056$. For this example, ignoring Q_E, Eq. (8.62) gives $Q'_T=1.172$ and $Q''_T=1.497$. To be more rigorous, Q_E must be evaluated and techniques are available for this purpose.

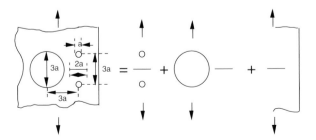

Figure 8.43 Example of the use of compounding to determine the stress intensity factor for a complex configuration. (Adapted from Parker, 1981, reproduced courtesy of Chapman and Hall Publishers, London, UK.)

8.8   **Indentation fracture**

Flaws in ceramics can be a result of contact events and there has been substantial effort in the last 20 years to understand the fracture behavior of these defects. The major scientific approach in simulating the size and nature of these flaws has been to study the morphology and growth behavior of the cracks formed by indenting the surface of a brittle material, usually with a Vickers hardness inden-ter. In this approach, two major types of cracks are formed: cracks perpendicular to the surface; and **lateral cracks** that are approximately parallel to the surface. The former cracks are termed **median–radial cracks** and they are approximately semi-circular. They are associated with the strength degradation caused by the contact event. The lateral cracks are associated with spalling and erosion that can occur during contact. Consider the behavior of the median-radial cracks after indentation and during subsequent loading (Fig. 8.44). At the contact impression, the material is deformed inelastically but outside of this field the material is elastic. This material cannot return to its initial position and, hence, the inelastic material 'wedges open' the radial indentation cracks with a residual stress.

After indentation, the stress intensity factor associated with the residual field can be written as $K_I = \chi P/c^{3/2}$, where P is the indentation load, c the crack size and χ a constant associated with the material and the indenter. If the effect of an applied stress (mode I loading) is considered, there will be two contributions to the total stress intensity factor, which is given by

$$K_I = \frac{\chi P}{c^{3/2}} + \sigma Y \sqrt{c} \tag{8.63}$$

The indentation crack growth behavior is summarized in Fig. 8.45. For $\sigma = 0$, i.e., immediately after indentation, K decreases with crack length and the crack will arrest when $K = T$, the fracture toughness, i.e.,

$$T = \frac{\chi P}{c_0^{3/2}} \tag{8.64}$$

Further crack growth will only occur if the material can undergo sub-critical crack growth (see Section 8.12). Now, consider the effect of the applied stress, as shown in Fig. 8.45. The additional term in Eq. (8.63) leads to a minimum in the total stress intensity factor. The indentation crack, therefore, undergoes stable

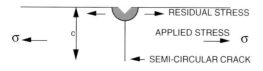

Figure 8.44 Schematic illustration of a semi-circular indentation crack under the action of residual and applied tensile stresses.

growth as the applied stress increases. Once the minimum reaches T, i.e., when c reaches c_m, the condition for final (unstable) failure is obtained. By setting $\sigma=\sigma_f$, $dK/dc=0$ and $K=T$, Eq. (8.63) can be written as

$$\sigma_f = \frac{0.47T^{4/3}}{Y(\chi P)^{1/3}} \tag{8.65}$$

and one can show that

$$c_m \sim 2.5c_0 \tag{8.66}$$

The localized residual stress field stabilizes crack growth prior to the final unstable failure. This contrasts strongly with the type of crack growth envisioned by Griffith (1920), in which the crack size does not change until the final unstable fracture. It has, however, strong similarities with the stable crack growth observed by Obreimoff (1930) (Section 8.3).

The development of indentation fracture mechanics has also allowed fracture toughness to be determined using indentation cracks. Indeed, the ease with which these cracks can be introduced and the simple specimen preparation involved has popularized this approach. Moreover, it has allowed crack behavior to be studied for cracks in a size range that is close to that found in practice. There are two main approaches for determining fracture toughness from indentation cracks. In the first approach, the size of the radial cracks that emanate from the hardness impression are measured. It is recognized that the parameter χ depends on the elastic properties of the indented material. It has been proposed that $\chi=\beta\sqrt{(E/H)}$, where β is a constant that depends only on the indenter geometry and H is hardness. Equation (8.64) can, therefore, be written as

$$T = \frac{\chi P}{c_0^{3/2}} = \beta\left(\frac{E}{H}\right)^{1/2}\left(\frac{P}{c_0^{3/2}}\right) \tag{8.67}$$

The parameter β has been determined from calibration tests and is usually considered to be ~ 0.016.

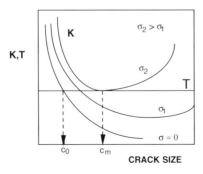

Figure 8.45 Influence of applied stress on growth of an arrested indentation crack, length c_0. The crack undergoes stable growth to a length c_m.

The second indentation approach to measure fracture toughness is to break indented specimens. The median–radial cracks are loaded in mode I and the indentation strength is determined. In some cases, more than one indentation crack is introduced. In this case, failure will occur from one indentation and the other indentations can be used to determine the extent of the stable growth. Equation (8.65) can be re-arranged to the form

$$T = \zeta \left(\frac{E}{H} \right)^{1/8} (\sigma_f P^{1/3})^{3/4} \tag{8.68}$$

The parameter $\zeta = (2.27 \, Y \beta^{1/3})^{3/4}$, which from calibration experiments was found to be ~ 0.59. Using Eq. (8.68), fracture toughness can be determined from the indentation strength, indentation load, Young's modulus and hardness.

8.9 *R* curves

To this point, it has been assumed that crack resistance R and fracture toughness T are independent of crack length. Recent work has, however, shown this is not necessarily the case. In some materials, the energy dissipation rate during fracture can increase as the crack extends. Thus, R and T are no longer constant but, rather, they may increase with increasing crack length. Consider the hypothetical *R* curve shown in Fig. 8.46 for a material loaded in mode I (uniaxial tension). The crack growth behavior becomes substantially more complex. For example, if the material contains a crack of length c_1, the value of $G(\sigma_1)$ is such that unstable crack extension occurs until $G(\sigma_1)$ becomes less than R at c_2. The crack then undergoes a period of stable crack growth as G is increased. This proceeds until at a value $G(\sigma_2)$ and crack length c_3, unstable fracture occurs because $G = R$ *and* $dG(\sigma_2)/dc = dR/dc$. If the material contains cracks smaller than c_0 or greater than c_3, failure will occur in an unstable manner. For the intermediate crack sizes, a period of stable growth occurs until unstable failure occurs at a single value of G and crack size. This implies that cracks with sizes in this range

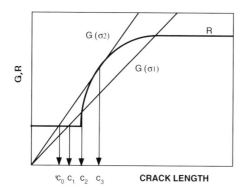

Figure 8.46 A rising crack resistance (R) curve gives rise to a region of stable crack growth and flaw insensitivity.

will all fail at the same stress, i.e., the strength becomes *independent* of initial crack size. This type of failure is clearly very different from that expressed by Griffith, in which strength varies inversely with $\sqrt{c}$. This **flaw insensitivity** is shown in Fig. 8.47, in which the effect of the rising R curve (Fig 8.46) on strength is shown. The R-curve behavior shown in Fig. 8.46 is associated with the absolute crack size. The rising R-curve effect could occur, for example, if a crack is initiated within a grain of a polycrystal but encounters greater crack resistance as it travels through the polycrystalline boundaries. In other materials, the rising R-curve behavior is associated with crack extension and with events occurring behind the crack tip. In order to measure R curves in materials, testing geometries in which G decreases with crack length are very important as this leads to stable crack growth. For example, if a fixed grip DCB test was used, one would be able to determine the complete R curve. In a uniaxial tension test, however, only a portion of the R curve would be measured on a single specimen, such as the region from c_2 to c_3 in Fig. 8.46.

Figure 8.48 shows an example of the importance of the shape of the R curve in determining the failure condition. One material has a higher fracture toughness than the other for long cracks but because the low-toughness material has an R curve that rises more steeply, failure occurs at a higher applied stress. For

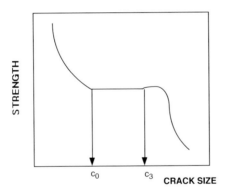

Figure 8.47 The rising R-curve behavior shown in Fig. 8.46 leads to a region in which strength is insensitive to crack size.

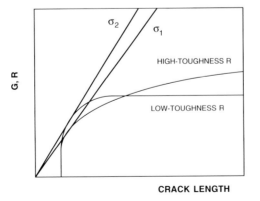

Figure 8.48 The rising R curve no longer implies there is necessarily a correlation between strength and toughness.

many years, it was assumed that increasing the toughness of a brittle material would increase strength if the flaw size remained unchanged. With this example, however, it is seen that the material with the lower toughness has the higher strength.

8.10 Mixed mode fracture

Brittle failure appears to occur almost exclusively in mode I loading but it is clear that cracks that initiate fracture may be oriented such that other loading modes are present. For example, consider the crack shown in Fig. 8.49. The applied stress can be resolved into the normal and shear components using Eqs. (2.42) and (2.44) with $\theta = \pi + \alpha$, thus

$$K_I = \sigma_1 \sqrt{\pi c} \sin^2\theta \text{ and } K_{II} = \sigma_1 \sqrt{\pi c} \sin\theta\cos\theta \tag{8.69}$$

Various theories have been put forward to define the failure condition and the direction of crack propagation. For example, if failure occurs at a critical value of G, one can use Eq. (8.20) (plane stress) to write

$$EG_C = K_I^2 + K_{II}^2 \tag{8.70}$$

Recognizing that at $\theta = 90°$, $K_{II} = 0$ and hence, $EG_C = K_{IC}^2$, Eq. (8.70) can be written as

$$\frac{K_I^2}{K_{IC}^2} + \frac{K_{II}^2}{K_{IC}^2} = 1 \tag{8.71}$$

For some materials, $K_{IC} \neq K_{IIC}$ and Eq. (8.71) is sometimes modified as

$$\frac{K_I^2}{K_{IC}^2} + \frac{K_{II}^2}{K_{IIC}^2} = 1 \tag{8.72}$$

The above approach does not consider the crack propagation direction. Observations have shown a mixed mode crack will *not* continue to propagate in the same plane. Indeed, a mixed mode crack is found to propagate at an angle to

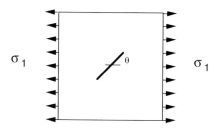

Figure 8.49 A crack under the action of mixed mode loading (I and II).

Table 8.1 *Some typical fracture toughness values for brittle materials*

Material	T (MPa $\sqrt{m}$)
Silicate glasses	0.6–0.8
Barium titanate	1
Lead zirconium titanate	1.5
Alumina	2–6
Silicon carbide	2–4
Silicon nitride	3–7
Zirconia ceramics	5–35
Tungsten carbide/cobalt	10–20

the crack plane and eventually the crack tip region becomes pure mode I loading. Two theories have been proposed to determine not only the failure condition but also the direction of crack propagation. These theories utilize the **maximum principal stress** and the **strain energy density** as their respective failure criterion. The two approaches give similar results and are often difficult to distinguish using experimental data.

The above discussion considered cracks from a macroscopic viewpoint but even for a pure mode I loaded crack, cracks are often deflected at a microstructural level. This crack deflection is a result of variations in the fracture toughness within real materials and the influence of localized stress fields. These observations imply that mixed mode effects can play a role even in a failure that appears to be a result of pure mode I loading, i.e., with the crack propagating normal to the maximum stress. Crack deflection on the microstructural scale are discussed further in the next section.

8.11   Microstructural aspects of crack propagation

It is essential to understand the mechanisms that can occur during fracture and relate these to the toughness of a material. Some of these mechanisms can involve the microstructure of a material. Thus, it follows that manipulation of the microstructure can be used to enhance toughness and develop new materials. The fracture toughness of ceramics was first evaluated in the late 1960s. Values of T were typically less than 5 MPa$\sqrt{m}$ but, as shown in Table 8.1, some materials have now been developed with much higher fracture toughness values. Toughening mechanisms in ceramics are sometimes classified into three groups, viz., crack tip interactions; crack tip shielding; and crack bridging, and these are discussed below.

8.11.1 Crack tip interactions

The primary aim of this type of toughening mechanism is to place obstacles in the crack path to impede crack motion. The obstacles could be second-phase particles, fibers, whiskers or, possibly, regions that are simply difficult to cleave. It is expected that stress concentrations or residual stresses associated with such obstacles would play a role in this process. There are two different consequences that can occur if a crack front is impeded by an array of obstacles. In one case, although the crack is pinned by the obstacles, it can by-pass them by a **crack bowing** process. In this process, shown schematically in Fig. 8.50, the cracks remains on virtually the same plane. An example of crack bowing is shown in Fig. 8.51. An alternative way to by-pass obstacles is by a **crack deflection** process. The deflection of the crack tip can be accomplished by tilting of the crack path or twisting of the crack front (Fig. 8.52). In a real situation, a combination of bowing and deflection may occur, but these will be discussed separately below. Figure 8.53 shows an example of crack deflection in a composite material.

(A) Crack bowing

This mechanism has been analyzed theoretically and various calculations have shown that crack bowing should lead to an increase in fracture toughness. For

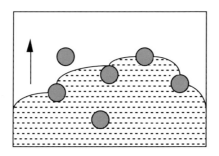

Figure 8.50 Schematic of crack front bowing, caused by interaction with tough obstacles. The shaded region is the cracked area.

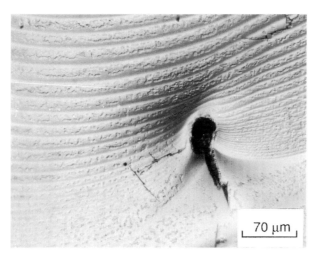

70 μm

Figure 8.51 Interaction of a crack front with an inclusion in glass showing crack bowing. The marks on the fracture surface were produced by ultrasonic fractography (frequency 1 MHz) and reflect successive crack front positions at 1 μs intervals.

example, Lange (1970) used a line tension analogy from dislocation theory to
suggest that

$$G_C = 2\left(\Gamma + \frac{T_L}{d}\right) \tag{8.73}$$

where Γ is the matrix fracture surface energy, d is the obstacle spacing and T_L is
the line tension of the crack front. The analogy between a dislocation line and a
crack front is tempting but, in reality, the stress fields have substantially different

a)

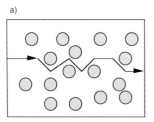

b)

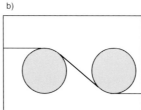

Figure 8.52 Schematic of crack deflection. In a) the crack path tilts to avoid obstacles and in b) the crack front twists to by-pass obstacles.

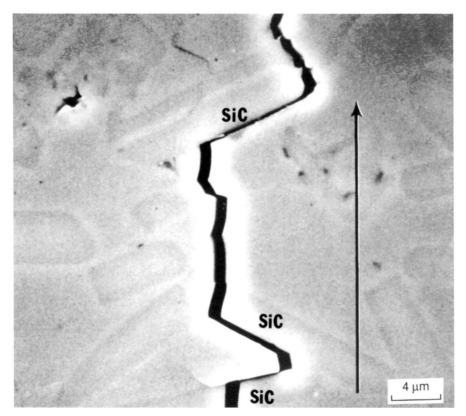

Figure 8.53 Crack deflection caused by SiC platelets in an alumina matrix; scanning electron micrograph. (From Chou and Green, 1993, reproduced courtesy of The American
Ceramic Society, Westerville, OH.)

forms and this makes the analogy rather tenuous. This analysis of the crack bowing mechanism assumed the obstacles were impenetrable but again, in reality, one would expect the strength and toughness of such obstacles to be a key issue. For example, the obstacle could fail before the bowing process is complete or the obstacles would be left behind as unbroken ligaments behind the crack tip. In this latter case, crack bowing becomes a precursor to the crack bridging mechanism (see Section 8.11.3). For the case of obstacle failure, the theoretical calculations will tend to overestimate the toughening contribution from crack bowing.

(B) Crack deflection

If a crack is deflected out of the plane that is normal to an applied uniaxial tensile stress, the crack is no longer loaded in a simple mode I and is therefore not subjected to the maximum tensile stress. Two types of deflection have been analyzed: tilting of the crack about an axis parallel to the crack front; and twisting about an axis normal to the crack front. The overall deflection process is manifested as roughness of the final fracture surface. The change in orientation of the crack plane during deflection leads to a reduction of the crack extension force. The crack deflection mechanism was analyzed by Faber and Evans (1983) by evaluating the mixed mode stress intensity factors on the deflected portion. The various components of K were then combined with a failure criterion and compared to the crack resistance of the matrix surrounding the obstacle. The fracture mechanics analysis demonstrated that the twist component in deflection contributes most to the fracture toughness. For a random array of obstacles, it was shown that the toughening increment depends on the volume fraction and shape of the particles. Figure 8.54 shows schematically the predicted differences between rod, disc and sphere-shaped obstacles. The toughening predicted for rod-shaped obstacles is shown in Fig. 8.55, indicating that rods with large aspect ratios impart maximum toughness. This aspect ratio effect is a result of the increase in twist angle that occurs with increasing aspect ratio. The majority of the toughening from crack deflection appears to develop for volume fractions of

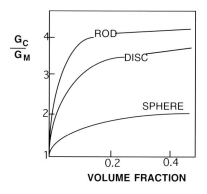

Figure 8.54 Toughening by crack deflection is estimated to increase as the obstacle shapes change from spheres, discs and rods (G_C is crack resistance force of composite, G_M is crack resistance force of matrix). (After Faber and Evans, 1983.)

obstacles <0.2. One of the difficulties with the crack deflection analysis is that it does not consider the process by which the deflection occurs. The local stress field at the obstacle will probably play a large role in the deflection process. Deflection could also be the result of the presence of a low-toughness interface or cleavage plane, an effect that is not included in the theoretical analysis. It is certainly dangerous to assume that an increase in crack deflection necessarily implies an increase in fracture toughness.

8.11.2 Crack tip shielding

The stresses near a crack tip in a linear elastic material are related to the applied stress intensity factor K_I^A. In some materials, non-linear deformation behavior may occur in the high-stress zone at the crack tip. The effect of this **process zone** will be to change the stresses at the crack tip. These stresses can often described by a *local* stress intensity factor K_I^L. If $K_I^L < K_I^A$, the stresses are reduced and the process zone is said to shield the crack tip from the applied loads. Figure 8.56 shows a process zone in which deformation is irreversible and, hence, remains in the wake of a moving crack. The change in stress intensity factor $\Delta K = K_I^L - K_I^A$ and, thus, shielding occurs if $\Delta K < 0$. Failure will occur when $K_I^L = T_0$, the fracture toughness of the material in the process zone. The measured toughness, however, is that determined by a critical value of the *applied* stress intensity factor, i.e., when $K_I^A = K_{IC}^A$. From the above discussion, it is clear that

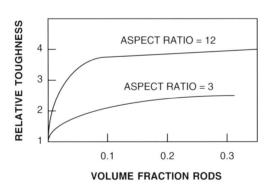

Figure 8.55 Toughening by crack deflection by rods is estimated to increase as the aspect ratio of the rods increases. (After Faber and Evans, 1983.)

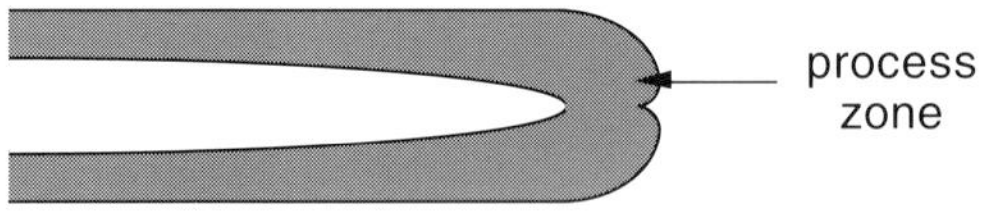

Figure 8.56 In crack tip shielding, various processes can be activated by the high stresses near the crack tip and this process zone can be left as a 'wake' behind the crack tip. If this process reduces the stresses at the crack tip, the zone 'shields' the crack tip.

$K_{IC}^A = T_0 + \Delta T$, where ΔT is the value of $-\Delta K$ at the critical condition. Two mechanisms that can give rise to crack tip shielding are **transformation toughening** and **stress-induced microcracking**.

(A) Transformation toughening

This mechanism could be exploited in many materials but the majority of the work has concentrated on using zirconia (ZrO_2) as the toughening agent. There is a well-known martensitic phase transformation in which tetragonal (t-) ZrO_2 transforms to monoclinic (m-) ZrO_2. The transformation (t-$ZrO_2 \rightarrow$ m-ZrO_2) involves a large shear strain ($\sim 7\%$) and a large volumetric strain ($\sim 4\%$). The tetragonal phase is usually found at high temperatures but, in some cases, it can be retained at low temperatures. The retention is primarily a result of the constraint on a zirconia particle that is produced by the surrounding material, as shown schematically in Fig. 8.57. If the transformation proceeds (a to b), a large amount of strain energy is produced in the surrounding material and this acts to 'oppose' the transformation. It has been shown that the zirconia particle must be below a *critical size* for retention to occur. The effect of particle size on the transformation temperature (during cooling) is shown schematically in Fig. 8.58. The ease of retention of the tetragonal phase is found to depend on the matrix surrounding the zirconia particles or precipitates. To be retained at room temperature, the zirconia particle size is typically $<0.5\ \mu m$.

Under the action of mechanical stresses, the transformation of the tetragonal phase can proceed (a to c in Fig. 8.57). Such a process can occur at crack tips in materials in which the t-ZrO_2 has been retained below the equilibrium transformation temperature. Indeed, using a variety of techniques, such as transmission electron microscopy (TEM), x-ray diffraction and Raman microprobe analysis, it has been confirmed that some or all of the t-ZrO_2 can be transformed to the monoclinic phase in the vicinity of the fracture surface. Figure 8.59 shows microscopic evidence of a transformation zone around a crack.

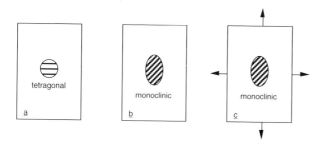

Figure 8.57 The transformation from tetragonal to monoclinic ZrO_2 involves large strains (a to b), which creates elastic energy that will oppose the transformation. This allows the tetragonal phase to be retained to temperatures below its 'unconstrained' transformation temperature. The transformation can be induced by the application of stress (a to c).

Several approaches have been taken to calculating the toughening associated with a stress-induced phase transformation and there is agreement between these various analyses. The simplest calculations assume the phase transformation is purely dilatational and that it is nucleated by a critical value of the mean stress at a crack tip. In a material that contains a crack, the transformation will occur in a frontal zone at the crack tip, as shown schematically in Fig. 8.60. For this case, the transformation is found *not* to increase toughness. Toughening only

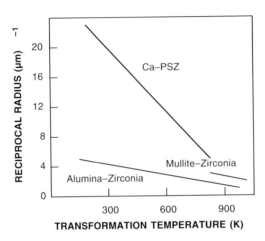

Figure 8.58 Critical particle size for the retention of tetragonal zirconia in various material systems. (PSZ is partially stabilized zirconia). (Adapted from Green *et al.*, 1989, reproduced courtesy of CRC Press, Boca Raton, FL.)

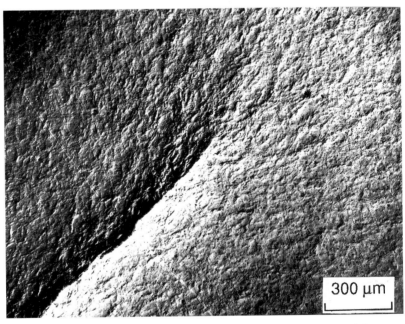

Figure 8.59 Optical micrograph showing the transformation zone around a crack in a partially stabilized zirconia. Nomarski interference is used to provide the contrast which is caused by the surface uplift of the transformed material. (From D. B. Marshall *et al.*, 1990, reproduced courtesy of The American Ceramic Society, Westerville, OH.)

occurs as the zone is left behind the crack tip (see Fig. 8.60), approaching an asymptotic limit, ΔT_{max}, as the crack extends. It follows, therefore, that the process of transformation toughening gives rise to a rising R curve (or T curve). The effect of the frontal zone on toughness is dependent on the particular zone shape. Other zone shapes may increase or decrease toughness and this will be related to parameters that allow the transformation to proceed. The process zone that is left in the crack wake is in a state of residual compression because the zone dilatation is constrained by the surrounding (non-transformed) material. The maximum toughness increment can be written as

$$\Delta T_{max} = \left(\frac{CEV_v^t e^t \sqrt{h}}{1-v}\right) \tag{8.74}$$

where C is a constant that depends on the assumptions in the analysis, E Young's modulus, V_v^t the volume fraction of zirconia that transforms, e^t the volumetric strain associated with the transformation, $2h$ the width of the transformation zone and v Poisson's ratio. Some examples of the calculated C values are given in Table 8.2. In the above discussion, it was assumed that the transformation was irreversible but it has been shown that toughening can occur, under some circumstances, even if the transformation is reversible.

Equation (8.74) is clearly very useful not only in understanding transformation toughening but also in designing materials for maximum toughness. The main difficulty with this approach is identifying the structural parameters that control zone size h. Moreover, there may be some interplay between the parameters in Eq. (8.74). For example, placing zirconia particles in a high-modulus matrix may increase E but decrease h, by making the transformation more difficult. It is worth considering further some of the parameters that control zone size.

The size of the zirconia particles is very important in controlling the extent of the stress-induced transformation and the effect is shown schematically in Fig. 8.61. At small particle sizes compared to the critical size d_c, a very high critical stress is needed to induce the transformation and thus h and the toughening are

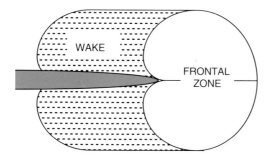

Figure 8.60 The zirconia transformation in a frontal zone does not lead to increases in fracture toughness if the transformation is controlled by the mean stress. The major toughening occurs as the transformation zone passes into the crack wake, with the zone being left in residual compression.

Table 8.2 *Results of steady-state toughening calculations*

(After Evans and Cannon, 1986, reproduced courtesy of Elsevier Press, Kidlington, UK.)

Net strain coupling	Zone shape	Toughness constant, C
Dilation	Hydrostatic	0.22
Dilation	Hydrostatic	0.21
Dilation	Shear band profile	0.38
Uniaxial dilation	Maximum principal stress	$0.55(1-v)$
Dilation	Relaxed shear front	0.22

small. As the particle size increases, the stress-induced transformation becomes easier and the toughness increases. The maximum toughness would occur if all the particles had a size just below d_c. Clearly, the size distribution of the zirconia particles plays a role in determining the volume fraction that transforms and a narrow size distribution will allow a greater number of particles to transform. Particles larger than d_c will have transformed prior to stressing and, thus, the amount of t-ZrO_2 available for the stress-induced transformation will be decreased.

Thermodynamic calculations indicate that the extent of transformation will increase as the amount of supercooling increases. This implies that the transformation stress will decrease at lower temperatures and, thus, the transformation zone size will be larger. It is, therefore, concluded that the fracture toughness of a transformation-toughened material is expected to *increase* with decreasing test temperature. At some critical temperature T_C, however, it may not be possible to retain the tetragonal phase and, then, the fracture toughness will pass through a maximum, as shown schematically in Fig. 8.62. The amount of undercooling can also be controlled by the presence of solute in the ZrO_2, as this reduces the normal (unconstrained) transformation temperature. Thus, at a given temperature, the amount of undercooling will be decreased as one adds the solute, decreasing the toughening increment. The benefit of adding solute is that it allows the t-ZrO_2 phase to be retained to lower temperatures for a given particle size.

Up to this point, it has been assumed that increasing zone size will be beneficial in terms of toughness. It must, however, be realized that there is a limiting zone size, above which the toughness must *decrease*. For example, if the critical transformation stress is reduced to a value such that the stress-induced phase transformation occurs throughout the specimen, the transformation zone will no longer be constrained. This implies that the zone will no longer be left in residual compression and the crack closure forces will vanish. The overall trend is shown schematically in Fig. 8.63. The stress required to induce the phase transformation decreases with increasing fracture toughness but this stress also acts to limit the maximum strength. If the applied stress approaches the transformation stress, the whole specimen will transform. Thus, strength and toughness values must be limited to the shaded region in Fig. 8.63. Within the shaded region, one expects the initial flaw size and the rising *R*-curve behavior to remain important in determining strength. This behavior is analogous with the transition from localized to general yielding in a metal.

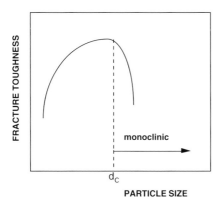

Figure 8.61 Influence of particle size on fracture toughness. The transformation zone increases with increasing particle size but above the critical particle size, it can no longer be retained in the tetragonal phase after processing.

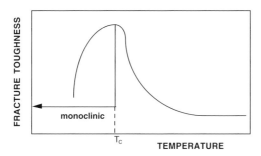

Figure 8.62 Influence of test temperature on fracture toughness. The transformation zone size increases with decreasing temperature, as the tetragonal particles become less stable. At some critical temperature, however, the transformation can become spontaneous.

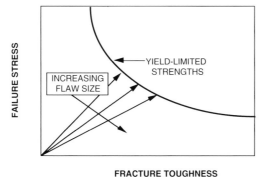

Figure 8.63 The fracture toughness of transformation-toughened ceramics will be limited by the transformation stress. In the shaded region, however, strength will be dependent on flaw size and the shape of the *R* curve.

(B) Microcrack toughening

Stress-induced microcracking has also been shown to give rise to crack tip shielding and this is usually termed **microcrack toughening.** Ceramics that contain localized residual stresses are known to be capable of microcracking. These residual stresses arise in ceramics as a result of phase transformations, thermal expansion anisotropy in single-phase materials and thermal expansion or elastic mismatch in multiphase materials. Regions of low toughness, such as grain boundaries, would also be expected to be attractive sites for such cracks. Microcracks can form spontaneously during the fabrication process if the grain or particle size is above a critical value. This effect was discussed earlier in the chapter (Section 8.7.4). Ceramics containing microcracks after fabrication have been associated with good thermal shock resistance but such materials are expected to have low strengths, as the microcracks are likely failure origins. A more attractive proposition, in terms of fracture toughness, is to fabricate materials with a particle size below that producing spontaneous microcracking but in which microcracks could be *stress-induced*. In terms of fracture, the micro-cracks would be expected to form in a zone around larger cracks, similar to a transformation zone. The creation of a microcrack zone around a propagating crack is expected to reduce the stresses near the crack tip, giving rise to shielding. An alternative view of the toughening phenomenon would be to consider the increased amount of fracture surface associated with the microcrack zone.

The shielding process associated with a microcrack zone is, in some ways, similar to that involved in transformation toughening. In the latter case, the dilatation of the process zone has been shown to be a primary factor in the toughening process. The formation of microcracks is generally associated with residual stress fields and when a crack forms in such a field it will give rise to a volume increase as the microcrack opens. Thus, if the volume increase associated with the microcracking could be predicted as a function of the microcrack density, the toughening increment could be determined from an equation with the same form as Eq. (8.74). There are, however, two other effects that are important in the microcracking process that are different from the transformation-toughening phenomenon. The first effect gives rise to additional shielding and is a result of the decrease in the elastic constants that occurs when a material microcracks. This 'modulus effect' reduces the stresses in the frontal microcracked zone and, thus, aids in the shielding process. The other effect is based on the realization that the microcracks must degrade the fracture toughness within the zone. This is because fracture surface (the microcracks) is already available for the propagation of the main crack. It is sometimes considered that the degradation and modulus effects approximately offset each other and, thus, that the toughening can be considered simply in terms of the dilatation. By analogy with Eq. (8.74), the maximum toughening increment from microcrack toughening can be written as

$$T_{max} = \left(\frac{C E e^M \sqrt{h}}{1-v} \right) \qquad (8.75)$$

where e^M is the zone volumetric strain associated with the microcracking process. The value e^M will depend on the details of the microcracking process but it is usually found to be less than the typical transformation strain. Thus, microcrack toughening is often considered less effective than transformation toughening. Rising R-curve behavior should occur in a microcracking material as $\Delta K \sim 0$ for a frontal zone but, as the microcracks enter the crack wake, dilatation toughening will occur. The relationship between zone size and the critical microcracking stress is, as yet, poorly understood but it will involve the flaw populations present in the residual stress fields and the mechanisms by which they nucleate and form the microcracks.

The overall trends in the fracture toughness increment due to microcracking are very similar to those of transformation toughening. For example, as shown in Fig. 8.64, the toughness is expected to increase with particle size up to the critical particle size for spontaneous microcracking. For larger particle sizes, the material is already microcracked prior to the application of stress and further toughening will only arise if one can increase the microcrack density. The temperature sensitivity could be somewhat different than discussed for transformation toughening, depending on the source of the residual stress. If the microcracks are a result of thermal expansion mismatch, one would expect that increasing temperature would result in a decrease in toughness, since the magnitude of the residual stress (and volumetric strain) will decrease. On the other hand, if the microcracks were formed by a phase transformation, the toughening effect would be relatively insensitive to temperature, unless the transformation reverses. Increasing the volume fraction of the microcracks is clearly attractive but, again, interaction effects may increase the likelihood of spontaneous microcracking. In addition, microcrack linking could be a prime source of pre-existing flaws in the material. The factors that control the size of the microcrack zone are expected to be similar to those that control spontaneous microcracking.

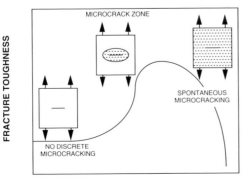

Figure 8.64 Fracture toughness increases with particle size in a microcrack-toughened ceramic unless the size exceeds the critical size for spontaneous microcracking.

These include the magnitude of the residual stress, the particle size and the size (and shape) of defects within the residually stressed regions. Similar to transformation toughening, a toughening limit is expected if the microcrack zone size approaches the specimen size and a 'yield-limit' to the strength will occur.

8.11.3 Crack bridging

In the discussion of crack tip interactions (Section 8.11.1), it was indicated that cracks may by-pass an obstacle, leaving it intact. In such cases, the obstacle is left as a ligament behind the crack tip. Figure 8.65 shows two examples of crack bridging. In one case, the bridges are being pulled out of the matrix. In the other, the bridges are ductile and fail by a plastic deformation process. Frictional contacts may also be formed by the mechanical interlocking of grains. Crack bridges will make it more difficult to open the crack at a given applied stress and will increase fracture toughness. A rising R-curve behavior is also expected if the bridging zone grows with crack extension. It is, however, expected that the bridging zone will reach a limiting size and will then move in conjunction with the crack tip. Some authors consider crack bridging to be a form of shielding process, as the bridging decreases the stress intensity factor at the crack tip. Crack bridging has been observed in frictionally bonded fiber composites, in large-grained polycrystals, in whisker-reinforced ceramics and in cermets.

To determine the effect of crack bridging on fracture toughness, it is necessary to know the stress–displacement relationships $p(u)$ for the bridging ligaments. For partially bridged cracks, the maximum toughening is given by

$$\Delta G_{\mathrm{C}} = 2V_{\mathrm{v}} \int_0^{u^*} p(u)\,\mathrm{d}u \tag{8.76}$$

where $2u^*$ is the critical crack opening displacement associated with bridge failure and V_{v} the volume fraction of bridges. Theoretical approaches similar to Eq. (8.76) are used to describe (atomistic) cohesive or yield zones at a crack tip (see Section 8.15).

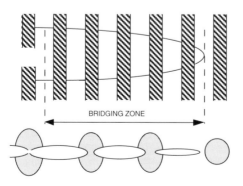

Figure 8.65 Crack bridging, in which unbroken ligaments are left behind the crack tip. The upper figure shows frictional bridges, while the lower figure shows ductile bridges.

BRIDGING ZONE

At a microstructural level, if the bridges are elastic, bridging must be accompanied by some debonding as the crack faces open. For this case, Eq. (8.76) can be written as

$$\Delta G_C = \frac{V_v d\sigma_b^2}{E} \tag{8.77}$$

where d is the obstacle debond length normal to the crack plane and σ_b the bridge strength. For ductile bridges, the toughening is usually expressed as

$$\Delta G_C = \chi V_v R \sigma_Y \tag{8.78}$$

where $2R$ is the bridge width, σ_Y the uniaxial yield stress and χ a 'work-of-rupture' parameter. This latter parameter is determined by the bridge ductility and the amount of debonding. If the bridges remain bonded, χ is approximately unity but for partially debonded bridges, χ may approach a value of 8. In the frictional bridging mechanism, energy must also be expended for debonding and frictional pull-out. Evans (1990) has suggested that Eq. (8.77) can be modified to account for these additional effects using

$$\Delta G_C = \frac{V_v d\sigma_b^2}{E} + \frac{4 V_v d \Gamma_i}{R(1 - V_v)} + \frac{2\tau V_v L^2}{R} \tag{8.79}$$

where Γ_i is the interfacial fracture energy, L the pull-out length and τ the shear resistance of the interface after debonding. Figures 8.66 and 8.67 show SiC whisker bridges in alumina. As the applied stress intensity factor is increased (Fig. 8.67) the bridge is pulled out of the matrix.

An extreme type of crack bridging is the fully bridged crack, shown schematically in Fig. 8.68, and this effect has been observed in some fiber-reinforced com-

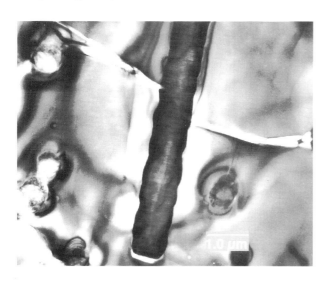

Figure 8.66 Transmission electron micrograph showing a silicon carbide whisker pull-out in alumina. (From P. Becher *et al.*, 1996, reproduced courtesy of The American Ceramic Society, Westerville, OH.)

posites. In this case, a crack passes through the matrix and leaves the fibers intact. This process can then be repeated and **multiple matrix cracking** will occur. For example, the addition of continuous carbon or SiC fibers to glasses or ceramics has been studied since about 1970. In some cases, the fiber composites are found *not* to undergo catastrophic failure in uniaxial tensile loading even though the matrix and the fibers are brittle. This 'ductile' type of behavior for a material composed of two brittle components is particularly attractive for structural applications. In optimum materials, as shown in Fig. 8.69, the tensile loading behavior is initially elastic until, at a particular stress, a crack passes through the matrix. This crack by-passes the fibers and leaves them available for load carrying so that the fibers *completely* bridge the crack. The by-pass process usually involves debonding of the fiber. Further loading causes the formation of regularly spaced matrix cracks until, at the peak load, the fibers fail. The ensuing failure, however, is not necessarily catastrophic as the fibers can continue to pull

2.6   5.4   5.9
Applied Stress Intensity, MPa√m

Figure 8.67 Series of scanning electron micrographs showing a silicon carbide whisker bridge being pulled out of an alumina matrix as the applied stress intensity factor is increased. (From P. Becher *et al.*, 1996, reproduced courtesy of The American Ceramic Society, Westerville, OH.)

a)

b)

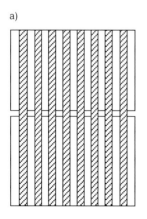
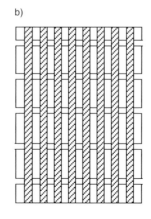

Figure 8.68 Schematic illustration of a fully bridged crack in which a crack passes through the matrix and not the fibers. If this process is repeated as shown in b), multiple matrix cracking is produced.

out of the matrix. In these materials, the final failure is not the result of the propagation of a single crack and, thus, a fracture toughness value cannot be defined. It is, however, possible to use fracture mechanics to describe the conditions at which the first crack passes through the matrix. In this analysis, the matrix cracking stress σ_{mc} is found to be *independent* of the initial flaw size and, thus, is a material property. The matrix cracking stress can be expressed as

$$\sigma_{mc} = \left(\frac{12\Gamma\tau V_f^2 E_f E_c^2}{(1-V_f)E_m^2 R}\right)^{1/3} \tag{8.80}$$

where E_f, E_m and E_c are the Young's moduli of the fiber, matrix and composite, Γ is the fracture energy of the matrix, R is the fiber radius, τ is the interfacial shear resistance and V_f is the volume fraction of fibers. Techniques to increase σ_{mc} are attractive but if this stress approaches that of the fibers then fiber failure will accompany matrix cracking. A transition to a brittle type of failure will then ensue, though it should still involve partial crack bridging. The overall effect of crack bridging in fiber composites can be shown in a normalized plot (Fig. 8.70). The normalization parameters take into account the specific material properties involved in the process. For a fully bridged crack, above a certain crack size the strength becomes independent of crack size, as discussed earlier. Clearly, the

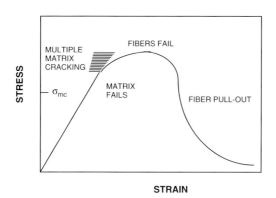

Figure 8.69 For a fully bridged crack, the matrix can undergo multiple cracking and the final failure can involve fiber pull-out. These effects give rise to non-linear stress–strain behavior even though both components are brittle. (σ_{mc} denotes the onset of matrix cracking.)

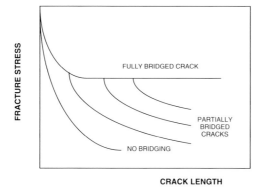

Figure 8.70 Crack bridging leads to a region in which the strength becomes insensitive to crack size.

insensitivity of strength to flaw size is very attractive in practical applications. For partially bridged cracks, the bridging increases strength for a given crack size but the strength insensitivity to crack size occurs over a smaller range. In the absence of bridging, the strength will follow the normal inverse dependence of the square root of crack size. Figure 8.71 shows a fully bridged crack in a glass–ceramic matrix composite.

8.12   Sub-critical crack growth

To this point, it has been assumed that failure occurs when $K_I = T$ (or $G = R$) but, in studies of fracture, it is sometimes found that crack growth can occur at lower values of K_I or G. Thus, kinetic effects must be included in any general formalism. There are several mechanisms that can give rise to sub-critical crack growth, but most attention has been directed to **stress corrosion**. This behavior has been extensively studied in silicate glasses but it can also occur in many polycrystalline ceramics. Figure 8.72 shows a typical response of ceramics to stress corrosion, with crack velocity v plotted as a function of K (or G). At low values of K, there often appears to be a **threshold value** K_I^0 of the stress intensity factor below which crack growth does not occur. At higher values of K, there is a strong sensitivity of v on the value of K. This region (Region I) is usually associated with chemically assisted crack growth and is found to depend on the concentration of the

Figure 8.71 Micrograph showing a fully-bridged matrix crack in a SiC fiber-reinforced glass-ceramic matrix composite; width of field 550 μm. (From D. B. Marshall and A. G. Evans, 1985, reproduced courtesy of The American Ceramic Society, Westerville, OH.)

environmental species that is 'aiding' the growth process. A plateau region (Region II) often follows and this is usually associated with the inability of the reacting species to keep pace with the crack tip motion (transport limited). The final region (Region III) relates to the crack growth behavior in vacuum and is usually associated with thermal activation of bond rupture. Figure 8.73 shows a schematic of a molecular species moving to the crack tip, then reacting and breaking the bonds at the crack tip. Various mechanisms have been analyzed to describe the Region I behavior, including chemical reaction kinetics and interfacial diffusion. At high temperatures, it is also found that **localized creep damage** can give rise to sub-critical crack growth.

For the engineering use of ceramics, the primary emphasis has been on the threshold and Region I behavior. This is because an engineering component will often spend most of its 'lifetime' in these regions. Indeed, it has been become popular to use an empirical power law to describe this behavior for $K \geq K_I^0$, with the crack velocity expressed as

$$v = v_0 \left(\frac{K}{K_{IC}} \right)^n \tag{8.81}$$

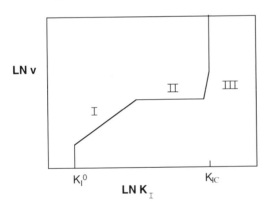

Figure 8.72 Crack velocity during sub-critical crack growth is a function of the stress intensity factor and is considered to show three different regions of behavior.

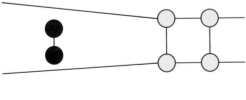

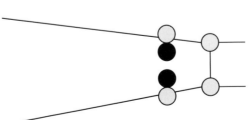

Figure 8.73 Schematic of environmental species reacting at the crack tip and leading to bond breakage.

where n and v_0 are the **sub-critical crack growth parameters**. These parameters will be dependent on the choice of material, environment and temperature. The use of this approach in predicting component lifetime and the evaluation of the sub-critical crack growth parameters will be discussed in the next chapter.

8.13   Fractography

The study of fracture surfaces, **fractography**, is very useful in identifying not only the processes that occur during fracture but also **failure origins**. In brittle materials, fracture surfaces have very distinctive features and the practiced observer can often obtain critical information by this process. Cracks initially propagate on a plane perpendicular to the maximum normal stress. If the failure is unstable, the crack can subsequently branch, as depicted in Fig. 8.74. This can give rise to a large number of fragments from the failure process but careful examination of the branching morphology can often pinpoint the region in which the failure initiated. It should be noted that the stress waves associated with the failure process can give rise to secondary fractures and surface markings, known as **Wallner lines**. Surface markings can also be produced artificially using a technique called ultrasonic fractography. An example of this technique was shown in Fig. 8.51.

Figure 8.75 shows an example of a failure origin in a brittle material. The region around the failure origin is initially rather smooth. In glasses, it is usually highly reflective and is known as the **mirror region**. The same region can be seen in polycrystalline materials but the granular nature of the surface reduces the reflectivity. As the crack progresses, the fracture surface increases in roughness and a less reflective region, known as the **mist region**, is formed. This region is difficult to discern in polycrystalline materials. Finally, the surface becomes extremely rough as the crack branches and striations, known as **hackle**, are formed. The hackle marks usually 'point back' to the failure origin and are, thus, useful in identifying the failure source. The boundaries between these various regions, e.g., mirror–mist, mist–hackle or branching can also be used to obtain *quantitative* information. If the distances from the failure origin to these bound-

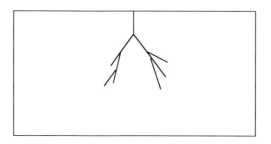

Figure 8.74 Crack initially propagates on a plane but, if failure is unstable, the crack will start branching.

Figure 8.75 Fracture surfaces in brittle materials generally show a smooth region that surrounds the failure origin (mirror region) but the surface increases in roughness as the crack accelerates (mist region) until crack branching occurs. The branched region contains 'ridges' known as hackle. (Optical micrograph courtesy of Matt Chou.)

aries are measured, e.g., the branching distance R_b, a simple relationship has been established between the branching radius and the fracture stress,[†] i.e.,

$$M_b = \sigma_f \sqrt{R_b} \tag{8.82}$$

where M_b is known as the **branching constant**, which is fixed for a given material and microstructure. This equation is particularly useful when the fracture stress is unknown, because knowledge of M_b and R_b allows one to estimate the fracture stress. An equation of the same form as Eq. (8.82) can be used for the mirror dimensions and one can define a **mirror constant** for the material.

Fracture surfaces are often described as to whether a crack passes through grains (**transgranular**) or between grains (**intergranular**). Failure ranges from being exclusively one type to a mixture. Intergranular failure is often associated with the presence of low-toughness grain boundaries or residual tensile stresses across these boundaries. In many cases, sub-critical crack growth is found to be intergranular but changing to transgranular once the crack becomes unstable.

Microscopic features on fracture surfaces are also often useful for determin-

---

[†] Sometimes written as a stress intensity factor, i.e., $K_b = \sigma_f Y \sqrt{R_b}$.

ing crack direction. For example, as a crack passes through a grain it may possess **cleavage steps**, with the crack front jogging from one plane to another. The cleavage steps run in the direction of crack propagation. As a cleavage crack crosses a grain boundary, the change in orientation often increases the density of the cleavage steps. The steps can, however, coalesce to form a **river pattern**, a term used because of the visual similarity with the coalescence of streams into rivers. As cracks by-pass inclusions or pores, a step is often formed at the rear of the obstacle, where the bowing segments re-join. Figure 8.51 shows an example of such a step after the crack has by-passed the inclusion.

It is often critical to determine the source of failure, as approaches can be developed to prevent or remove the source. The failure origin should be located in the center of the mirror region. **Processing defects** are common failure origins and several example are shown in Figs. 8.76 to 8.81. For example, Figs. 8.76, 8.77 and 8.79(a) show pores that acted as failure origins. These pores are sometimes associated with organic impurities that are thermally pyrolyzed in the fabrication process, i.e., during sintering. In two-phase materials, poor mixing of the two components can cause agglomerates to act as failure origins. Two examples, showing failure from alumina agglomerates in alumina–zirconia composites, are given in Fig. 8.78. An example of failure from a zirconia agglomerate was given earlier in Fig. 1.11. Clearly, inclusions picked up as impurities in the processing can also act as failure origins. An example is given in Fig. 8.80, in which a silicon carbide impurity inclusion in silicon nitride acted as a failure origin. For this case, it was concluded that the inclusion would be in residual tension at low tem-

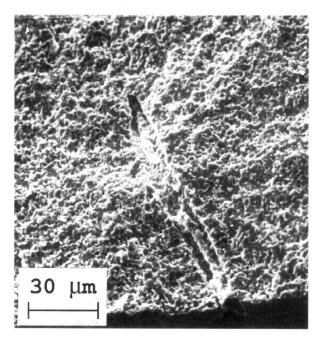

Figure 8.76 Lenticular void that acted as a fracture origin in an alumina-zirconia composite; scanning electron micrograph. (From Green *et al.*, 1989, reproduced courtesy of CRC Press, Boca Raton, FL.)

30 μm

Figure 8.77 Spherical void that acted as a fracture origin in an open-cell vitreous carbon; scanning electron micrograph. (Courtesy of Rasto Brezny.)

peratures and these stresses would aid in crack initiation. **Contact-induced flaws** are also common in brittle materials and usually result from machining damage, impact or simply surface contact. The origin in Fig. 8.42 was found to be a result of surface contact and Fig. 8.79(b) shows failure from a surface flaw in a sapphire (single-crystal alumina) fiber. Chemical surface reactions during service can also lead to defects such as oxidation pits. Figure 8.81 is an example in which the presence of the hackle helped to identify the failure origin. The failure in the glaze was caused by impact but the cracking initiated on the side of the glaze opposite the impact site. This implied the glaze was failing under the action of bending stresses.

8.14   Contact-damage processes

Ceramics can often be damaged under low contact loads and, thus, it is important to understand these damage mechanisms. These processes not only give rise to strength degradation but are linked to wear and erosion. Indentation fracture mechanics has been found to be a very useful approach in understanding these contact-damage processes (see Section 8.8).

a)

b)

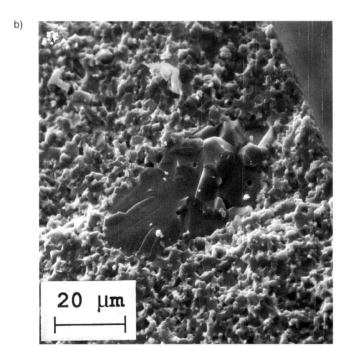

Figure 8.78 Scanning electron micrographs showing processing defects that acted as fracture origins in an alumina–zirconia composites: a) porous alumina agglomerate; b) large alumina grain. (From Green *et al.*, 1989, reproduced courtesy of CRC Press, Boca Raton, FL.)

a)

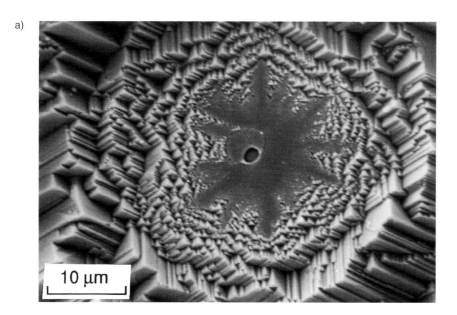

10 μm

b)

40 μm

Figure 8.79 Scanning electron micrographs showing failure origins in single-crystal alumina (sapphire) fibers: a) internal pore; b) surface flaw. (Courtesy of Paul Heydt.)

Figure 8.80 Silicon carbide inclusion that acted as a fracture origin in silicon nitride; scanning electron micrograph.

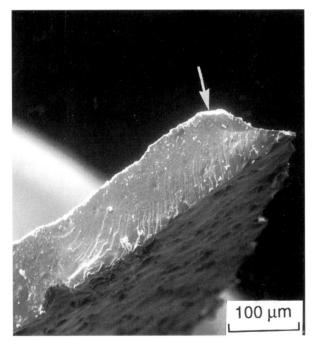

Figure 8.81 Impact failure origin in glaze coating from a space shuttle tile; scanning electron micrograph. (Reproduced courtesy of The American Ceramic Society, Westerville, OH.)

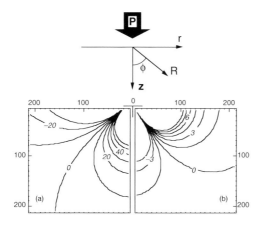

Figure 8.82 Contours of the stress, acting normal to the median plane ($\varphi=0$), generated by a Vickers indentation: (a) at the maximum load (9.8 N); (b) after the indenter is removed. Lengths and stresses are expressed in μm and MPa, respectively. (From Sglavo and Green, 1995, reproduced courtesy of The American Ceramic Society, Westerville, OH.)

8.14.1 Point-force solution

The elastic problem for an elastic half-space contacted by a normal point force P was solved by Boussinesq in 1885.[†] The stress field is axisymmetric around the force direction and has the general form, in spherical coordinates,

$$\sigma_{ij}=\frac{P}{\pi R^2}f_{ij}(\varphi) \tag{8.83}$$

where φ is the angle between the radial vector R and the surface normal. The stresses fall off rapidly with distance from the contact point. This highly localized stress field leads to crack stability effects, similar to those observed in the vicinity of residually stressed inclusions. Equation (8.83) indicates a stress singularity at $R=0$ and, thus, one expects inelastic deformation near the contact point. The complete stress field can be written

$$\sigma_{rr}=\frac{P}{\pi R^2}\left(\frac{1-2v}{4}\sec^2\left(\frac{\varphi}{2}\right)-\frac{3}{2}\cos\varphi\sin^2\varphi\right) \tag{8.84}$$

$$\sigma_{\theta\theta}=\frac{P}{\pi R^2}\left(\frac{1-2v}{2}\right)\left[\cos\varphi-\frac{1}{2}\sec^2\left(\frac{\varphi}{2}\right)\right] \tag{8.85}$$

$$\sigma_{zz}=-\frac{3P\cos^3\varphi}{2\pi R^2} \tag{8.86}$$

$$\sigma_{rz}=-\frac{3P\cos^2\varphi\sin\varphi}{2\pi R^2}; \quad \sigma_{r\theta}=\sigma_{\theta z}=0 \tag{8.87}$$

Figure 8.82(a) shows the nature of this stress field across a median plane. A tensile region is found below the contact point and, thus, cracks may be expected to form in this region. The presence of such cracks have been confirmed during

[†] See Timoshenko and Goodier, 1970.

the loading of sharp indenters and these are termed **median cracks**. There are, however, other crack geometries found at sharp indenters and these are related to the residual stress field that forms during unloading. As pointed out earlier, inelastic deformation occurs near the contact point and, thus, the surrounding elastic material cannot return to zero displacement. The elastic-plastic field, therefore, leaves residual stresses during unloading. The nature of the residual field is shown in Fig. 8.82(b) and the tensile region now moves to the surface region outside the contact. One group of cracks associated with the residual field are the **lateral cracks**, which are approximately parallel to the surface. The formation process for median and lateral cracks is shown schematically in Fig. 8.83. The median cracks form and grow in the loading phase, while the lateral cracks form during unloading (as load approaches zero). Markings left by the sub-critical growth of a median crack on a glass fracture surface are shown in Fig. 8.84. Fractography allows the crack shape to be identified and the growth process to be studied. Median cracks often extend across the surface during

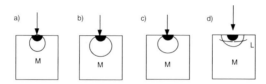

Figure 8.83 Schematic of median (M) and lateral (L) crack formation: a) median crack forms during loading; b) increases in size with increasing load; c) closes slightly during unloading; and d) lateral cracks form just prior to complete unloading, while the median crack extends across the surface.

Figure 8.84 Fracture surface marking shows outline of median crack front at a Vickers indentation in soda-lime-silica glass. (Optical micrograph courtesy of Vincenzo Sglavo.)

unloading, driven by residual stresses. This geometry is sometimes termed the **median–radial crack system**, with the crack shape being semi-elliptical (Fig. 8.85). Even in the absence of a median crack, cracks perpendicular to the surface can form during unloading, as shown schematically in Fig. 8.86. These are termed **radial cracks** and their formation is driven by the residual stresses that *increase* during unloading. It is important to note that radial cracks often form even in the absence of a median crack. It should be noted that median and radial cracks are associated with strength degradation, while lateral cracks are linked to erosion and wear by their tendency to spall. Figure 8.87 shows the indentation cracking pattern at a Vickers indentation in glass.

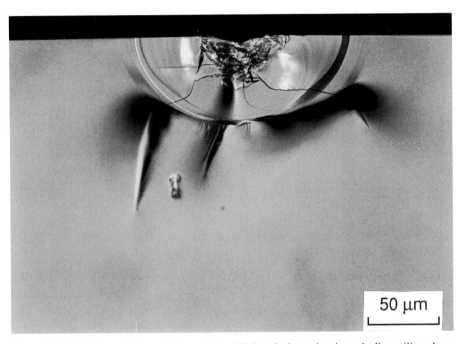

Figure 8.85 Median-radial crack formed at a Vickers indentation in soda-lime-silica glass; the crack shape is semi-elliptical. (Optical micrograph courtesy of Vincenzo Sglavo.)

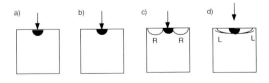

Figure 8.86 Schematic illustration of radial and lateral crack formation: a) and b) no median crack nucleated; c) radial cracks (R) form during unloading; and d) lateral cracks (L) form close to complete unload.

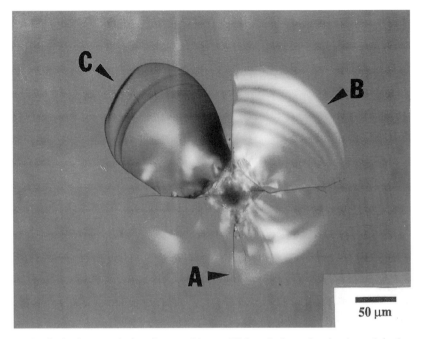

Figure 8.87 Optical micrograph showing cracking at Vickers indentation in glass: A is the radial crack surface trace; B shows an interference pattern from the lateral crack below the surface; and C is a spall produced by a lateral crack intersecting the surface. (Courtesy of Matt Chou.)

8.14.2 Blunt contacts

The elastic problem of contact between spherical particles was solved by Hertz in 1881.[†] Of interest here is the limiting case of the contact of a sphere, radius R, with a flat surface (blunt contact). The spherical indenter forms a circular contact, radius a, with the surface. The contact area increases with size as the load increases. It can be shown that

$$a^3 = \frac{4kPR}{3E_S} \tag{8.88}$$

where

$$k = \frac{9[(1-v_S^2)E_I + (1-v_I^2)E_S]}{16E_I} \tag{8.89}$$

where E is Young's modulus and v is Poisson's ratio. The subscripts I and S refer to the indenter and surface materials, respectively. It can also be shown that the distance of mutual approach δ is given by

---

[†] See Timoshenko and Goodier, 1970.

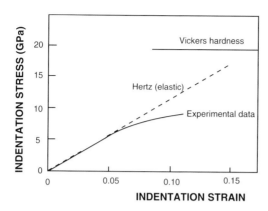

$$\delta^3 = \frac{16k^2P^2}{9E_S^2 R} \tag{8.90}$$

For this loading geometry, the maximum tensile stress σ_{max} occurs on the surface, just outside the contact area and is given by

$$\sigma_{max} = \frac{P(1-2\upsilon_S)}{2\pi a^2} \tag{8.91}$$

Outside the contact area, the tensile stresses fall off as r^{-2}, where r is the radial distance from the center of the contact area. Inside the contact area, the stresses are compressive and form a hemispherical distribution. It is also found that the region below the indenter is compressive to a depth $\sim 2a$, in contrast to the tensile stress field found for a point indentation. At larger distances, the stress field (necessarily) converges to that for the point-force problem. From Eq. (8.90), it is clear, even for elastic contacts, that the load–deflection behavior is non-linear. This is a reflection of the growth in the contact area as the load increases. If the contact area can be measured, it is sometimes useful to rearrange Eq. (8.88) into the (linear) form

$$\frac{P}{\pi a^2} = \left(\frac{3E_S}{4\pi k}\right)\left(\frac{a}{R}\right) \tag{8.92}$$

where the left-hand side is considered the indentation stress and a/R the indentation strain. Figure 8.88 shows some recent data on polycrystalline alumina with three different grain sizes (3, 23 and 48 μm). The data exhibit elastic behavior to about 5 GPa but deviate at higher stresses due to microfracture below the indenter. For other materials, deviation could be due to plastic deformation processes.

If the stresses are high enough, cracking occurs at blunt indenters but, in this case, the crack is initially circular (**ring crack**) but then extends into a cone geometry (Fig. 8.89). Experimentally, it is found that the load P_c required to initiate a **cone crack** is given by

$$P_c = \beta R \tag{8.93}$$

which is known as **Auerbach's Law**. Lawn (1993) has shown that Auerbach's Law can be derived using fracture mechanics arguments. For anomalous (densifying) glasses, such as fused silica, cone-cracking has been observed even at sharp indentations.

8.15   J-integral

The fracture mechanics developed earlier were based on linear elasticity. It is, however, unreasonable, even in brittle materials, to expect such behavior in the highly stressed region near a crack tip. If the non-linear region is small then one can still use linear elastic fracture mechanics (LEFM). Various criteria have been developed to determine the amount of non-linearity (e.g., inelastic zone size) that will allow the G and K measurements still to be valid. At some point, however, a new approach is needed, especially for materials that exhibit significant inelastic behavior. For this reason, the J-integral has been developed as a fracture criterion for non-linear *elastic* materials. This assumption implies that the deformation should be reversible but it can be applied to inelastic materials if no unloading occurs. For example, it has been applied to materials that undergo plastic deformation or other irreversible processes (e.g., metals, transformation-toughened materials, etc.). Indeed, standard procedures are available to measure a fracture criterion based on the J-integral. Another important aspect of the J-integral approach is that it allows the continuum view of fracture to be linked to thermodynamics and structural effects, whether they occur at the atomistic or microstructural level.

For a cracked, non-linear elastic body one can define a parameter J that represents the mechanical energy release rate, i.e., $J = -\mathrm{d}U_M/\mathrm{d}c$. In this sense, J is analogous to the LEFM parameter G but for a non-linear rather than a linear elastic material. Rice (1968) has shown that J can be written as the line integral

$$J = \int_C (U_D \mathrm{d}y - T_i \frac{\partial u_i}{\partial x} \mathrm{d}s) \tag{8.94}$$

where U_D is the strain energy density, x and y are rectangular coordinates normal to the crack front, $\mathrm{d}s$ is an increment along the integration contour C, T_i is the

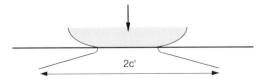

Figure 8.89 Schematic illustration of cone-crack formation at a blunt indentation.

2c'

stress vector acting on the contour and u_i is the displacement vector (see Fig. 8.90). For the case where the loading at the outer boundary is linear elastic, $J=G$. An important property of the J-integral is that it is zero when the contour is closed. Consider the closed contour around a crack tip, as shown in Fig. 8.91. The segments of the contour along the crack face must have zero tractions (at free surfaces) and thus $J_O+J_I=0$. If the integration direction is reversed for one of the contours, one obtains $J_O=J_I$ (reverse). The power of this approach is that the outer contour can be taken as the specimen perimeter where (linear elastic) loading and displacements can be measured, while the inner contour can surround the non-linear zone. This allows the energetics to be linked to the micromechanics in the non-linear region. For example, consider a zone at a crack tip that is characterized by a stress function $p(u)$ as shown in Fig. 8.92. The integral is similar to that given in Eq. (8.76) for crack bridging and can be written

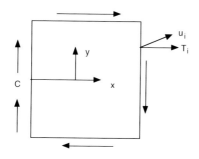

Figure 8.90 A cracked body loaded by a traction, T_i with displacement u_i.

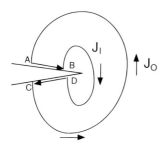

Figure 8.91 A closed contour for the J-integral around a crack tip. It can be shown that the inner and outer integrals are equivalent.

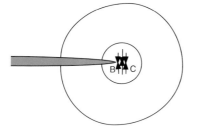

Figure 8.92 J-integral contours around a crack tip for a non-linear zone, BC.

$$J_C = \int_0^{u*} 2p(u)\, du \tag{8.95}$$

where $u*$ is the maximum crack opening displacement at B. If the non-linear zone can be considered in terms of an average stress p, then $J = 2pu*$. For perfectly brittle materials, one can equate p to the theoretical cleavage stress and $u*$ will represent the range of the intrinsic cohesion. For this case, the integral will also be equal to the surface energy γ associated with each fracture surface (2γ) and the Griffith failure criterion can be retrieved. For example, using values of $p = 10$ GPa and $2u* = 1$ nm, one finds $J = 10$ J/m^2, similar to the expected value of 2γ. For materials showing limited ductility, p can be equated to the yield stress and $2u*$ becomes the crack tip opening displacement caused by crack tip blunting.

Problems

8.1   A notched beam of yttria-stabilized ZrO$_2$ containing a through-thickness surface crack is broken in four-point bending (mode I), beam depth 10 mm. The fracture stress was 100 MPa and the critical crack size (c) was 1 mm. The Young's modulus and Poisson's ratio of the ZrO$_2$ are 200 GPa and 0.30, respectively.
   a) Determine the fracture surface energy, the fracture toughness and the crack resistance force.
   b) If the ZrO$_2$, when unnotched, breaks from semi-circular surface cracks, determine the fracture stress for a material that contains a 100 μm radius crack.
   c) If the ZrO$_2$ was indented with a Vickers hardness indenter, what radial crack size would you expect for a 20 kg load? Determine the strength of the indented material and the critical crack size. Why is the critical crack size greater than the indentation crack size?

8.2   A notched thick plate of Al$_2$O$_3$, containing a through-thickness internal crack is broken in uniform tension (mode I). The fracture stress was 12 MPa and the critical crack size $(2c)$ was 60 mm. The Young's modulus and Poisson's ratio of Al$_2$O$_3$ are 400 GPa and 0.26, respectively.
   a) Determine the fracture surface energy, the fracture toughness and the crack resistance force.
   b) Plot the total energy of the system as a function of crack length at failure.
   c) A broken Al$_2$O$_3$ component has a branching radius of 1 mm. If the associated branching constant is 20 MPa $\sqrt{m}$, determine the frac-

ture stress and the critical crack size. Suggest two reasons why the initial crack size could be smaller than the critical size.

8.3   The following data were gathered on Y_2O_3-stabilized ZrO_2:

Crystal structure: cubic fluorite
Lattice parameter, nm: 0.5144
Thermodynamic surface energy, J/m2: $\gamma=1.4$ ({100} planes)
Elastic stiffness constants, GPa: $c_{11}=403.5$; $c_{12}=102.4$; $c_{44}=59.9$
Hardness, GPa: $H=11.4$
Branching constant, MPa $\sqrt{m}$: $M_b=12.8$

a) Calculate the theoretical cleavage strength for {100} cleavage.
b) A large polycrystalline ZrO_2 plate is being stressed in tension. It has a mode I, through-thickness internal crack, length $2c=20$ mm. Calculate K_I and G for a stress of 10 MPa. In addition, determine the stresses (σ_{11}, σ_{12} and σ_{22}) in the crack plane ($\theta=0$) at distances of 1 μm and 10 μm from the crack tip. When the stress was increased to 20 MPa, the plate failed. Calculate the fracture toughness T, the crack resistance R and the fracture surface energy, Γ.
c) Plot G and R as a function of crack length for stresses of 10 and 20 MPa. Determine the largest internal crack size that could exist at 10 MPa.
d) The same ZrO_2 material was broken in uniaxial tension at a stress of 300 MPa. The failure origins were spherical voids. Determine the void size that caused failure. (Assume the void is equivalent to a circular crack with the same radius.)
e) Look in some research journals to see if you can find a microstructural modification that has been used to toughen cubic ZrO_2. Briefly explain the toughening approach. If you cannot find an approach, suggest one.

8.4   Suppose a material has a rising R curve that can be described by $R=R_0$ for $c\le\delta$ and $R=R_0+\Delta R[1-(\delta/c)^2]$ for $c\ge\delta$. Describe the crack growth behavior one would expect to see for such a material. Assume the loading geometry is uniaxial tension ($G=\pi\sigma^2c/E$). Draw a graph for the strength of the material as a function of crack size. Derive equations for the strength values in the different regions. How does the strength behavior compare to a material in which the fracture resistance is constant $=R_0$?

8.5   A beam is being loaded in pure bending with a nominal applied maximum stress σ and it contains a mode I edge crack. Using a Green's function approach, determine the equation for K_1, assuming the surface correction factor is 1.12 and the beam can be considered a semi-infinite body. Compare the answer with solutions in which one considers the crack being subjected to a uniform stress (maximum, mean and crack tip methods). This should be performed by plotting $K_1/(\sigma\sqrt{c})$ as a function of c/h (for $c/h=0$ to 0.5), where h is the beam depth and c is crack length. Finally, compare your answer to a solution found in a stress intensity factor handbook.

8.6   The fracture toughness of Al_2O_3 (4.0 MPa $\sqrt{m}$) is to be increased by adding 30 vol.% of the following materials:
   a) ZrO_2 particles using the transformation-toughening mechanism (transformation zone size, $2h=100$ μm).
   b) SiC whiskers using crack bridging. (Assume the whisker strength is 10 GPa, the debond length is approximately the whisker diameter (5 μm) and no pull-out occurs.)
   c) Al particles using crack bridging. (Assume the aluminum bridges are 50 μm in diameter.)
   Estimate the increase in fracture toughness that is expected from each mechanism. Indicate any assumptions in your analysis (try to be realistic). For part c), estimate the pull-out length required for the pull-out contribution to be equivalent to that of bridging.

   Data: E; and v(ZrO_2: 200 GPa; 0.31), (Al_2O_3: 400 GPa; 0.23), (SiC: 430 GPa; 0.20). Yield strength of Al=100 MPa.

8.7   The fracture toughness of α-SiC (4.0 MPa $\sqrt{m}$) is to be increased by adding 30 vol.% of TiB_2 particles to induce microcrack toughening. Estimate the increase in fracture toughness that is expected from this mechanism, assuming the microcrack zone size $2h=100$ μm.

   Data: E; and v (TiB_2: 520 GPa; 0.32); (SiC: 430 GPa; 0.20). Thermal expansion coefficient (SiC 5.0×10^{-6}/K), (TiB_2 8.2×10^{-6}/K).

8.8   a) A large plate under tension contains an internal, cylindrical hole and radial cracks emanating from the hole. The crack planes are perpendicular to the applied stress (worst case). Using a Green's function approach determine the equation for K_1. Compare the answer with solutions discussed in connection with Fig. 8.30.
   b) A pin-loading arrangement is to be used as a gripping approach for a tensile test on SiC ($K_{1C}=3$ MPa $\sqrt{m}$) as shown in Fig. 8.93. In machining the hole, cracks of 100 μm depth can be introduced. Determine the maximum stress that can be applied to the grip.

8.9   Derive the expression given as Eq. (8.65).

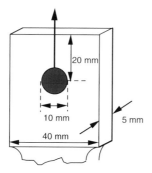

Figure 8.93 Pin-loaded grip in a uniaxial tensile specimen.

References

P. Becher, C.-H. Hseuh, K. B. Alexander and E. Y. Sun, Influence of reinforcement content and diameter on the *R*-curve response in SiC-whisker-reinforced alumina, *J. Am. Ceram. Soc.*, **79** (1996) 298–304.

D. Broek, *Elementary Engineering Fracture Mechanics*, Sitjhoff and Noordhoff, The Netherlands, 1978.

B. Budiansky, J. W. Hutchinson and J. C. Lambropoulus, Continuum theory of dilatant transformation toughening in ceramics, *Int. J. Solids Struct.*, **19** (1983) 337–55.

Y-S Chou and D. J. Green, SiC platelet-reinforced alumina; III Toughening mechanisms, *J. Am. Ceram. Soc.*, **76** (1993) 1985–92.

D. R. Clarke and K. T. Faber, Fracture of ceramics and glasses, *J. Phys. Chem. Solids*, **48** (1987) 1115–57.

R. F. Cook and G. M. Pharr, Direct observation and analysis of indentation cracking in glasses and ceramics, *J. Am. Ceram. Soc.*, **73** (1990) 787–817.

R. W. Davidge, *Mechanical Behaviour of Ceramics*, Cambridge University Press, Cambridge, UK, 1979.

A. de S. Jayatilaka, *Fracture of Engineering Brittle Materials*, Applied Science Publishers, 1979.

G. de Portu, *Introduction to Mechanical Behaviour of Ceramics*, CNR/IRTEC, Faenza, Italy, 1992.

A. G. Evans, Structural reliability: A processing-dependent phenomenon, *J. Am. Ceram. Soc.*, **65** (1982) 127–37.

A. G. Evans, Perspective on the development of high-toughness ceramics, *J. Am. Ceram. Soc.*, **73** (1990) 187–206.

A. G. Evans and R. M. Cannon, Toughening of brittle solids using martensitic transformation, *Acta Metall.*, **34** (1986) 761–800.

A. G. Evans and K. T. Faber, On the crack growth resistance of microcracking brittle materials, *J. Am. Ceram. Soc.*, **67** (1984) 255–60.

H. L. Ewald and R. J. H. Wanhill, *Fracture Mechanics*, Edward Arnold, London, 1984.

K. T. Faber and A. G. Evans, Crack deflection processes: I. Theory; II. Experiment, *Acta Metall.*, **31** (1983) 565–76, 577–84.

D. J. Green, Stress intensity factor estimates for annular cracks at spherical voids, *J. Am. Ceram. Soc.*, **63** (1980) 342.

D. J. Green, Microcracking mechanisms in ceramics, pp. 457–78 in *Fracture Mechanics of Ceramics*, Vol. 5, edited by R. C. Bradt *et al.*, Plenum Press, 1983.

D. J. Green, R. H. J. Hannink and M. V. Swain, *Transformation Toughening of Ceramics,* CRC Press, Boca Raton, 1989.

A. A. Griffith, The phenomenon of rupture and flow in solids, *Phil. Trans. Roy. Soc., Lond.,* **A221** (1920) 163–98.

F. Guiberteau, N. P. Padture and B. R. Lawn, Effect of grain size on Hertzian contact damage in alumina, *J. Am. Ceram. Soc.,* **77** (1994) 1825–31.

R. W. Hertzberg, *Deformation and Fracture of Engineering Materials,* J. Wiley and Sons, 1989.

C. E. Inglis, Stresses in a plate due to the presence of cracks and sharp corners, *Trans. Inst. Naval Archit.,* **A127** (1913) 219–30.

G. R. Irwin, Fracture, pp. 551–89 in *Handbuch der Physik,* Vol. 6, Springer-Verlag, Berlin, 1958.

J. F. Knott, *Fundamentals of Fracture Mechanics,* Butterworth Press, London, 1979.

F. F. Lange, Interaction of a crack front with a second phase dispersion, *Phil. Mag.,* **22** (1970) 983.

B. R. Lawn, *Fracture of Brittle Solids –* Second Edition, Cambridge University Press, Cambridge, UK, 1993.

D. B. Marshall and A. G. Evans, Failure mechanisms in ceramic fiber/ceramic matrix composites, *J. Am. Ceram. Soc.,* **68** (1985) 225–31.

D. B. Marshall and J. E. Ritter, Jr, Reliability of advanced structural ceramics and ceramic matrix composites, *Bull. Am. Ceram. Soc.,* **66** (1987) 309–17.

D. B. Marshall, M. C. Shaw, R. H. Dauskardt, R. O. Ritchie, M. J. Readey and A. H. Heuer, Crack-tip transformation zones in toughened zirconia, *J. Am. Ceram. Soc.,* **73** (1990) 2659–66.

J. W. Obreimoff, The splitting strength of mica, *Proc. Roy. Soc. Lond.,* **A127** (1930) 290–97.

A. P. Parker, *The Mechanics of Fracture and Fatigue: An Introduction,* E. & F.N. Spon, London, 1981.

I. S. Raju and J. C. Newman, Jr, Stress intensity factors for a wide range of semi-elliptical surface cracks in finite thickness plates, *Engg. Frac. Mech.,* **11** (1979) 817–29.

J. R. Rice, A path-independent integral and the approximate analysis of strain concentration by notches and cracks, *J. Appl. Mech.,* **35** (1968) 379–86.

D. P. Rooke, F. I. Baratta and D. J. Cartwright, Simple methods of determining stress intensity factors, *Engg. Frac. Mech.,* **14** (1981) 397–425.

V. M. Sglavo and D. J. Green, Subcritical growth of indentation median cracks, *J. Am. Ceram. Soc.,* **78** [3] (1995) 650–56.

H. Tada, P. C. Paris and G. R. Irwin, *The Stress Analysis of Cracks Handbook,* Del Research Corporation, St. Louis, MO, 1985.

A. S. Tetelman and A. J. McEvily, Jr, *Fracture of Structural Materials,* J. Wiley and Sons, 1967.

S. P. Timoshenko and J. N. Goodier, *Theory of Elasticity,* McGraw-Hill, New York, 1970.

S. M. Wiederhorn, Influence of water vapor on crack propagation in soda–lime–silica glass, *J. Am. Ceram. Soc.,* **53** (1967) 407–14.

Chapter 9

Strength and engineering design

To ensure the use of mechanically reliable ceramic components in technological applications, it is critical to establish an approach that can be incorporated into the engineering design process. In this chapter, the emphasis will be on the use of strength data in designing reliable ceramic components. After briefly describing strength measurement techniques, the use of failure statistics will be introduced. Finally, the time dependence of strength will be considered. As seen in the previous chapter, cracks can grow sub-critically in ceramics, causing strength to decrease with service time.

9.1 Strength testing

A common technique for the strength determination of a ceramic component is the **bend (flexure) test**. This approach has been popular as it involves simple specimen shapes. This is particularly useful when the specimen is machined from larger production units. The loading configuration is usually either three- or four-point bending and American Society for Testing and Materials (ASTM) standards are available for both approaches (ASTM C 1161, 1990). Assuming the material fails in tension, the bend strength is determined from the maximum applied tensile stress, using Eq. (4.6). The bending configuration is statically indeterminate and, thus, the stress equations assume that the material is linearly elastic. Four-point is often preferred over three-point bending as the specimen has a larger region under the maximum stress. Moreover, there are no shear stresses in this region (see Fig. 4.7). The specimens used in bend tests usually have a rectangular cross-section but circular rods are sometimes used. For the former specimen type, the corners of the specimens are often rounded or beveled, to minimize failure from corner cracks. Such cracks are often produced in the machining process and are more deleterious to strength than surface cracks (see Section 8.6.3). In bend tests, substantial attention must be given to the accuracy of the loading geometry, specimen dimensions, friction effects at the loading points and alignment. This topic has been reviewed recently by Quinn and Morrell (1991).

Biaxial flexure is another approach that uses a bending configuration. In this test, the specimen is supported on a ring or three balls and the load is applied in the center of the opposite face. The specimens for this test are often circular plates but rectangular plates can also be utilized.

Uniaxial tension is finding increased use in the strength testing of ceramic components. Specimens usually have a dog-boned shape with a circular or rectangular cross-section. The specimen shape involves more extensive machining than a bend test specimen, especially in the transition region between the gage section and the gripping area. It is critical that specimens fail in the gage section, where the stress is well defined, and not in the transition region. Alignment is extremely critical in this test as any bending stresses will lead to premature failure. Gripping must also be accomplished so that unwanted stress concentrations do not occur. The brittle nature of many ceramics does not allow the gripping stresses to relax, enhancing the possibility of grip failure.

In all the above testing approaches, it is critical for the details of the testing procedure to be reported. As shown later, the strengths of ceramics are *dependent* on the loading geometry and specimen dimensions. It is particularly important to understand these effects, when comparing strength data from various sources.

9.2   Failure statistics

As indicated in Chapter 8, ceramics contain flaws that can vary substantially in size and type, causing strength to vary significantly from sample to sample. This variability in strength is often expressed in terms of a **failure probability**. To describe a strength distribution at least two parameters are needed, to measure the width and magnitude of the distribution. The difficulty encountered is that the form of this distribution is not known a priori. For this reason, an empirical distribution, first suggested by Weibull (1951) is often used. Once the strength of a material is fitted to this distribution, the failure probability can be predicted for any applied stress distribution. If this failure probability is too high it will clearly impact safety, and the design needs to be changed or the material improved. Clearly, it is preferable to reduce the failure probability until it is neg-

ligible but this may impact the cost. There is often, therefore, an interplay
between safety and economics in the overall design process.

9.2.1 Weibull approach

A common empirical approach to describing the strength distribution of a brittle
material is the **Weibull approach**. It is a 'weakest link' approach, in that the
strength of a body involves the *products* of the survival probabilities for the individual volume elements. This is analogous to the strength of a chain, in which
strength is determined by the weakest link. Once the chain is broken, the next
weakest link will determine the strength of the remaining parts and so on. The
Weibull approach assumes a simple power-law stress function for the survival of
the elements, which is integrated over the body volume. For example, the three-
parameter Weibull distribution for a body failing under the action of a tensile
stress σ can be written as

$$F = 1 - \exp\left[-\int_V \left(\frac{\sigma - \sigma_{min}}{\sigma_0} \right)^m dV \right]$$ (9.1)

where F is the failure probability, m the **Weibull modulus**, σ_0 the **characteristic
strength** and σ_{min} the **minimum strength**. The Weibull modulus describes the
width of the strength distribution and the characteristic strength 'locates' the distribution in stress space. The higher the value of the Weibull modulus, the lower
is the strength variability. Values of m for ceramics are often in the range 5–20,
though there has been a recent impetus to reduce strength variability (i.e.,
increase m) and higher values are now being reported.

In many cases, a two-parameter Weibull distribution is assumed, where
$\sigma_{min} = 0$. For this case, Eq. (9.1) can be simplified as follows

$$\ln\left(\frac{1}{1-F} \right) = L_F V \left(\frac{\sigma_{max}}{\sigma_0} \right)^m = \left(\frac{\sigma_{max}}{\sigma^*_0} \right)^m$$ (9.2)

where σ_{max} is the maximum applied stress, V the stressed volume and L_F the
loading factor. The loading factor is a reflection of the stress distribution in the
body; it has a value of unity for uniaxial tension. From this equation, σ_0 can be
interpreted as the uniaxial strength of a body with unit volume at a failure probability of 0.632. It is often convenient to use the parameter, $\sigma^*_0 = \sigma_0 (L_F V)^{1/m}$. This
value represents the specific characteristic strength of a body at $F = 0.632$ for
particular values of V and L_F. The loading factor is a measure of how effectively
the body is stressed compared to uniaxial tension. The product $(L_F V)$ is often
termed the **effective volume** as it indicates how 'effectively' the body is being
stressed. Some examples of loading factors for common testing techniques are
given in Table 9.1. The above analysis assumes that the flaws are distributed

Table 9.1 *Examples of loading factors (volume flaws)*

Geometry	Loading factor, L_F
Uniaxial tension	1
Flexure testing	
pure bending	$\dfrac{1}{2(m+1)}$
3–point bending	$\dfrac{1}{2(m+1)^2}$
4–point bending	$\dfrac{mL_i+L_o}{2L_o(m+1)^2}$
(L_i=inner span, L_o=outer span)	

throughout the volume of the material. A similar analysis can be used for bodies that contain only surface flaws, by replacing the volume terms by area terms.

To analyze strength data, Eq. (9.2) is usually written

$$\ln \ln\left(\frac{1}{1-F}\right)=m\ln\sigma_{max}-m\ln\sigma_0^* \tag{9.3}$$

Thus, a plot of the left side of Eq. (9.3) as a function of the natural logarithm of strength should yield a straight line. In such a procedure, a failure probability is needed for each test specimen. This is usually estimated using

$$F=\frac{n-0.5}{N} \tag{9.4}$$

The strength data of N specimens are organized from weakest to strongest and given a rank n with $n=1$ being the weakest specimen. Equation (9.4) attempts to account for the finite number of test specimens. Other forms of this equation have been suggested, e.g., $F=n/(N+1)$ but Eq. (9.4) is now widely accepted. A plot based on Eq. (9.3) will give the Weibull modulus m from the slope and σ_0^* from the intercept ($=-m\ln\sigma_0^*$). The fitting of a straight line is often done using linear regression. Some workers have, however, suggested that other fitting procedures, such as the maximum likelihood estimator, may be preferable. The parameter σ_0^* is related to the mean strength σ_{av} by $\sigma_{av}=\sigma_0^*\Gamma(1+1/m)$, where $\Gamma(\ )$ is the Gamma Function, which is found in mathematical tables.

An important feature of the above strength analysis is that a large number of test specimens need to be broken before the Weibull parameters can be known with any accuracy. For example, one often needs to break at least 30 specimens before m is known within 20% and σ_0^* within 5%. It is, therefore, not unusual to require 30 to 50 specimens in a strength testing program.

9.2.2   Effect of specimen (component) size and loading mode

An important feature of the Weibull approach is that the strength of a brittle material depends on component size and the loading configuration (Eq. (9.2)). For example, the strengths of two bodies with different volumes but broken in the same loading geometry are shown schematically in Fig. 9.1(a). The specimen with the larger volume is predicted to possess the lower strength because there is an increased probability of 'finding' a larger flaw in a larger body. For specimens of equivalent volumes, the type of loading geometry becomes important. From Table 9.1 and the associated stress distributions, one concludes that bending tests stress the volume less effectively than uniaxial tension. Thus, for a given specimen size, strength values from bend tests are higher than those from uniaxial tension (e.g., see Fig. 9.1(b)).

To calculate the relationship between the strength (σ_1) for one loading geometry (L_{1F}) and specimen size (V_1) and another (σ_2, L_{2F}, V_2), Eq. (9.2) can be written for fixed failure probability as

$$\left(\frac{\sigma_1}{\sigma_2}\right) = \left(\frac{L_{2F}V_2}{L_{1F}V_1}\right)^{1/m} \tag{9.5}$$

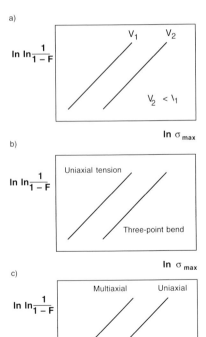

Figure 9.1 Effect of various factors on strength distribution: a) specimen size; b) loading geometry; and c) multiaxial loading.

Multiaxial loading changes the form of Eq. (9.1) and, in some cases, can promote fracture. In biaxial stress fields, one can continue with Eq. (9.1), using the maximum principal stress (σ_1^*) for the applied stress as long as the other principal stress (σ_2^*) does not exceed 0.8 σ_1^*. In triaxial fields, the same approach can be taken as long as $(\sigma_2^*/\sigma_1^*)(\sigma_3^*/\sigma_1^*)<0.5$. Above these ratios, the multiaxial nature of the field will significantly decrease strength (see Fig. 9.1(c)) and this effect becomes more pronounced as the Weibull modulus increases. In other words, multiaxial stress fields allow a greater fraction of cracks to be in an orientation likely to cause failure. This increases failure probability and, thereby, decreases strength. For multiaxial stress problems, it is often assumed that the survival probability or reliability R in each principal stress direction is independent. Hence, the total survival probability R_T is the product of the survival probabilities for each principal stress $(R_T = R_1 R_2 R_3)$. An expression for the loading factor is obtained from Eqs. (9.1) and (9.2), i.e.,

$$L_F = \int_V \left(\frac{\sigma}{\sigma_0}\right)^m \frac{dV}{V}$$

(9.6)

Thus, if the analytical stress distribution is known, the loading factor for each principal stress can be calculated. In this process, compressive stresses are usually neglected. In many realistic component geometries, the stresses need to be determined by numerical techniques, such as finite-element analysis. From Eq. (9.1) with $\sigma_{min}=0$, the survival probability for a single element R_{ij} under the action of a single principal stress can be written as

$$R_{ij} = \exp\left[\left(-\frac{\sigma_{ij}}{\sigma_0}\right)^m V_j\right]$$

(9.7)

where i and j represent the principal stress and the element number, respectively. The total survival probability is obtained by taking the product of this expression over all the elements and the three principal stresses. This approach has now been incorporated into a special software package (CARES, published by the National Aeronautic and Space Agency, NASA) and it is a powerful tool in determining how changes in design will affect failure probability.

When performing strength testing, it is critical that the flaws leading to failure in the test specimens are the same as those that cause failure in service. Spurious results will be obtained, for example, if flaws are produced in the machining of test specimens. Additionally, flaws in the component may appear during service (e.g., by oxidation or corrosion) and these may be different than those in test specimens. In some cases, several types of flaws may be present in the same specimen and this leads to non-linearities in the Weibull plots. Techniques have been put forward for extracting the Weibull parameters from such plots. In many situations, Weibull plots need to be extrapolated to failure probability values well below those obtained in the laboratory and this can lead to large uncertainties. It should also be remembered that the Weibull analysis is empirical and, thus, there is no guarantee that it is the best approach. Finally, as will be seen in the next section, strength can degrade during service and this effect must also be incorporated into the design process.

9.3   Time dependence of strength

In Section 8.12, the idea of sub-critical crack growth was introduced. This discussion indicated that cracks can grow under the application of stress in service and, thus, strength can *decrease* with use. In an active environment for sub-critical crack growth, it becomes important to define a reference state for strength. This is called the **inert strength** and it defines the strength of a specimen or component if tested in an inert environment or at a very fast stressing rate, i.e., where no sub-critical crack growth occurs. For example, consider the hypothetical situation shown in Fig. 9.2, where it is shown that the inert strength S is greater than the **dynamic strength** σ_f measured in the active environment. Both the active and inert strengths are distributed values, and the initial ranges are shown as the shaded regions in Fig. 9.2. It is, however, possible to define the strength degradation in terms of changes in the inert strength S. Two inert strength 'trajectories' are shown and these define the strength degradation. For one trajectory, the dominant crack becomes critical and failure occurs at the maximum applied stress. For the second trajectory, the material survives the stress application but under some circumstances, such as that shown in the figure, the strength can be less than the maximum applied stress. This effect is a result of the strength degradation that occurs during unloading.

To analyze the strength degradation, a simple approach will be taken in which it will be assumed that only Region I sub-critical crack growth occurs (see Fig. 8.72) with no threshold. This may seem rather restrictive but, in many situations, cracks often spend most time in this region. To define the effect of sub-critical crack growth on strength, three items are needed: the initial inert strength S_I; the crack velocity equation; and the failure condition ($K_{IC}=T$). Using Eq. (8.81) as

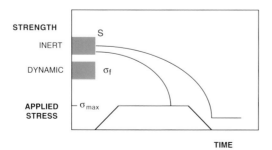

Figure 9.2 The degradation in strength that occurs under stress can be followed by considering the inert strength as a reference state.

the crack growth law, in conjunction with $K_I = \sigma Y\sqrt{c}$ and $K_{IC} = S_I Y\sqrt{c}$, one can show for an arbitrary stress cycle $\sigma(t)$ that

$$S_F^{n-2} - S_I^{n-2} = -\frac{1}{B}\int_0^{t_f} [\sigma(t)]^n dt \tag{9.8}$$

where S_F is the inert strength after the stress application and

$$B = \frac{2K_{IC}^2}{(n-2)Y^2 v_0} \tag{9.9}$$

It is useful to analyze two situations that are associated with strength degradation. If a constant stress σ_A is applied to a material, the component will fail after a given time period. This process is termed **static fatigue** and, if one assumes $S_F^{n-2} \ll S_I^{n-2}$, Eq. (9.8) can be easily integrated to give the failure time (lifetime) t_f of the material as

$$t_f = \frac{BS_I^{n-2}}{\sigma_A^n} \tag{9.10}$$

Figure 9.3 shows a (logarithmic) plot of lifetime in terms of the applied stress. The second situation of interest is when a constant stress rate ($d\sigma/dt$) is applied to a material; this is termed **dynamic fatigue**. Integration of Eq. (9.8) for this situation and assuming $S_F^{n-2} \ll S_I^{n-2}$ gives the **dynamic strength** σ_f as

$$\sigma_f^{n+1} = B(n+1)S_I^{n-2}\left(\frac{d\sigma}{dt}\right) \tag{9.11}$$

Figure 9.4 (ln–ln plot) shows that dynamic strength increases with increasing stress rate, reflecting the reduced time for crack growth at the higher stress rates.

Another well-known source of damage is **cyclic fatigue** and Eq. (9.8) can be used to predict strength degradation for this case. For some ceramics, however, additional sources of damage can occur during cyclic fatigue. Toughening mechanisms such as crack shielding and crack bridging can be degraded by a cyclic stress. For example, frictional crack bridges can lose their effectiveness by wear processes. Under these conditions, Eq. (9.8) is no longer applicable and the prediction of strength degradation becomes more complex.

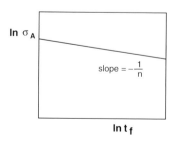

Figure 9.3 Schematic of expected static fatigue behavior for cracks experiencing Region I sub-critical crack growth.

9.4 Determination of sub-critical crack growth parameters

To include the effect of sub-critical crack growth into design, it is necessary to determine experimentally the values of B and n, the sub-critical crack growth parameters. These parameters define the behavior of a material in a given active environment. Three types of approach are commonly used for this purpose and these will be discussed below.

9.4.1 Direct techniques

One approach is to directly view the crack growth and to measure the crack velocity as a function of the applied stress intensity factor. The values of v_0 and n are determined by fitting the data to Eq. (8.81), which requires a value of K_{IC}. In order to use Eq. (9.8) to predict strength degradation a crack geometry needs to be assumed for Y to obtain the value of B from Eq. (9.9). In the direct approach, fracture mechanics test geometries are often used and the artificially introduced cracks are larger than 'natural' flaws. Recently, it has been shown that indentation cracks may be used to characterize sub-critical crack growth (Dwivedi and Green, 1995). Figure 9.5 shows a series of crack-front markings that occurred during sub-critical crack growth of an indentation crack. The aspect ratio of the crack was found to change during growth and this effect was incorporated in the analysis.

9.4.2 Dynamic fatigue

An alternative approach, termed dynamic fatigue, is to measure strength as a function of stressing rate. Equation (9.11) can be written as

$$\ln \sigma_f = \frac{1}{n+1}\left\{\ln[B(n+1)S_1^{n-2}]+\ln\left(\frac{d\sigma}{dt}\right)\right\} \tag{9.12}$$

From a logarithmic plot of the dynamic strength versus stressing rate, the value of n is obtained from the slope $[=1/(n+1)]$ and B is obtained from the intercept. An attractive feature of this approach is that failure occurs from 'natural' cracks.

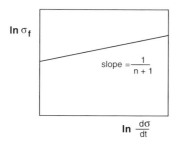

Figure 9.4 Schematic of expected dynamic fatigue behavior for cracks subjected to Region I sub-critical crack growth.

A difficulty with this approach is the large scatter in the data at each stressing rate, requiring a large number of specimens to characterize the sub-critical crack growth parameters. This problem can be overcome by using indentation cracks to characterize dynamic fatigue. The data are analyzed using an equation similar to Eq. (9.12) (see Marshall and Lawn, 1980).

9.4.3   Static fatigue

The final approach is called static fatigue, in which lifetime for different levels of constant applied stress are determined. Writing Eq. (9.10) as

$$\ln \sigma_A = \frac{1}{n}\{\ln[BS_1^{n-2}] - \ln t_f\}$$
(9.13)

allows the data to be fitted to a linear function on a ln-ln plot, with n being obtained from the slope $(= -1/n)$ and B from the intercept. This approach is not as common as dynamic fatigue, because it usually involves longer testing times.

In the above characterizations, it was assumed that there was no threshold stress intensity factor below which sub-critical crack growth did not occur. In some circumstances, however, a threshold does exist and this effect needs to incorporated into the design process. Several techniques have been put forward to determine the threshold stress intensity factor. For example, Sglavo and Green (1995) have recently analyzed the **interrupted static fatigue test**, in which specimens are held at a constant stress for different times and, then, the surviving specimens are broken. A comparison of the strength after this test with the initial strength distribution allows the threshold to be estimated. Alternatively, crack growth behavior near the threshold can be measured using fracture mechanics specimens (see Lawn, 1993).

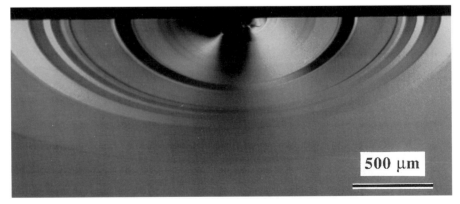

Figure 9.5 Crack front markings showing the position of an indentation crack as it undergoes sub-critical crack growth; optical micrograph. (From Dwivedi and Green, 1995, reproduced courtesy of The American Ceramic Society, Westerville, OH.)

9.5 SPT diagrams

It is useful to introduce failure statistics into the time dependence of strength and this can be accomplished by stating the initial inert strength in a Weibull form, i.e,

$$\ln\frac{1}{1-F}=\left(\frac{S_1}{S_0^*}\right)^m \tag{9.14}$$

By substituting Eq. (9.14) into Eq. (9.8), it is possible to construct **stress–probability–time (SPT) diagrams**. For example, SPT diagrams showing lifetime under constant stress for different levels of failure probability can be formulated, as shown schematically in Fig. 9.6.

An alternative approach has been put forward by Davidge (1979), in which dynamic fatigue data is transformed into an SPT diagram. This process involves combining Eqs. (9.12) and (9.13), eliminating the dependence on B and the inert strength. The steps in this approach are

1. Determine the Weibull parameters in the active environment at one stressing rate. (This involves breaking at least 20 specimens.)
2. Determine the value of n from dynamic fatigue tests (several stressing rates).
3. Determine the values of F and failure time t_f for each specimen tested in step 1.
4. Transform the failure times in step 3, assuming the failure stress was constant (i.e., not a constant stress rate). From Eqs. (9.12) and (9.13), one can show that the equivalent failure time t_0 under constant stress is given by

$$t_0=\frac{t_f}{n+1} \tag{9.15}$$

5. Calculate the (constant) applied stress values that would be needed to produce a single failure time. From Eq. (9.10), this is accomplished for each failure time (and F value) using

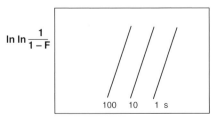

$$\text{In In}\frac{1}{1-F}$$

100   10   1 s

In σ_A

Figure 9.6 Schematic of stress–probability–time (SPT) diagram for a body under constant stress.

$$\left(\frac{\sigma_{A2}}{\sigma_{A1}}\right)^n = \frac{t_{01}}{t_{02}} \tag{9.16}$$

For example, the value of t_{01} obtained for each applied stress (σ_{A1}) could be used to find the stress σ_{A2} for a failure time of 1 s ($t_{02}=1$ s). A line for a 1 s lifetime can then be fitted to this data. Lines for other failure times are then obtained using Eq. (9.16).

It is important to be able to generalize the above approach to more complex stress distributions. For example, the body can be broken down into volume elements in which the principal stresses can be assumed constant. Equation (9.10) can then be used to describe the strength degradation of each element. Ritter (1992) has shown that the fatigue survival probability, R_{ij}^F, for each element can be written

$$\ln R_{ij}^F = \ln R_{ij} \left(\frac{\sigma_{ij}^2 t_f}{B}\right)_{n-2}^m \tag{9.17}$$

where R_{ij} was defined in Eq. (9.7). Taking the product for each element and principal stress direction, the overall fatigue survival probability is obtained. Clearly, this approach is easily adapted into numerical stress analysis techniques and as such, it has been incorporated into some software packages (e.g., NASA *CARES/Life*).

9.6   Improving strength and reliability

In the above testing methodology outlined for use in engineering design, it is sometimes found that the failure probability associated with the design stress is unacceptably high. It is, therefore, worth considering the steps that can be taken to overcome this problem. If one considers the basic equation between strength and fracture toughness,

$$\sigma_f = \frac{K_{IC}}{Y\sqrt{c}} \tag{9.18}$$

several alternatives can be identified, as follows.

9.6.1   Decrease flaw size

One approach is to decrease the critical crack size associated with fracture, without changing the fracture toughness. This could be accomplished by 1) improved processing of the ceramic material, i.e., removing the dominant failure sources; 2) improved finishing operations (removing machining flaws); 3) eliminating flaws produced during service; 4) **non-destructive evaluation** to identify components that contain large flaws (often with ultrasonic waves); or 5) **proof testing**. With this last technique, components are over-stressed prior to use and those that fail or give an indication of failure (e.g., by acoustic emission) are discarded. It is preferable to conduct proof testing in an inert environment. This ensures no component with a strength less than the maximum proof stress will survive. As shown earlier in Fig. 9.2, stressing in an active environment can lead to strengths less than the maximum stress. After proof testing in an inert environment, the minimum strength of the components is then equivalent to the maximum proof stress σ_p. Calculations can be performed using this value to determine the minimum failure time for the components. For the two-parameter Weibull distribution, the effect of proof testing in an inert environment is to produce a three-parameter distribution with $\sigma_{min} = \sigma_p$, as shown in Fig. 9.7. In some cases, it is not possible to perform proof testing in an inert environment and, in those cases, it is critical to demonstrate that the strength distribution is properly truncated at the design stress. For example, proof testing was performed on the thermal protection tiles of the space shuttle before its maiden voyage. Figure 9.8 compares the initial strength distribution of the tiles to that after proof testing. It is evident that proof testing caused truncation of the after-proof

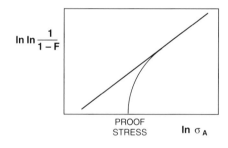

Figure 9.7 Schematic of effect of proof test on a two-parameter Weibull distribution (inert environment).

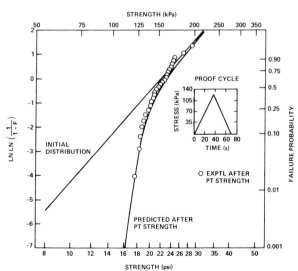

Figure 9.8 Effect of proof testing on space shuttle tile materials. Compared to the initial strength distribution, the strength of the surviving specimens was found to be truncated close to the maximum proof stress. (From Green et al., 1982, reproduced courtesy of The American Ceramic Society, Westerville, OH.)

strength data near the maximum proof stress. This behavior was in good agreement with the predicted behavior from Eq. (9.8) (Green *et al.*, 1982).

9.6.2   Improve fracture toughness or flaw insensitivity

Returning to Eq. (9.18), an alternative approach to improving the reliability of a ceramic is to introduce new toughening mechanisms which will increase strength even if there is no change in the critical flaw size. A particularly attractive approach is to introduce rising *R*-curve behavior into a material (see Section 8.9). This has the effect of decreasing the strength variability by making strength less sensitive to flaw size. A decrease in strength variability can have a large effect on the expected failure probability at the design stress.

9.6.3   Decrease applied stresses

The final possibility for increasing reliability is to decrease the applied stress. For example, a change in a component geometry may remove unwanted stress concentrations, thereby decreasing the design stress. A less obvious way to decrease stress is to introduce residual compression in the region where cracks initiate. For example, if surface cracks are a dominant flaw, **residual surface compression** will decrease the applied stress in the surface region, making the propagation of these cracks more difficult.

9.7   Temperature dependence of strength

Many ceramics being considered for structural applications are covalent materials which are brittle at room temperature, failing from the most severe flaws. This behavior persists to temperatures up to 1000 °C.[†] Above this temperature, creep and other inelastic deformation mechanisms are activated, such that strength decreases, as shown schematically Fig. 9.9. In this region, the strength approaches the yield stress and thermal activation assists the inelastic processes (see Chapters 6 and 7). For ceramics with a greater degree of ionic bonding, this transition to semi-brittle behavior occurs at a lower temperature. For most polycrystalline ceramics there are usually insufficient slip systems to allow complete ductility.

9.8   Thermal stresses and thermal shock

The thermal expansion of ceramics can lead to significant stresses during use. In earlier chapters, residual stresses from thermal expansion mismatch or

---

[†] Transformation-toughened ceramics exhibit decreasing strength with increasing temperature (see Section 8.11).

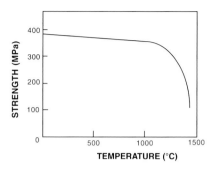

Figure 9.9 The strength of many structural ceramics is insensitive to temperature below 1000 °C but is degraded at the higher temperatures, usually by creep-related mechanisms. The trend shown here is typical for the flexural strengths of a dense high-purity alumina.

anisotropy have been discussed. Similar principles are involved in **thermal stresses** which are also related to dimensional changes. These stresses can be high enough to damage or fracture a component (**thermal shock**) and are often a key aspect of material selection, especially for ceramics.

If a component is at a different temperature than its surrounding attachments then stresses will develop. For example, a rod attached to a rigid constraint will be placed under a thermal stress if it is at a different temperature than the constraint. If the constraint is at temperature T_0 and the rod at temperature T_1, a strain $\alpha(T_0 - T_1)$ will develop in the rod, where α is the thermal expansion coefficient of the rod. If the rod is linearly elastic, the thermal stress σ_T will be given by $\sigma_T = E\alpha(T_0 - T_1)$, where E is Young's modulus. Clearly, the situation is more complex if the rod can creep, as these stresses may relax over time. For this chapter, it will be assumed that the ceramics are linearly elastic and isotropic, in order to set out the basic principles.

Although external constraint can lead to thermal stresses, a more common source is due to **temperature gradients** that occur *within* a material. For example, if the surface of a material is cooler than the interior, the surface will attempt to contract but this will be constrained by the internal material. The surface will be placed in tension and the interior in compression. For brittle materials, the emphasis is usually placed on the tensile stresses as these may be sufficient to extend cracks, causing damage or fracture.

Consider a large plate that possesses a thermal gradient across its thickness (Fig. 9.10). The simplest situation is when the temperature gradient is symmetric with respect to the x_1 axis.[†] For the symmetric case, the thermal stresses at any point are given by

$$\sigma_1 = \sigma_3 = -\frac{\alpha TE}{1-\upsilon} + \frac{1}{2h(1-\upsilon)} \int_{-h}^{h} \alpha TE \, dx_2 \qquad (9.19)$$

where T is the temperature at that point and υ is Poisson's ratio. The stresses are equibiaxial because the plate is free to expand or contract in the x_3 direction. If

---

[†] An asymmetric temperature gradient gives rise to a more complex situation as the plate will bend to relax some of the stresses.

the plate is being cooled from both sides, the surface region will be in tension and there will be compensating compression inside to produce the force balance. The temperature integral in Eq. (9.19) is sometimes termed the average temperature T_{av}, so the equation can be written

$$\sigma_1 = \sigma_3 = \frac{\alpha E(T_{av} - T)}{1 - v} \tag{9.20}$$

For the case of rapid cooling from a temperature T_0, only the surface temperature T_S will change and the surface stress will be the maximum possible with $T_{av} = T_0$, i.e.,

$$\sigma_{max} = \frac{\alpha E(T_0 - T_S)}{1 - v} \tag{9.21}$$

In most situations, the cooling is not so rapid and the thermal stresses are usually less than this maximum value. For example, slower cooling may give rise to a parabolic temperature gradient, i.e.,

$$T = T_0 \left[1 - \left(\frac{x_2}{h} \right)^2 \right] \tag{9.22}$$

The thermal stresses are obtained by substituting Eq. (9.22) into Eq. (9.19) and integrating

$$\sigma_1 = \sigma_3 = \frac{\alpha E T_0}{1 - v} \left[\frac{2}{3} - \left(1 - \frac{x_2^2}{h^2} \right) \right] \tag{9.23}$$

This stress distribution is shown in Fig. 9.11 and the surface stress ($x_2 = h$) in this case equals $2\alpha E T_0 / [3(1 - v)]$.

A more general approach to the thermal stress problem is to consider the parameters that control the temperature gradients in a material. For these prob-

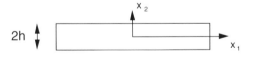

Figure 9.10 Geometry of a large plate being subjected to a thermal gradient across its thickness (2h).

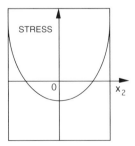

Figure 9.11 Thermal stress distribution in a thick plate subjected to a parabolic temperature profile.

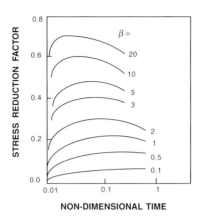

Figure 9.12 Stress reduction factor as a function of non-dimensional time for various values of Biot's modulus, β.

lems, a non-dimensional parameter, termed the Biot's modulus β is introduced. This parameter is defined as

$$\beta = \frac{ht_S}{k} \tag{9.24}$$

where $2h$ is a body dimension, t_S the surface heat transfer coefficient and k the thermal conductivity of the material. The parameter t_S depends both on the quenched material and the surrounding medium, while k is a material property. The thermal stresses that develop are determined by the rate at which heat can enter or be removed from the surface and the rate at which it 'moves' within the material. The surface stresses that develop in a cooled body are now uniquely related to β and can be written

$$\sigma_{max} = \frac{\phi\alpha E\Delta T}{1 - v} \tag{9.25}$$

where ΔT is the difference between the original and new surface temperature and ϕ is termed the **stress reduction factor**. The parameter ϕ lies between zero and unity, so that for high β, $\phi = 1$ and Eq. (9.25) becomes equivalent to Eq. (9.21). Figure 9.12 shows the stress reduction factor plotted as a function of non-dimensional time[†] for different β values. The thermal stresses are found to decrease as β decreases and the maximum stress in a body does not occur instantaneously but at some time after the thermal transient.

9.9   Thermal shock resistance parameters

It is useful to define parameters that indicate the resistance of a material to thermal shock. The difficulty with this approach is that the thermal stresses

[†] Defined as $\kappa t/h^2$, where κ is the thermal diffusivity, h is a characteristic component dimension of the body and t is time.

depend on the particular details of the heat transfer process. With recent computational innovations it is now possible to determine thermal stresses using finite-element analysis. This can be combined with failure statistics to determine the overall failure probability of a component, as outlined in Section 9.2. The **thermal shock resistance (TSR) parameters** can still, however, be useful in the initial phase of material selection. A large number of TSR parameters have been formulated and these define either the conditions for crack propagation or for crack arrest under the action of a thermal stress.

9.9.1   Fracture initiation

For extremely rapid cooling, the maximum stress from Eq. (9.21) can be set equal to the strength σ_f of the material and a TSR parameter $\mathcal{R}$ can be defined as the critical temperature difference that causes failure, i.e.,

$$\mathcal{R}=\Delta T_c=\frac{\sigma_f(1-v)}{\alpha E} \tag{9.26}$$

It is not immediately clear which type of testing geometry should be associated with the strength value but the biaxial nature of the stress field would suggest that biaxial flexure values might be the most appropriate estimate.

The very rapid cooling rates needed to define $\mathcal{R}$ values do not occur often and, thus, a second TSR parameter $\mathcal{R}'$ is sometimes used. This approach introduces the other important material parameter, thermal conductivity. For slow cooling rates, it is found that the stress reduction factor is $\sim 0.31\beta$ and is, thus, inversely dependent on thermal conductivity. If this condition is introduced, $\mathcal{R}'$ is defined as

$$\mathcal{R}'=\frac{\sigma_f k(1-v)}{\alpha E} \tag{9.27}$$

and the critical temperature difference for thermal shock damage is given by

$$\Delta T_c=\frac{\mathcal{R}'}{0.31ht_s} \tag{9.28}$$

Table 9.2 gives some examples of $\mathcal{R}$ and $\mathcal{R}'$ values for various brittle materials, with high values being associated with the best thermal shock resistance. It is important to remember that both the thermal conductivity and the thermal expansion coefficients of ceramics can be sensitive to temperature (see Fig. 9.13). For the $\mathcal{R}$ parameters, Young's modulus values and the thermal expansion coefficients play a major role, with low values being preferred. For example, even though pyrex glass has a low strength, it is predicted to be more resistant to thermal shock than alumina. For the $\mathcal{R}'$ parameters, however, the thermal conductivity is now included and this approach can rate materials much

Table 9.2 *Typical thermal shock resistance parameters for engineering ceramics* (Adapted from Davidge, 1979.)

Material	Strength (MPa)	E (GPa)	υ	α (10^{-6}/K)	k (W/m K)	$\mathcal{R}$ (K)	$\mathcal{R}'$ (kW/m)
HP Si$_3$N$_4$	850	310	0.27	3.2	17	625	11
RB Si$_3$N$_4$	240	220	0.27	3.2	15	250	3.7
RB SiC	500	410	0.24	4.3	84	215	18
HP Al$_2$O$_3$	500	400	0.27	9.0	8	100	0.8
WC (6% Co)	1400	600	0.26	4.9	86	350	30
Pyrex glass	100	70	0.25	3.0	1.1	357	0.4

Note:
HP – hot-pressed; RB – reaction-bonded

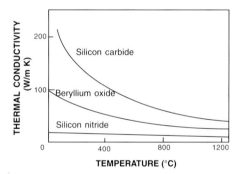

Figure 9.13 The variation in the thermal conductivity of ceramics with temperature can be an important factor in the determination of thermal shock resistance.

differently. For example, pyrex is now ranked lower than alumina with respect to thermal shock. Of the two TSR parameters, $\mathcal{R}'$ is often preferred as it relates to more realistic thermal transients.

9.9.2 Degree of damage

The above approach only accounts for fracture initiation but, clearly, crack arrest can occur in materials with steep temperature gradients. The crack arrest process is enhanced for materials with high fracture toughness and rising R-curve behavior. Materials with low elastic moduli can also be useful as these materials limit the thermal stresses and the amount of stored elastic energy. For this reason, porous and microcracked materials are often found to show good thermal shock resistance with the degree of damage being limited.

Two types of strength degradation are found during thermal shock. For materials in which crack propagation is unstable, there is a catastrophic decrease in strength at the critical temperature difference (Fig. 9.14). This effect becomes less pronounced for materials with low Weibull modulus values. For materials in which crack growth is stable, the strength decreases in a more gradual fashion with increasing temperature change (Fig. 9.15). Some strength data for a porous ceramic foam that shows this gradual decrease in strength is given as Fig. 9.16. The damage in these materials can also be quantified by the measurement of the elastic modulus, which also decreases gradually with increasing thermal shock, Fig. 9.17.

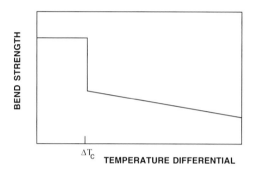

Figure 9.14 As the temperature from which quenching is performed is increased, there is often a pronounced strength degradation associated with a critical temperature differential.

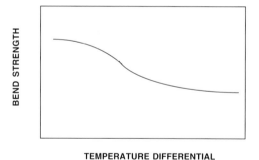

Figure 9.15 Some ceramics do not show a sudden strength loss with quenching from high temperatures. In these materials, the thermal stresses often lead to a gradual increase in the degree of damage.

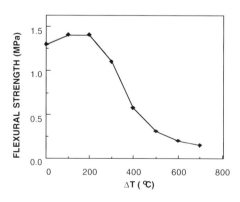

Figure 9.16 Flexural strength degradation with increasing quench temperature differential for a porous alumina foam. The shaded region is the as-received strengths. (From Orenstein and Green, 1992, reproduced courtesy of The American Ceramic Society, Westerville, OH.)

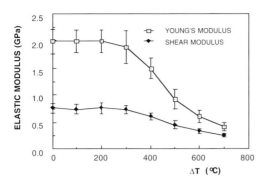

Figure 9.17 Elastic constant degradation with increasing quench temperature differential for a porous alumina foam. (From Orenstein and Green, 1992, reproduced courtesy of The American Ceramic Society, Westerville, OH.)

9.10   Residual stresses

Residual stresses are analogous to thermal stresses, with the linear thermal strain ($\alpha \Delta T$) being replaced by the linear strain associated with the size mismatch ($\Delta V/V$)/3. Some residual stress solutions were discussed in Section 4.8 but it is useful to mention some other examples here.

9.10.1   Coatings and glazes

When coatings or glazes are applied to a ceramic body, residual stresses develop during cooling after fabrication if there is a thermal expansion mismatch. It is often critical to 'fit' the coating by matching the thermal expansion coefficients. In some cases, coatings are left in residual compression and this can contribute to strength.

9.10.2   Composition gradients

Various processes, such as chemical reactions, phase transformations, crystallization, diffusion, drying, oxidation, etc., can give rise to compositional gradients in a material. If this leads to a gradient in the thermal expansion coefficient or to shrinkage, residual stresses will arise. If the stresses are tensile, this can lead to crack formation but if compressive, they can again be used for strengthening. For example, surface ion exchange is often used to strengthen glass articles (chemical tempering).

9.10.3   Thermal tempering

Thermal stresses have been used for some time to temper glass and this places the surface in residual compression. Thermal tempering depends on the ability of materials to relax stress. In the cooling of the glass, the temperature of the interior of the body can be above the glass transition temperature allowing the stresses to relax. This gives rise to a temperature gradient in the material but with low thermal stress. Once the body cools to an equilibrium temperature, however, the inside will need to contract more than the exterior and, thus, the surface will be placed in residual compression.

Problems

9.1   The following data were measured for the uniaxial tensile strength of SiC fibers in air. The fibers were 100 mm long and 10 μm in diameter (stress rate, 1 GPa/s).

Strength values, GPa: 5.15, 6.21, 6.71, 5.61, 5.44, 5.65, 6.23, 4.87, 6.64, 4.07, 5.54, 6.02, 6.27, 5.33, 5.20, 5.02, 5.47, 5.54.

These fibers also showed a dependence of strength on stress rate such that the average strengths when tested at 100 MPa/s, 10 MPa/s and 1 MPa/s were 4.93, 4.58 and 4.25 GPa, respectively.

a) Using the two-parameter Weibull distribution, determine the Weibull parameters m and σ_0^*, assuming volume flaws.

b) Estimate the maximum tensile stress that can be applied to the fibers for a failure probability of <0.001.

c) If 1000 of these fibers are used in a unidirectional fiber composite (10 vol.% fibers in an alumina matrix), estimate the applied stress for the first fiber to break, assuming the rest of the composite remains intact (i.e., no matrix microcracking and well-bonded interface). Hint: find the relationship between the applied stress and the stress on the fibers from the constitutive relationship for the elastic properties.

d) Assuming the flaw population does not change, what would the average strength be for fibers 5 μm in diameter?

e) Construct an SPT diagram for failure of SiC fibers for the same environment under a constant stress such that the failure times are 1, 100 and 10 000 s.

9.2   Uniaxial tension tests were performed on polycrystalline zirconia specimens (specimen volume = 10 000 mm³).

Strength values, MPa: 220, 270, 298, 320, 330, 334, 340, 356, 362, 370, 372, 386, 390, 390, 398, 402, 402, 412, 426, 430, 444, 448, 462.

a) Calculate the Weibull parameters m and σ_0^* using linear regression (assume volume flaws).

b) From the strength data, estimate the stress for a failure probability of 1%. How would this stress change if the specimens had twice the volume but were still broken in uniaxial tension?

c) How would the strength change if the volume remained the same as the original tests but three-point bending was used instead of uniaxial tension ($F=1\%$)?

d) Suppose the strength data were obtained in air (stress rate 100 MPa/s) and the material was subject to sub-critical crack growth ($n=30$). Determine the average strength at a stress rate of 10 MPa/s.

e) How long would a specimen (of average strength) survive at a constant stress of 360 MPa?

9.3 The following data were gathered on the flexural strength of soda-lime-silica glass.

Stress rate (MPa/s)	Strength (MPa)
0.285	63.3, 35.9, 46.0, 64.7, 46.9, 52.3, 42.5, 40.3, 30.7, 58.2.
1.51	78.8, 50.8, 78.8, 44.0, 25.9, 66.3, 48.4, 58.0, 49.3, 37.5
3.16	66.6, 67.8, 41.5, 42.2, 56.7, 59.6, 61.5, 42.2, 65.7, 70.1
7.55	70.3, 92.4, 53.6, 59.9, 48.5, 51.4, 70.3, 63.3, 77.8, 58.5

Construct an SPT diagram for failure of soda-lime-silica glass with the same loading geometry and environment under a constant stress such that the failure times are 1, 100 and 10 000 s.

9.4 The following data were measured for the tensile strength of a space shuttle tile material (assume the specimens to be unit volume).

Strength values, kPa: 565, 623, 487, 664, 407, 554, 602, 627, 533, 520, 515, 621, 671, 561, 544, 502, 547, 554.

a) Using the two-parameter Weibull distribution, determine the Weibull parameters m and σ_0^*.

b) Estimate the maximum tensile stress that can be applied to the tiles so that the failure probability does not exceed 0.001.

9.5 The following strength data (three-point bending) were gathered on glass specimens (beam width=16 mm, depth=1 mm) in an inert environment.

Strength values, MPa: 110, 149, 224, 186, 165, 195, 170, 222, 193, 178, 201, 231, 178, 213, 185, 206, 135, 201, 181, 160, 195, 199, 213, 206, 167.

a) Determine the Weibull parameters m and σ_0^*.

b) Determine the mean strength and the median strength.
c) Estimate the stresses that 99.99%, 99.9%, 99% and 90% of the same glass specimens can withstand in three-point bending (same geometry).
d) Estimate the stress that 99.99% of the same glass specimens can withstand in uniaxial tension, assuming failure occurs from surface flaws.

9.6   The same glass specimens as problem 9.5 were tested in water using stressing rates of 100, 10, 1, 0.1 and 0.01 MPa/s. The median strength values were found to be 92.8, 82.3, 72.2, 64.1 and 57.4 MPa, respectively.
a) Determine the sub-critical crack growth parameters B and n for this glass in water.
b) Draw an SPT diagram for this glass being subjected to a constant stress. Indicate the stresses and failure probability for lifetimes of 1, 10, 100 and 1000 s.
c) Determine the stress that such a glass could withstand for a period of 1 year in water and have a 99% chance of surviving.

9.7   For the data presented below, determine the Weibull modulus m and the characteristic strength σ_0^* for each set of data. Are the two sets of data consistent with a single flaw distribution? The effective volumes are $L_F V = 30$ mm^3 for the four-point flexure specimens and $L_F V = 0.1$ m^3 for the internally pressurized cylindrical tubes.

Four-point flexure (strength values, MPa)
204, 223, 224, 229, 236, 237, 237, 241, 243, 246, 247, 248, 249
249, 249, 250, 251, 254, 254, 255, 255, 257, 258, 260, 262, 268

Internally pressurized cylindrical tubes (strength values, MPa)
49, 63, 68, 71, 71, 75, 75, 80, 93, 100

9.8   Glass discs are considered as prime candidates for substrates in computer hard disc drives. A key question is whether the drives are mechanically reliable. Suppose you are involved in the materials aspect of such a design and a manufacturer offers to sell you circular glass discs (density 2200 kg/m^3) that contain a central hole (for mounting). The outer diameter is 130 mm, the inner diameter is 40 mm and the disc thickness is 1.9 mm. The manufacturer also has information on the inert strength distribution from spin testing: average strength=83.2 MPa; Weibull modulus, $m=14.0$; the number of specimens tested was 100. Complete the following tasks, clearly indicating assumptions, test conditions, references, etc.

a) Estimate the inert strength of the weakest and strongest samples in a batch of 100 discs. If K_{IC} of the glass is 0.8 MPa $\sqrt{m}$, estimate critical crack size associated with these two strengths.

b) Determine the failure probability at 20 000 rpm for the inert strengths.

c) Design a proof test that would allow you to reduce the failure probability at 20 000 rpm by a factor of 10.

d) Suppose the discs (non proof-tested) were used at a constant speed of 20 000 rpm, draw an SPT diagram in terms of rpm, failure probability and lifetime. Is the predicted lifetime at 20 000 rpm reasonable for the consumer? Tests performed previously in your laboratory had shown that the sub-critical crack growth parameters for this glass are $n=18$ and $\ln B = -5.395$ (B units: MPa^2 s).

e) Suppose the glass was not proof-tested but was ion-exchanged instead to give residual surface compression. Assuming the surface compression (-600 MPa) was uniform to a depth of 200 μm, plot the strength distribution of the ion-exchanged glasses and compare it with the original distribution. How would this change the answer to part b)?

f) In use, the discs are subjected to thermal stresses as the outside surface cools and heats. Assuming Young's modulus $=70$ GPa, Poisson's ratio $=0.2$, thermal expansion coefficient $=5\times10^{-6}/°C$, estimate these thermal stresses if the temperature changes are ~ 50 °C. The thermal analysis expert indicates the Biot's modulus should be ~ 5. Would these stresses change your answer to part b)?

9.9   ZnS is to be used as an infra-red window. The final component is circular with a diameter of 100 mm and is held in place around its circumference. The inside of the system is under vacuum while the outside is at atmospheric pressure (air). In the initial development, a series of tests was performed and the results were as follows:

- *Microstructure.* Grain size was from 50 to 300 μm (average 150 μm)
- *Strength tests.* The strength was measured in uniaxial tension in an inert environment (stress rate 1 MPa/s). The surface area of the gage section of the tensile specimen was 600 mm^2.
- *Strength values,* MPa. 51.7, 17.3, 23.8, 39.9, 30.7, 37.4, 32.6, 46.9, 35.6, 46.8, 26.3, 36.7.
- *Fractography.* All specimens failed from semi-circular machining flaws.
- *Time-dependent strength.* Subsequent strength data were gathered in uniaxial tension at stressing rates of 1, 0.1, 0.01, 0.001 and 0.0001 MPa/s in air. The average strengths were found to be 32.5, 27.3, 24.1, 21.7 and 18.9 MPa, respectively.

• *Fracture toughness.* K_{IC} was determined using a chevron-notched specimen and found to be 1 MPa $\sqrt{m}$.

In the subsequent manufacturing process, it was decided to proof test component windows in a biaxial flexure test in air to eliminate specimens that had flaws bigger than 500 μm. Windows that survived were given an additional polish so that the machining flaws were reduced to values between 1 and 10 μm. Complete the following tasks, explaining your approach and indicating clearly any assumptions you have made (justify).

a) Plot the inert strength distribution and calculate the Weibull parameters m and σ_0^*. Show on the graph the expected tensile strength distribution if the specimens had been tested in air at 0.01 MPa/s.

b) Determine the minimum thickness required for the window to ensure it would last a year with a failure probability of 0.01. (For a discussion of the loading factor see Shetty *et al.*, *J. Am. Ceram. Soc.*, **66** (1983) 36.)

c) Design an appropriate scheme for the proof-testing in air and plot the before and after proof-test strength distributions. What recommendations would you make regarding the proof-testing procedure?

d) Suppose the ZnS was strengthened by surface compression to a depth of 1 mm with a uniform stress of -10 MPa instead of being proof-tested and polished. How long would the window survive with a failure probability ≥ 0.01? (use same thickness as part b).

e) In use, the windows are subjected to some thermal stresses, as the outside surface cools and heats. Assuming Young's modulus $=90$ GPa, Poisson's ratio $=0.3$, thermal expansion coefficient $= 10 \times 10^{-6}/°C$, estimate these thermal stresses if the temperature changes are $\sim 50\ °C$. The thermal analysis expert indicates that the Biot's modulus should be ~ 5. Would these stresses change the answer to part b)?

9.10   The inert uniaxial tensile strength distribution of some glass rods is given by $\ln [1/(1-F)] = (S_I/100)^{10}$ where S_I is in MPa.

a) If these rods were to be used at a constant tensile stress of 20 MPa in a stress corrosion environment, how many would fail in 1 year? The values of the stress corrosion parameters n and B are 18 and 0.16 MPa^2 s.

b) If the rods were subjected to a proof test consisting of increasing the stress rate at 4 MPa/s for 10 s, holding at 40 MPa for 10 s and unloading instantaneously (in the stress corrosion environment), how many would fail?

c) If the proof-tested rods were used at 20 MPa, calculate the minimum lifetime.

9.11   A thin-walled, cylindrical, gas pressure vessel, radius (r) 100 mm and wall thickness (t) 2 mm, must withstand internal pressures (P) up to 50 MPa. Cracks on the inside surface perpendicular to the hoop stress ($=Pr/t$) are the normal source of failure.

a) Use linear elastic fracture mechanics techniques to determine the maximum size of crack that can be tolerated in this material without vessel failure ($K_{IC}=10$ MPa$\sqrt{m}$).

b) What is the strength of a material containing the maximum crack size?

9.12   The following data were measured for the uniaxial tensile strength of soda–lime–silica glass rods.

Strength values, MPa: 55.0, 51.5, 53.3, 59.4, 57.2, 48.7, 58.2, 54.1, 56.1, 55.4.

a) Using the two-parameter Weibull distribution, determine m and σ_0^*.

b) Estimate the maximum tensile stress that can be applied to the rods for a failure probability of 0.001.

c) If the flaw population controlling strength does not change, what would the strength be at $F=0.5$, if the samples (same test volume) were tested in three-point bending.

9.13   The strength distribution of a ceramic tube in a particular application is described by $\ln [1/(1-F)]=(\sigma/500)^{10}$, where σ is the applied stress in MPa.

a) What is the maximum stress that could be applied to the tube such that the survival probability is $\geq 99\%$?

b) What would happen to the strength at this survival probability if the diameter was increased by a factor of 10 (assume volume flaws)?

c) How would the above Weibull equation be changed if specimens with strengths less than 100 MPa could be detected non-destructively and not used?

9.14   The median strengths of a glass tested in water at stressing rates of 100, 10, 1, 0.1 and 0.01 MPa/s were found to be 92.8, 82.3, 72.2, 64.1 and 57.4 MPa, respectively. In addition, the inert strength S_1 of the glass was found to fit the two-parameter Weibull equation: $\ln \ln [1/(1-F)]=7.36$ $\ln(S_1/199)$, where S_1 is in MPa. Determine the lifetime of similar glass specimens in water under a constant stress of 20 MPa such that 50% of specimens would survive. (Assume that only Region I sub-critical crack growth occurs.)

9.15   The following strength data were obtained on glass-bonded 85 wt% alumina using three- and four-point bending (tested in air). The specimen dimensions were identical in the two types of bend tests ($4 \times 3 \times 45$ mm). For the four-point bending, the outer and inner spans were 38 mm and 22 mm. For the three-point bending the loading span was 38 mm. The stress rate was 50 MPa/s.

Four-point bend strength values, MPa: 274.6, 252.5, 286.0, 308.5, 241.2, 267.7, 273.6, 287.1, 277.8, 270.2, 270.1, 268.5.

Three-point bend strength values, MPa: 297.8, 288.5, 291.6, 284.1, 320.3, 308.6, 315.4, 314.0, 309.7, 320.4, 285.9, 314.6.

a) Determine the Weibull modulus m and the characteristic strength σ_0^* for the material. Show the data in a Weibull plot. Are the two sets of data consistent with a single flaw Weibull distribution for a beam under differing loading modes?
b) Estimate the median uniaxial tensile strength for a cylindrical rod of this material with length 200 mm and radius 5 mm.

9.16   The glass-bonded alumina described in Problem 9.15 is to be used as a grinding ball for ball-milling operations. In the initial development, a series of tests was performed and the results were as follows:

- *Microstructure.* Grain size was between 2 and 5 μm (average 3.0 μm) and the glassy phase was 25 vol.%.
- *Fractography.* The specimens failed from pores.
- *Time-dependent strength.* Subsequent strength data were gathered in four-point bending at stress rates of 500, 5, 0.5 and 0.05 MPa/s in air and the average strengths were found to be 325, 241, 217 and 189 MPa, respectively. The average inert strength was 355 MPa.
- *Fracture toughness.* K_{IC}, determined using an indentation crack length approach, was found to be 3.6 MPa $\sqrt{m}$.
- *Bulk density.* 3.450 Mg/m^3.
- *Vickers hardness.* 9.7 GPa (at 10 kg indentation load).

Complete the following tasks, explaining your approach and indicating clearly any assumptions you have made (justify).
a) Draw an SPT diagram for this material when subjected to a constant stress in air in the four-point bend test geometry. Use lifetimes of 1, 10, 100, 1000 and 10000 s.

b) Choose one of the specimens broken in four-point bending (50 MPa/s) and calculate the mirror-to-initial flaw size ratio (mirror constant,[†] $M_m = 12$ MPa $\sqrt{m}$). How would this ratio change if the specimen had been tested in an inert environment?

c) Estimate the stress associated with a failure probability of 0.001 in four-point bending at 1 MPa/s.

d) Design an appropriate proof-testing scheme to reduce the failure probability to zero at a stress of 250 MPa (four-point bending at 1 MPa/s). Sketch the strength distribution before and after the proof test.

e) In use, the balls are subjected to some thermal stresses, as the outside surfaces cool and heat. Assuming the thermal expansion coefficient of the balls is $8 \times 10^{-6}/°C$, estimate these thermal stresses if the temperature changes are ~ 50 °C. The thermal analysis expert indicates that the Biot's modulus may be as high as unity under some circumstances. Are the thermal stresses significant enough to be incorporated into the design stress?

9.17   A ceramic rod, diameter 10 mm, is to be used as a valve lifter in an internal combustion engine. The maximum thermal stress occurs in a cooling cycle when the temperature drops by 300 °C. For this situation, the surface heat transfer coefficient was assumed to be 1000 W/m² K. The following data was gathered on the two candidate materials.

Property	Reaction-bonded SiC	Hot-pressed Si_3N_4
Weibull σ_0^* (MPa)	400	800
Weibull modulus m	8	10
Young's modulus E (GPa)	410	310
Poisson's ratio v	0.24	0.27
Thermal expansion coefficient		
α (10^{-6}/K)	4.3	3.2
Thermal conductivity k		
(W/m K)	50	5
Density ρ (Mg/m³)	3.10	3.18
Specific heat (J/g K)	0.71	0.83
Hardness (GPa)	24.0	18.5
Fracture toughness K_{IC}		
(MPa $\sqrt{m}$)	3.0	4.5

a) Calculate the thermal shock resistance parameters $\mathcal{R}$ and $\mathcal{R}'$ based on a failure probability of 0.1%.

[†] Derived as an analog of Eq. (8.82), $M_m = \sigma_f \sqrt{R_m}$, where R_m is the mirror radius.

b) Determine the failure probability for the two types of material during the actual thermal transient

c) Determine the time after the start of the thermal transient that the maximum stress is reached.

d) Based on the above information which material would you choose.

e) It is suggested that the lifter is 'over-designed' and the diameter could be reduced to 5 mm. Would this change your answer to part d)?

f) What other tests would you recommend to ensure the reliability of the lifter?

9.18   The surface of transformation-toughened ceramics can be transformed by grinding. For a single-phase, yttria-stabilized zirconia,

a) Estimate the surface stress if the surface of a 10 mm thick plate is ground and the material completely transforms to the monoclinic phase to a depth of 30 μm.

b) Derive an expression for the surface stress if the monoclinic composition gradient is linear, changing from 100% at the surface to 0% at a depth of 30 μm.

9.19   A plate of alumina is to be coated with a glaze of thickness 1 mm. If the thermal expansion coefficient of the glaze is $5 \times 10^{-6}/°C$, estimate the stress in the glaze assuming the glass transition temperature of the glaze is 800 °C.

9.20   If the surface sodium ions in a soda-lime-silica glass are ion-exchanged with potassium ions, estimate the surface stress. Assume the exchange is performed below the glass transition temperature and the percentage ionic volume increase is estimated to be 4.2% (sodium to potassium).

9.21   Derive an expression for residual stresses in a rod which possesses a second-phase composition gradient $C(r)$, where r is the radial distance from the center of the rod. The second phase has a different molar volume than the parent material.

References

F. I. Baratta, Requirements for flexure testing of brittle materials, pp. 194–222 in *Methods for Assessing the Structural Reliability of Brittle Materials*, ASTM STP 844, American Society for Testing and Materials, Philadelphia, PA, 1984.

F. I. Baratta, G. D. Quinn and W. T. Mathews, Errors associated with

flexure testing of brittle materials, *US Army Report No. 87–35*, US Army Materials Technology Laboratory, Watertown, MA, 1987.

B. A. Boley and J. H. Weiner, *Theory of Thermal Stresses*, J. Wiley and Sons, New York, 1960.

R. W. Davidge, *Mechanical Behaviour of Ceramics*, Cambridge University Press, Cambridge, UK, 1979.

D. G. S. Davies, The statistical approach to engineering design of ceramics, *Proc. Brit. Ceram. Soc.*, **22** (1973) 429–52.

G. J. DeSalvo, *Theory and Structural Design Applications of Weibull Statistics*, Report WANL-TME-2688, Westinghouse Electric Corporation, 1970.

P. J. Dwivedi and D. J. Green, Determination of sub-critical crack growth parameters by *in-situ* observation of indentation cracks, *J. Am. Ceram. Soc.*, **78** [8] (1995) 2122–8.

D. J. Green, J. E. Ritter, Jr, and F. F. Lange, Fracture behavior of low density ceramics, *J. Am. Ceram. Soc.*, **65** [3] (1982) 141–6.

R. G. Hoagland, C. W. Marshall and W. H. Duckworth, Reduction of errors in ceramic bend tests, *J. Am. Ceram. Soc.*, **59** (1976) 189–92.

W. D. Kingery, H. K. Bowen and D. R. Uhlmann, *Introduction to Ceramics*, J. Wiley and Sons, New York, 1976.

B. R. Lawn, *Fracture of Brittle Solids: Second Edition*, Cambridge University Press, Cambridge, UK, 1993.

A. F. Maclean and D. L. Hartsock, Design with structural ceramics, pp. 27–97 in *Structural Ceramics, Treatise on Materials Science and Technology, Vol. 29*, edited by J. B. Wachtman, Jr, Academic Press, San Diego, CA, 1989.

C. W. Marshall and A. Rudnik, Conventional strength testing of ceramics, *Fracture Mechanics of Ceramics, Vol. 1*, pp. 69–92, edited by R. C. Bradt *et al.*, Plenum Press, 1973.

D. B. Marshall and B. R. Lawn, Flaw characteristics in dynamic fatigue: the influence of residual contact stresses, *J. Am. Ceram. Soc.*, **63** (1980) 532–6.

D. B. Marshall and J. E. Ritter, Jr, Reliability of advanced structural ceramics and ceramic matrix composites, *Bull. Am. Ceram. Soc.*, **66** (1987) 309–17.

R. C. Newnham, Strength tests for brittle materials, *Proc. Brit. Ceram. Soc.*, **24** (1975) 281–93.

R. M. Orenstein and D. J. Green, Thermal shock behavior of open-cell ceramic foams, *J. Am. Ceram. Soc.*, **75** (1992) 1899–1905.

G. D. Quinn and R. Morrell, Design data for engineering ceramics; a review of the flexure test, *J. Am. Ceram. Soc.*, **74** (1991) 2037–66.

J. E. Ritter, Jr, Engineering design and fatigue failure of brittle materials, pp. 667–86 in *Fracture Mechanics of Ceramics, Vol. 4*, edited by R. C. Bradt *et al.*, Plenum Press, New York, 1978.

J. E. Ritter, Jr, Assessment of reliability of ceramic materials, pp. 227–51 in *Fracture Mechanics of Ceramics, Vol. 5*, edited by R. C. Bradt *et al.*, Plenum Press, New York, 1985.

J. E. Ritter, Jr, Strength and reliability of ceramics, pp. 105–21 in *Introduction to Mechanical Behaviour of Ceramics*, edited by G. de Portu, CNR/IRTEC, Faenza, Italy, 1992.

V. M. Sglavo and D. J. Green, The interrupted static fatigue test for evaluating threshold stress intensity factor for ceramic materials: a numerical analysis, *J. Eur. Ceram. Soc.*, **15** (1995) 777–86.

S. P. Timoshenko and J. N. Goodier, *Theory of Elasticity*, McGraw-Hill, New York, 1970.

W. Weibull, A statistical distribution function of wide applicability, *J. Appl. Mech.*, **18** (1951) 293–7.

Comprehension exercises

Chapter 1

1   Give an example of a structural use of ceramics.
2   Materials scientists relate the properties of materials to the
    _____ and _____ of a material.
3   Name a mechanism that has been used to toughen alumina.
4   Give three examples of the non-structural use of ceramics.
5   Structures that consist of one type of material between two layers of
    another material are called _____.

Chapter 2

1   What is the definition of an elastic material?
2   How does the stiffness of an atomic bond relate to the interatomic
    potential?
3   Give an example of an inelastic deformation mechanism.
4   Name the four engineering elastic constants for isotropic materials.
5   Suggest a typical value of Young's modulus for an oxide ceramic.
6   The energy required to fracture an elastic brittle ceramic is similar to the
    atomic bond energy. True or False?
7   The continuum definition of strain at a point involves the interrelation-
    ship between two vectors. Name the vectors.
8   Why is $e_{ij} = \partial u_i / \partial x_j$ inappropriate as a definition of strain?

9   How many components are there in a fourth-order tensor? How many components would be independent if the tensor was symmetric?

10   Some combinations of stress or strain components are called invariants because they do not vary with _____.

11   What is the name of the relationships that ensure the strain components give rise to continuity of displacement in a body?

12   To separate volume change and shear distortion, the strain components are often split into _____.

13   In the subscript notation, the order of a tensor can be obtained by _____.

14   Thermal expansion anisotropy exists in polycrystalline ceramics with _____ crystal structures.

15   Spontaneous microcracking can be eliminated in ceramics with thermal expansion anisotropy if _____.

16   Use a sketch to show simple shear is equivalent to a pure shear and an equal rotation.

17   Define principal strains.

18   There are a set of relationships for the components of a stress tensor that restrict their variation with position in a body. These are called _____ .

19   The general version of Hooke's Law involves either the elastic compliance or elastic _____ constants. This array of elastic constants is a _____-order tensor.

20   Using the general version of Hooke's Law, $\sigma_i = c_{ij}\epsilon_j$, write the equation for σ_2.

21   How could you recognize from the array of elastic constants that a single crystal would produce a shear strain in response to a normal stress?

22   The Zener ratio is the ratio of the shear moduli for deformation on which crystallographic planes in a cubic crystal (include direction of displacement)?

23   For cubic crystals, Young's modulus is a maximum along <100> if the Zener ratio Z is less than unity. For $Z>1$, Young's modulus is a maximum along _____ .

24   If the Zener ratio is unity, the shape of the Young's modulus representational surface for a cubic crystal is _____ .

25   A low value of Poisson's ratio implies a high resistance to shear deformation. True or False?

26   Young's modulus is a measure of the resistance of a material to dilatation only. True or False?

27   Do you expect the adiabatic bulk modulus of a material to be greater or less than the isothermal value?

28   Some elastic materials possess a component of strain that is time

dependent. This is termed _____ and results from internal friction in the material.

29   The engineering elastic constants of isotropic materials can be measured from static loading. Name two other general measurement techniques.

30   The transverse elastic wave velocity through a covalently bonded ceramic is higher than that through an ionically bonded ceramic. True or False?

31   Three types of sonic waves propagate from the site of an earthquake: longitudinal; transverse; and surface waves. Which type arrives first? Which type creates the most devastation (i.e., largest amplitude)?

32   The Young's modulus usually increases with increasing temperature. True or False?

33   To obtain the specific Young's modulus of a material, the Young's modulus is divided by _____.

34   Name the three basic vibration modes in dynamic resonance.

Chapter 3

1   What is the role of the Madelung constant in the interatomic potential?

2   Covalently bonded ceramics are expected to possess higher melting points, lower thermal expansion coefficients and higher elastic moduli than ionically bonded ceramics. True or False?

3   The bulk modulus of the alkali halides decreases with decreasing lattice parameter. True or False?

4   Based on the anion size, which of the following would you expect to possess the highest bulk modulus? CaF_2, SrF_2, BaF_2 or PbF_2.

5   Which has a higher bulk modulus in this series? NaF, NaCl, NaBr or NaI.

6   The Young's modulus of an aluminosilicate glass is expected to increase as the number of non-bridging oxygens increases. True or False?

7   The Zener ratio is less than unity for most rock-salt structures. The exceptions usually arise because _____.

8   For systems with two sets of springs (stiff and compliant), systems with the springs in series will be stiffer than those where the springs are in parallel. True or False?

9   Why do pores and microcracks decrease the elastic moduli of a polycrystalline ceramic?

10   Name one approach for obtaining the elastic constants of a random (single-phase) polycrystal from the single-crystal elastic constants of the same phase.

11   Name one of the theories that is used to predict the effect of porosity on the elastic constants.

12   Give an example of a ceramic in which one might expect microcracking to occur (caused by residual stresses). What phenomenon led to the microcracking?

13   The single-crystal thermal expansion coefficients for mullite are
$$\alpha_{11}=3.9\times10^{-6}/°C, \ \alpha_{22}=7.0\times10^{-6}/°C, \ \alpha_{33}=5.8\times10^{-6}/°C.$$
Estimate the thermal expansion coefficient of polycrystalline mullite.

14   The maximum shrinkage on cooling in a mullite single crystal will occur along [010]. True or False?

15   Name the two bounding approaches used to describe the constitutive elastic behavior of particulate composites. What assumption needs to be made about the geometry of the second phase in these approaches?

16   The presence of pores in a polycrystalline ceramic will not change the thermal expansion coefficient. True or False?

17   Why can microcracking in non-cubic polycrystalline materials reduce the thermal expansion coefficient?

Chapter 4

1   What is the name of the principle that allows one to ignore complex loading in a region of a material at large distances from that region?

2   Which of the following stress fields is (are) statically determinate? i) Uniaxial tension, ii) Three-point bending, iii) Thin-walled pressure vessel.

3   Sketch a beam with a circular cross-section and show the region in which the normal stress is a maximum for four-point bending. Is this region a point, a line, a surface or a volume?

4   What measurements must be made to determine Young's modulus from a beam subjected to three-point bending?

5   Why does a column undergo Euler collapse at a critical load?

6   How many components of strain (out of nine) must be zero in plane strain?

7   Two-dimensional elastic problems can be solved by finding solutions to the _____ equation and using the appropriate boundary conditions. This equation involves the use of a parameter called the _____ stress function.

8   For a cylindrical pressure vessel, will the maximum (tensile) stress change significantly ($>10\%$) if the outer radius is increased from ten times the inner radius to twenty times?

9   Describe the stress state in an inclusion that has a higher thermal expansion coefficient than surrounding material. Assume the material was stress-free at the fabrication temperature and was then cooled to room temperature.

10   For the surrounding material in the previous exercise, are the radial and tangential stresses tensile or compressive?

11   If cracks form at the inclusion in Exercise 9, describe their geometry.

12   For a small circular hole in a large plate under uniaxial compression, what is the stress concentration factor (maximum stress divided by the applied stress)?

13   For a small circular hole in a large plate under equibiaxial tension, what is the stress concentration factor?

14   Define elastic instability.

15   Why does a compressive point-force often lead to a residual stress after unloading?

Chapter 5

1   Define viscosity for a Newtonian viscous material (in words or with equation). What are the units (SI)?

2   Name two approaches that are used to describe the temperature dependence of viscosity. Mention the key concept behind the approach.

3   Describe a technique that can be used to measure viscosity. Indicate what is measured in the experiment.

4   ι dilatant material is one in which the viscosity increases with increasing shear rate. True or False?

5   Define thixotropy.

6   Using springs and dashpots, draw the standard linear solid model.

7   Anelasticity in a material gives rise to permanent deformation. True or False?

8   For uniaxial tension performed at constant strain rate, how would the stiffness of a standard linear solid change with decreasing strain rate?

9   Describe a mechanism that can give rise to anelasticity in polycrystalline ceramics.

10   Using springs and dashpots, draw a model that gives rise to anelasticity.

Chapter 6

1   Theoretical shear strength decreases with increasing unit slip distance and decreasing interplanar spacing. True or False?

2   Indicate whether the Burgers vector is parallel or perpendicular to the
     edge and screw dislocation lines.

3   Screw dislocations move in same direction as their Burgers vector. True
     or False?

4   Dislocations are expected to be wider in ionically bonded ceramics than
     in those which are covalently bonded. True or False?

5   What type of elastic field exists around a screw dislocation?

6   Sketch an edge dislocation and indicate the region(s) that is (are) in
     tension?

7   Would a dislocation be attracted to, or repelled from, a free surface?

8   What effect limits the maximum velocity of a dislocation?

9   Based on the elastic field, the energy per unit length of a dislocation is
     primarily dependent on _____ .

10   In rock-salt structures, why are the primary slip planes {110} and not
     {100}? Which set of planes has the largest interplanar spacing?

11   Give an example of the Burgers vector for the smallest unit dislocation
     in the rock-salt structure.

12   In many crystals, plastic deformation occurs by movement of partial
     dislocations. What defect arises from this phenomenon?

13   Why does the smallest unit dislocation in alumina not lie in the close-
     packed oxygen direction?

14   Sketch the dependence of yield stress on temperature for a polycrys-
     talline material.

15   Plastic deformation in a polycrystal is usually easier than in a single
     crystal of the same material. True or False?

16   Name two strengthening mechanism for ductile polycrystalline materi-
     als.

17   The strength of a ductile polycrystal usually increases with decreasing
     grain size. True or False?

18   Name two yield criteria for materials that undergo plastic deformation.

19   Name two tests that are used for measuring hardness and describe the
     indentor geometry.

Chapter 7

1   Name two mechanisms that give rise to diffusional creep in polycrys-
     talline ceramics.

2   In primary creep, the strain rate is found to decrease with time. True or
     False?

3   What processes are usually occuring in tertiary creep?

4   Give an example of a dislocation creep mechanism.

5   Define steady-state creep.
6   What tests would you perform to determine the grain size exponent for creep?
7   Name a creep mechanism that obeys Newton's Law of viscosity.
8   In diffusional creep of polycrystals, vacancies flow from grain boundaries under tension to those under compression. True or False?
9   Why, for ceramics, is the creep rate in tension different to that in compression?
10   Describe a mechanism by which a grain boundary glassy phase can lead to creep.

Chapter 8

1   Suggest a typical value for the theoretical cleavage strength of a crystalline oxide.
2   In the Griffith approach, the failure condition (uniaxial tension) is defined in terms of the energy U of a system and the crack size c. Write down a mathematical expression for this failure criterion.
3   Give two examples of microstructural features that could lead to the initiation of a crack.
4   Explain in words or with an equation the parameter G that is used in linear elastic fracture mechanics.
5   Define the parameter K that is used in linear elastic fracture mechanics.
6   Crack stability is often defined in terms of the crack size dependence of G and R, the crack extension and resistance forces. What is the condition for stable crack growth?
7   Give two reasons why one cannot use reversible thermodynamics to describe the fracture process.
8   Name three non-indentation loading geometries that are used in practice to determine fracture toughness.
9   Fracture toughness can be measured from indentation cracks by measuring either their length or _____ .
10   Why do indentation cracks undergo stable growth prior to failure?
11   Why does crack bowing increase fracture toughness?
12   What happens to a crack-path geometry during the crack deflection toughening process?
13   Give an example of a loading geometry that gives constant K.
14   For mode I loading of an elliptically shaped crack by a uniform tensile stress, the stress intensity factor varies with position along the crack front and is a maximum at the ends of the minor axis. True or False?

15   Crack deflection theory suggests that, for a given volume fraction, rod-shaped obstacles will increase toughness more than disc-shaped obstacles. True or False?

16   What is the main microstructural control used to retain zirconia in the tetragonal phase?

17   The unconstrained transformation from monoclinic to tetragonal zirconia gives rise to a large shear and volume strain. True or False? Does the volume increase or decrease (monoclinic to tetragonal)?

18   Other than dilatation in the process zone, microcracks give rise to two other effects that influence toughening. Name one of them.

19   Name two microstructural features that can be used for crack bridging.

20   Some fiber composites can have fully bridged matrix cracks. How does this influence the stress–strain behavior?

21   What is the threshold stress intensity factor in sub-critical crack growth?

22   Name four features that are seen on brittle fracture surfaces.

Chapter 9

1   A high value of the Weibull modulus indicates a narrow strength distribution. True or False?

2   For equal test (stressed) volumes would you expect the strength in three-point bending to be higher or lower than that in four-point bending?

3   For equal test (stressed) volumes would you expect the strength in three-point bending to be higher or lower than that in biaxial flexure?

4   Specimens for strength testing are often smaller than a final engineered component. Will the strength of the component be higher or lower than the test specimen?

5   Various assumptions are used in applying the Weibull analysis to strength prediction. Name one.

6   Name two approaches for determining the sub-critical crack growth parameters.

7   For a material that undergoes sub-critical crack growth, the dynamic strength decreases with increasing stress rate. True or False?

8   A cyclic stress is applied to a material that undergoes sub-critical crack growth. Will the number of cycles to failure increase or decrease with increasing frequency?

9   If the strength of a material is unacceptably low, name two methods that can be used to increase the mechanical reliability of that material.

10   The thermal stress resistance parameter $\mathcal{R}$ is based on the maximum stress that arises during a thermal transient. The stresses are usually less

than the maximum and a stress reduction factor ϕ is introduced. The magnitude of ϕ is dependent on a normalized thermal property parameter, known as the _____.

11   Name two material properties that control the magnitude of the thermal stresses in a material.

12   The thermal conductivity of most (electrically insulating) ceramics increases with increasing temperature. True or False?

Appendix 1 Explicit relations between the stiffness and compliance constants for selected crystal classes

The elastic compliance constants are related to the elastic stiffness constants by the following equations.

ISOTROPIC

$$s_{11} = \frac{c_{11}+c_{12}}{(c_{11}-c_{12})(c_{11}+2c_{12})}; \quad s_{12} = \frac{-c_{12}}{(c_{11}-c_{12})(c_{11}+2c_{12})}; \quad s_{44} = \frac{2}{(c_{11}-c_{12})}$$

CUBIC

$$s_{11} = \frac{c_{11}+c_{12}}{(c_{11}-c_{12})(c_{11}+2c_{12})}; \quad s_{12} = \frac{-c_{12}}{(c_{11}-c_{12})(c_{11}+2c_{12})}; \quad s_{44} = \frac{1}{c_{44}}$$

HEXAGONAL

$$s_{11} = s_{12}\frac{c_{33}}{c}; \quad s_{11} = s_{12} = \frac{1}{(c_{11}-c_{12})}; \quad s_{13} = \frac{-c_{13}}{c}; \quad s_{33} = \frac{c_{11}+c_{12}}{c}; \quad s_{44} = \frac{1}{c_{44}}$$

where $c = c_{33}(c_{11}+c_{12}) - 2c_{13}^2$

TETRAGONAL

$$s_{11}+s_{12} = \frac{c_{33}}{c}; \quad s_{11}-s_{12} = \frac{1}{(c_{11}-c_{12})}; \quad s_{13} = \frac{-c_{13}}{c}; \quad s_{33} = \frac{c_{11}+c_{12}}{c}; \quad s_{44} = \frac{1}{c_{44}}; \quad s_{66} = \frac{1}{c_{66}}$$

where $c = c_{33}(c_{11}+c_{12}) - 2c_{13}^2$

TRIGONAL

$$s_{11}+s_{12} = \frac{c_{33}}{c}; \quad s_{11}-s_{12} = \frac{c_{44}}{c'}; \quad s_{13} = \frac{-c_{13}}{c}; \quad s_{14} = \frac{-c_{14}}{c'}; \quad s_{33} = \frac{c_{11}+c_{12}}{c}; \quad s_{44} = \frac{c_{11}-c_{12}}{c'};$$

$$s_{66} = \frac{1}{c_{66}}$$

where $c = c_{33}(c_{11}+c_{12}) - 2c_{13}^2$ and $c' = c_{44}(c_{11}+c_{12}) - 2c_{14}^2$

Appendix 2 Young's modulus as a function of direction for various single crystals

The Young's modulus representational surface for the various crystal classes can be found using the following equations.

Crystal class	Equation for $1/E$
CUBIC	$s_{11} - 2(s_{11} - s_{12} - s_{44}/2)(a_1^2 a_2^2 + a_2^2 a_3^2 + a_1^2 a_3^2)$
TETRAGONAL	$s_{11}(a_1^4 + a_2^4) + s_{33}a_3^4 + (2s_{12} + s_{66})(a_1^2 a_2^2) + (2s_{13} + s_{44})(a_3^2 - a_3^4)$ $+ 2a_1 a_2 (a_1^2 - a_2^2)s_{16}$
HEXAGONAL	$s_{11}(1 - a_3^2) + s_{33}a_3^4 + (2s_{13} + s_{44})(a_3^2 - a_3^4)$
ORTHORHOMBIC	$s_{11}a_1^4 + 2s_{12}a_1^2 a_2^2 + 2s_{13}a_1^2 a_3^2 + s_{22}a_2^4 + 2s_{23}a_2^2 a_3^2 + s_{33}a_3^4 + s_{44}a_2^2 a_3^2$ $+ s_{55}a_1^2 a_3^2 + s_{66}a_1^2 a_2^2$
TRIGONAL	$s_{11}(1 - a_3^2)^2 + s_{33}a_3^4 + (2s_{13} + s_{44})(a_3^2 - a_3^4) + 2s_{14}a_2 a_3 (3a_1^2 - a_2^2)$ $+ 2s_{25}a_1 a_3 (3a_2^2 a_1^2)$
MONOCLINIC	$s_{11}a_1^4 + 2s_{12}a_1^2 a_2^2 + 2s_{13}a_1^2 a_3^2 + 2s_{15}a_1^3 a_3 + s_{22}a_2^4$ $+ 2s_{23}a_2^2 a_3^2 + 2s_{25}a_1 a_2^2 a_3 + s_{33}a_3^4 + 2s_{35}a_1 a_3^3$ $+ s_{44}a_2^2 a_3^2 + s_{46}a_1 a_2^2 a_3 + s_{55}a_1^2 a_3^2 + s_{66}a_1^2 a_2^2$
TRICLINIC	$s_{11}a_1^4 + 2s_{12}a_1^2 a_2^2 + 2s_{13}a_1^2 a_3^2 + 2s_{15}a_1^3 a_3 + s_{22}a_2^4$ $+ 2s_{23}a_2^2 a_3^2 + 2s_{25}a_1 a_2^2 a_3 + s_{33}a_3^4 + 2s_{35}a_1 a_3^3 + s_{44}a_2^2 a_3^2$ $+ s_{46}a_1 a_2^2 a_3 + s_{55}a_1^2 a_3^2 + s_{66}a_1^2 a_2^2 + 2s_{14}a_1^2 a_2 a_3 + 2s_{16}a_1^3 a_2$ $+ 2s_{24}a_2^3 a_3 + 2s_{26}a_1 a_2^3 + 2s_{34}a_2 a_3^3 + 2s_{36}a_1 a_2 a_3^3$ $+ 2s_{45}a_1 a_2 a_3^2 + 2s_{56}a_1^2 a_2 a_3$

Note:
The direction of interest $[hkl]$ makes the angle cosines a_1, a_2 and a_3 with the x, y and z axes, respectively.

Appendix 3 Relationships between engineering elastic constants for isotropic materials

As only two elastic constants are needed to describe the linear elastic behavior of isotropic materials, there are various relationships between the four engineering elastic constants. Some of these are given below.

$$E=3B(1-2v)=2\mu(1+v)=\frac{9\mu B}{3B+\mu}$$

$$\mu=\frac{3EB}{9B-E}=\frac{E}{2(1+v)}=\frac{3B(1-2v)}{2(1+v)}$$

$$v=\frac{3B-E}{6B}=\frac{E-2\mu}{2\mu}=\frac{3B-2\mu}{2(3B+\mu)}$$

$$B=\frac{\mu E}{3(3\mu-E)}=\frac{2\mu(1+v)}{3(1-2v)}=\frac{E}{3(1-2v)}$$

Appendix 4 Madelung constants for various crystal types

The Madelung constant represents the interactions between ions in the attractive component of the interionic potential for ionic crystals.

Crystal	Madelung constant
Sodium chloride	1.747 56
Cesium chloride	1.762 67
Calcium chloride	2.365
Calcium fluoride	2.519 39
Cadmium iodide	2.355
Cuprous oxide (cuprite)	2.221 24
Zinc oxide	1.498 5
Zinc sulphide (zinc blende)	1.638 06
Zinc sulphide (wurtsite)	1.641 32
β-Quartz	2.219 7
Corundum, α-alumina	4.171 9

Appendix 5 Stress and deflection for common strength testing geometries

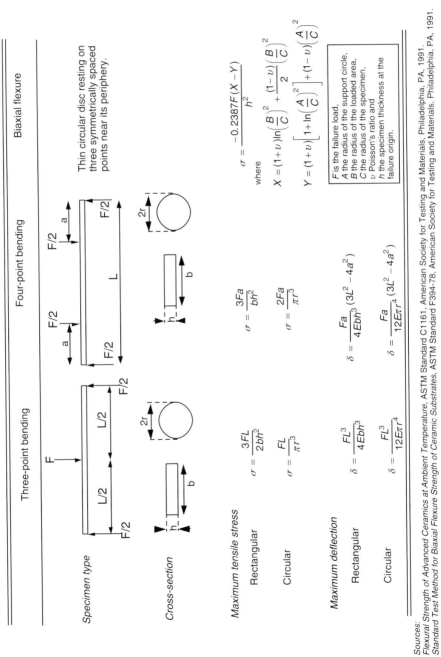

	Three-point bending	Four-point bending	Biaxial flexure

Specimen type

Thin circular disc resting on three symmetrically spaced points near its periphery.

Cross-section

Maximum tensile stress

Rectangular

$$\sigma = \frac{3FL}{2bh^2}$$

$$\sigma = \frac{3Fa}{bh^2}$$

$$\sigma = \frac{-0.2387F(X-Y)}{h^2}$$

where

$$X = (1+v)\ln\left(\frac{B}{C}\right)^2 + \frac{(1-v)}{2}\left(\frac{B}{C}\right)^2$$

$$Y = (1+v)\left[1+\ln\left(\frac{A}{C}\right)^2\right] + (1-v)\left(\frac{A}{C}\right)^2$$

Circular

$$\sigma = \frac{FL}{\pi r^3}$$

$$\sigma = \frac{2Fa}{\pi r^3}$$

F is the failure load,
A the radius of the support circle,
B the radius of the loaded area,
C the radius of the specimen,
v Poisson's ratio and
h the specimen thickness at the failure origin.

Maximum deflection

Rectangular

$$\delta = \frac{FL^3}{4Ebh^3}$$

$$\delta = \frac{Fa}{4Ebh^3}(3L^2 - 4a^2)$$

Circular

$$\delta = \frac{FL^3}{12E\pi r^4}$$

$$\delta = \frac{Fa}{12E\pi r^4}(3L^2 - 4a^2)$$

Sources:
Flexural Strength of Advanced Ceramics at Ambient Temperature, ASTM Standard C1161, American Society for Testing and Materials, Philadelphia, PA, 1991.
Standard Test Method for Biaxial Flexure Strength of Ceramic Substrates, ASTM Standard F394-78, American Society for Testing and Materials, Philadelphia, PA, 1991.

Index

Adams–Gibbs model 138
Airy stress function 116, 133
anelasticity 57–8, 148, 152–8
Auerbach's Law 278

beams
  bending 108–12, 329
  buckling 113
  chevron-notched 228
  deflection 111
  double cantilever 226–7
  cantilever 111
  composite 112
  sandwich 5, 7–8, 98, 112, 123
  single edge-notched 228
bending
  bending moment diagrams 109
  deflection 111
  first moment of area 110
  four-point 109, 285–6, 329
  non-uniform bending 110
  pure bending 109
  second moment of area 110
  shearing-force diagrams 109
  three-point 108, 285–6, 329
biharmonic equation 116, 118
Bingham flow 146
Biot's modulus 301
body forces 13, 44

boundary strengthening 185
Boussinesq (elastic) solution 273
branching constant 267
brittle–ductile transition 181
buckling 113
Burgers vector 164–5, 174
Burgers (viscoelastic) model 156

Cauchy relation 53
cavitation 144, 194, 199
ceramics
  applications 1–3
  structural 1–3
  functions 2–3
chemical strengthening 186
chevron-notched beam 228
cleavage steps 268
colloids 146
compatibility equations 35, 116
complex
  bulk modulus 156
  compliance 154
  modulus 154
  shear modulus 154
composites
  fiber 5, 120–2, 261–4
  elastic constitutive relations 78–88
  particulate 78–9, 249–59
  thermal expansion 85

constitutive (elastic) relations
   fiber composites 85–7
   microcracked materials 92–3
   particulate composites 78–84
   porous materials 88–92
   polycrystals 87–9
continuity equation 142
contact(s)
   blunt 276–8
   damage 216, 269, 273–8
   sharp 127, 273–5
   stresses 127, 216, 273, 277
crack(s)
   arrest 216
   bowing 249–51
   branching 224, 266–7
   bridging 7, 260–4
   circular 226
   cone 277
   critical size 214
   deflection 249, 250–1
   elliptical 230
   extension force 218–19
   formation 216–17
   lateral 243, 274
   median 243, 274–5
   multiple 262–3
   nucleation 180, 216–17
   radial 243, 275
   resistance force 222
   ring 277
   stability 223–4
   through–thickness 225
crack growth
   stability 223–4, 245
   sub-critical 229, 264–6, 291–6
crack tip
   interactions 249–52
   shielding 252
creep
   cavitation 194, 199
   Coble 196–7, 203
   compliance 152
   damage 265
   deformation mechanism maps 201–2, 208
   dislocation 195
   diffusional 195–7
   grain boundary sliding 197–8
   mechanisms 199–200, 203
   Nabarro–Herring 195–6, 203
   primary 193

rupture 201
   secondary 193–4
   solution–reprecipitation 197
   stages (of) 193–4
   steady state 193–4
   stress asymmetry 199
   tertiary 193–4
   viscoelastic 149
critical grain (particle) size
   microcracking 39, 217, 238–40, 258–9
   phase retention 253, 256–7
critical resolved shear stress 179
cross-slip 169
cylindrical polar coordinates 117–18

damping
   capacity 154
   linear 157
   thermoelastic 157
deflocculation 147
degree of damage 303–4
differential method 84
dilatant viscosity 145
diffusional creep 195–7
dislocation(s)
   climb 171, 195
   conservative motion 171
   creep 195
   edge 164
   energy 168–9
   Frank–Read source 171
   friction stress 166
   glide 162, 166, 171, 179
   jogs 170
   kinks 170
   line tension 169
   locks 170
   partial 176–8
   pinning 170
   screw 164–5
   sessile 170
   sources 171
   stress fields 166–8
   tilt boundary 169
   velocity 172
   width 165
dispersion strengthening 185–6
double cantilever beam 226–7
double integration approach 111
double torsion specimen 229
dynamic fatigue 291–4,

dynamic resonance 62–5

easy glide 179
effective medium 83
elastic anisotropy factor 52
elastic behavior
   anisotropic 47–54, 75–8
   atomic structure 14–17, 70–4
   chemical bonding 70–1
   constitutive relations 78–88
   elastic strain energy 22, 47
   isotropic 55–6
   orthotropic 47
   porous materials 88–92
   pressure effects 57, 77
   stability 113–14
   strain energy density 22,
   temperature effects 57, 77–8
elastic constants
   adiabatic 58
   bulk modulus 24
   compliance constants 46, 325
   crystal structure 46–9, 73
   engineering 18–24, 327
   glasses 74
   higher-order 57
   isothermal 58
   Lamé constants 55
   measurement techniques 62–5
   microcracking 92–4
   modulus defect 58
   Poisson's ratio 21
   shear modulus 23
   specific 56
   stiffness constants 45, 54, 325
   Young's modulus 20
elastic solutions
   Boussinesq 273
   circular hole 124–5
   contact 127, 273, 276–8
   cracks 131, 220
   dislocations 166–8
   elliptical hole 126
   Hertz 276–8
   spherical hole 126
elastic stability 113
elastic waves
   elastic constants measurement 62–3
   longitudinal or irrotational 59
   surface or Rayleigh 60
   transverse or equivoluminal 60

   velocities 59–60
electro-rheological solids 147
engineering design 5–6, 285
equilibrium equations 44, 116
Euler's formula 114

failure
   origins 10, 266–72
   probability 286–8
   statistics 286–91
fatigue
   cyclic 292
   dynamic 291–4
   static 292–4
finite-element analysis 128, 290
flaw
   insensitivity 246
   populations 216
flexural rigidity 111, 112
flexural strength 110, 285–6, 329
fluidity 135
fractography 240, 266–72
fracture
   Griffith concept 213–15
   indentation 243–5
   intergranular 267
   mirror 266–7
   mixed mode 247–8
   Obreimoff experiment 215–16
   origins 216–17, 266–69
   transgranular 267
fracture toughness 222–3,
   measurement 229–30, 244–5
Frank–Read source 171
free volume 138

gelation 146
glass transition temperature 136
grain boundary sliding 197–8
Green's functions 236–8
griffith approach 213–16

hackle 266, 269
Hall–Petch relation 185
Halpin–Tsai equations 87
hardness 188–9
   Brinell 188
   Knoop 189
   Meyer 189
   Vickers 189
Hashin–Shtrikman bounds 79–82

Hertz (elastic) solution 276–7
Hooke's Law 17, 44–7

I-beams 111
indentation fracture 243–5
internal friction 57, 148, 157
isoelectric temperature 174

J-integral 278–80

Kronecker delta
    definition 35

laminates 7
linear differential time operator 151
linear elastic fracture mechanics 218–23
loading factors 287–8
loss
    compliance 154
    factor 155
    modulus 153
    tangent 154

Madelung constant 72, 328
magneto-rheological solids 147
Maxwell (viscoelastic) model 150, 156
material selection maps 56
metal-matrix composites 186
microcrack(s)
    critical grain size 39, 217, 238–40, 258–60
    elastic behavior 92–4
    toughening 258–60
mirror constant 267
modulus defect 58, 152
modulus strengthening 112, 186
Mohr's circle 33
Mohr–Coulomb yield criterion 187
Mori–Tanaka methods 85
multiple cracking 262–4

Navier–Stokes equation 143
necking 135, 182
neutral surface 110,
non-destructive evaluation 240,

Obreimoff experiment 215–16, 244

particulate composites
    elastic behavior 78–83
    fracture 248–60
Peierls–Nabarro stress 166

permutation symbol 40
photoelasticity 128
plane strain 114–15
plane stress 114–15
plastic deformation
    slip (glide) 162
    twinning 162
point force (elastic) contacts 273–5
porosity
    stress concentration 124–6
    thermal expansion 97
processing defects 11, 216, 268
precipitation strengthening 185–6
pressure vessels
    thick-walled 118–20
    thin-walled 107–8
principle of superposition 17, 105
processes
    independent and sequential 202
pseudoplasticity 145

quality factor 155

R curves 245–7, 255, 257, 259, 260
relaxation time 151, 156
relaxed modulus 152
residual stress 37, 95, 120–3, 274, 305
retardation time 152, 156
river patterns 268

sandwich structures 5, 7–8, 112
    elastic behavior 98,
    residual stress 123
self-consistent (elastic) solutions 83
    generalized 84
single edge-notched beam 228
slip systems 172–7, 181
    alumina 175–6
    ceramics 177
    geometry 172–6
    independent 181
    rock salt 173–4
solid solution strengthening 181, 183–4
sols
    lyophillic 146
    lyophobic 146
    viscosity 147
sound waves
    see vibrations *or* elastic waves
spring–dashpot models 148–56
SPT diagrams 295–6

stacking fault 178
St Venant's Principle 105
standard linear solid 152–5
static fatigue 292–4
statically determinate 106
statically indeterminate 107
Stokes' Law 148
storage compliance 154
storage modulus 153
strain(s)
  compatibility relations 35, 116
  definition 20, 24–9
  deviatoric 35,
  dilatational 34,
  engineering 21
  invariants 33, 40
  normal 26
  principal 29, 33–4
  shear 26
  transformation 30–4
  true 21–2
strain energy release rate 218–19
strength
  characteristic 287
  dynamic 291–2
  improving 296–8
  inert 291
  loading geometry 289–91, 329
  minimum 287
  multiaxial 290
  specimen size 289–91
  testing 285–6, 329
  variability 246, 286–8
stress(es)
  concentration 124–6, 212
  contact 127, 216, 273, 277
  contours 128
  corrosion 264
  cyclic 153
  definition 20, 40–1
  deviatoric 41
  equilibrium equations 44, 116
  hardening 172
  hydrostatic 41
  hysteresis 155
  invariants 43
  on a plane 41–3
  principal 43
  reduction factor 301
  relaxation 149–50
  residual 37, 95, 120–3, 274, 305

resolved 41–3, 179
  softening 179
  thermal 216, 298–305
  true 22
stress intensity factor
  compounding 242
  definition 220–3
  Green's functions 236–8
  solutions 224–42
  stress concentration approach 233–4
  stress distribution approach 234–5
  superposition 232–3,
  surface correction factor 225
  threshold 264
  weight functions 240–2
structural relaxation 139
sub-critical crack growth 229, 264–6, 291–6
  parameters 265–6, 292–4
  power law 265, 291
subscript notation 24, 27, 36
surface force 13
survival probability 290

tensors
  antisymmetric 39
  order 36
  symmetric 39
theoretical cleavage strength 210–11
theoretical shear strength 162–4
thermal expansion coefficients 36–9, 94–8
  anisotropic 37
  definition 37
  isotropic 37
  composites 85
  microcracking 96
  temperature effects 96
thermal stress (shock) 216, 298–305
thermal shock resistance parameters 301–3
thermoelastic effect 57
toughening mechanisms
  crack bowing 249–51
  crack bridging 7, 260–4
  crack deflection 249, 251–2
  microcrack 258–60
  self-reinforcement 7
  transformation toughening 6, 253–7
transverse isotropy 47
Tresca yield criterion 187
triboluminescence 216

ultimate tensile strength 182

unrelaxed modulus 152

vibrations
    modes 61, 63
    resonant 61
viscoelastic models
    Burgers 156
    Kelvin 151
    Maxwell 150, 156
    standard linear solid 152–5
    Voigt 151, 155
    Zener 152–5
viscosity
    definition 134–5
    linear 134–6
    Navier–Stokes equation 143
    non-linear 145–6
    silicate glasses 135–6
    temperature dependence 137–9
viscous behavior
    adhesion 144
    Bingham flow 146
    creep 197–8
    dilatant 145
    Einstein's relation 147
    Newtonian 134–6
    pastes 147
    pseudoplastic 145
    slip casting 147

viscous flow solutions
    parallel plates 140
    concentric cylinders 141–2
    cylindrical tube 140–1
    liquid between parallel plates 144
Vogel–Fulcher–Tammann equation 137
Voigt–Reuss bounds 79–82
Voigt viscoelastic model 151, 155
von Mises yield criterion 187

Wallner lines 266
Weibull approach 286–91
Weibull modulus 287
weight functions 240–2
Williams–Landel–Ferry relation 138
work hardening 170, 184

yield
    criteria 186–8
    drop 180
    offset stress 182
    stress 179–82
Young's modulus
    definition 20
    representational surface 50–1, 326

Zener (viscoelastic) model 152–5
Zener ratio 52, 75–6,

Date Due

JUL 5 1999			

BRODART, CO.   Cat. No. 23-233-003   Printed in U.S.A.